HP QuickTest Professional
Version 10.00

Advanced Keyword-Driven Testing

WORK SHOP SERIES

LEVEL 1

Author : **Ananth Rao**

HP QuickTest Professional WorkShop Series : Level 1
HP Quicktest

V2.0

Outskirts Press, Inc.
http://www.outskirtspress.com

ISBN: 978-1-4327-5340-5

PRINTED IN THE UNITED STATES OF AMERICA

HP QUICKTEST PROFESSIONAL (Level 1)

PRE-REQUISITES

- Introduction To Computers , Use of Keyboard & Mouse
- Windows 7, VISTA, XP, 2000 or higher
- DOS Commands (Minimum knowledge)
- Concepts of Drives, Folders (*Directories*) and Files
- Windows Explorer & Mapping to a network drive
- Searching for Files & Folders
- Copying, Editing & Saving Files
- **Testing Process, Quality Assurance Terminology, Methods & Procedures**
- **Test Cases, CheckPoints, Volume Testing**
- **ODBC** (Creating ***DSN's***) , **SQL Queries and MS SQL Server Utility (BCP)**
- Basic Programming concepts *(variables, arrays, functions, methods, modules, libraries & output procedures)*
- Visit the vendors site to see if they offer a limited trial : **www.hp.com** → Search ***HP QuickTest Pro***

LAB REQUIREMENTS

1. **HP QuickTest Professional** Functional Testing Software
2. **HP QuickTest Sample Application – Flight4A**
3. **MS SQL Server** Database Software plus the utility **bcp.exe**
4. **MS EXCEL and MS WORD**

Dedication

This book is dedicated to the memory of my late father **Dr. H.V Ramakrishna Rao, PhD .** He was a *Scientist*, *Researcher* and one of the ***"Best"*** teachers I have ever known in my life. From a young age, my father exposed my brother, sister and myself to *Universities*, *Colleges* & and teaching institutions where he was employed. He would keep us ***kids*** busy by giving us **Experiments** to perform by *studying, observing* & *documenting* the results. My dad was a stronger believer in that ***Learning*** did not end in the classroom, but rather that **knowledge** is acquired in every day of life. He had his own unique technique & style of encouraging students to be motivated to learn without belittling or berating them. But rather, he would ask students basic questions that they could answer and slowly build up their confidence & capability in understanding as to complete their work. This same methodology was used constantly in my journey thru life and I thank my father every day.

Acknowledgments

I want to thank my wife, ***Sandhya A Rao*** for having the patience and perseverance in helping me writing this book. She has spent many, many *nights* and *weekends* allowing me to concentrate and work on publishing a book that would help others. In addition she has been the ***catalyst*** in motivating & inspiring me every day in my life. We both have always believed in helping *other*s (as passed on by our ***parents***) whether it be in the classroom or in life.

Who is This Book For ?

This book is targeted towards an audience who are new to Software Automation and wishes to Learn **HP QuickTest Professional** from the beginning. In addition, the book can help you refresh & brush up on some of the topics you may have forgotten. The book is a ***Step-By-Step*** set of instructions that walks you through on using HP QTP. The book assumes you have some basic knowledge in the following :

a) Test Cases

b) Concepts of *Verifications* & *Checkpoints*

c) Manual Testing

d) SQL Basic knowledge

e) Basic Concepts of Programming : (*Variables, Loops, Conditional Statements*, etc …)

f) MS Excel

g) Using the Windows *Explorer* and working with *Folders* & *Files*

All of the automated tests will refer to the sample *Flight Reservation* application provided by HP. The Work Shop Series has been designed to help the novice - quickly & easily learn the topics and gain confidence in building & testing automation scripts.

GAINING CONFIDENCE (3X Method)

- Learning (any subject) for an average Adult typically needs to be performed three (3) times !
- Follow and execute the 3X (times) technique in this Book :

1. *Reading* & *Perform* the Labs & Mini-Projects (Follow the Steps in Workbook)
2. Re-Execute the Labs on your own by Referring to the WorkBook 50% of the time
3. Re-Execute the Labs on your own (Avoid the WorkBook)

HP QUICKTEST PROFESSIONAL (Level 1)

Chapter 1 QTP Testing Process, Object Model & Repository
LAB 1A → LAB1D

Object Spy & Object Repository
LAB 2A , LAB2B, LAB2C

Analog & Low-Level Recording
LAB 2D , LAB2E

TO DO Pane & Comments

Optional Labs

Chapter 2 Object Recognition, Synchronization, Checkpoints
Maintenance Wizard, Run Update mode

LAB 3A → LAB3C , LAB 4A → LAB4E

Chapter 3 Exporting Objects, Basic Programming, Retrieving Properties, Debugging
LAB 5A → LAB5E

DataTables, Global & Local Sheets, Data Driven Tests
LAB 6A → LAB6D

Chapter 4 Data Driver Wizard, Basic Validations & Intro to Multiple Action Files
LAB 7A - Data Driver Wizard
LAB 7B - Create Orders (Basic Validation)

LAB 7C - Import Data from Database
LAB 7D - Creating Multiple Action Files
LAB 7E - Running Batch Tests

LAB 7F - Mini Project 1
LAB 7G - Mini Project 2
LAB 7H - Mini Project 3

CHAPTER 1

Objectives of this section

- QTP *Testing* Process
- Test Cases
- Quick-Test Components
- Creating Basic Test (QTP Script)
- Main Window
- LAB 1A → LAB 1D
- Object Model
- Object Spy & Repository
- LAB 2A , LAB 2B, LAB2C
- Analog & Low-Level
- LAB 2D , LAB 2E
- Optional Labs

Window Objects *(Re-cap)*

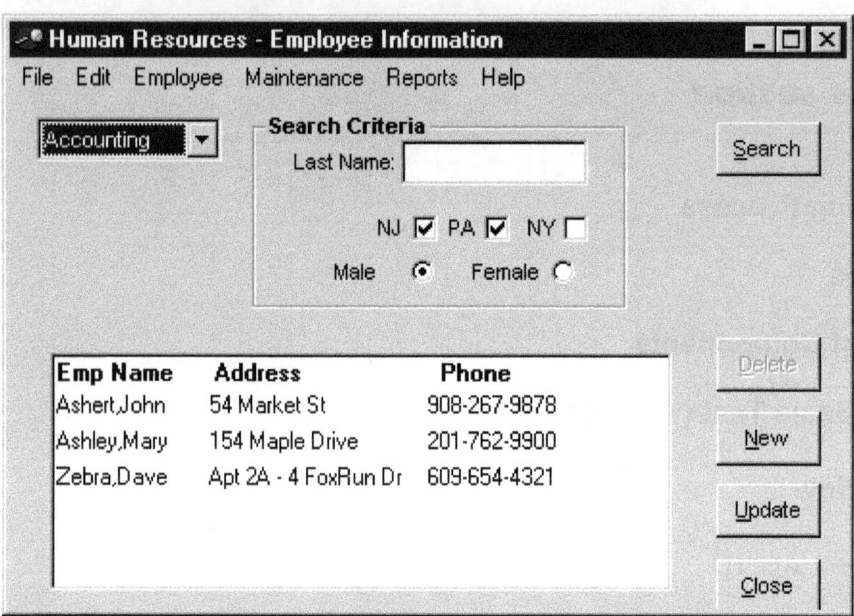

Common Window Objects: (*There may be more …*)

- Title Bar: Located at the top of the window: **Human Resources – Employee Information**
- Radio Buttons - Single Choice (**Male** or **Female**)
- Check Boxes - Multiple Choices (**NJ, PA, NY**)
- Combo Box - Selection List (**Accounting**, Marketing, Sales, etc …)
- Datawindow / RDO, ADO, DAO, DBGrids - Information from a Database allowing you to perform: *inquiry, update*, *delete* or modify data
- Command Buttons - Action taken by the user to execute an event (**Search, Delete, Update**, etc …)
- Static Text - Information displayed on the screen (does not change by the user)
- Input Fields - User enters data from the keyboard or mouse (**Last Name, Male, Female**, etc …)
- Menu Bar: - Optional choices located below the title bar :

  (**File, Edit, Employee, Payroll, Insurance, Maintenance, Reports, Help**)
- Tool Bar: - Icons providing a *shortcut* to sub-menu options:

TESTING PROCESS (Summary)

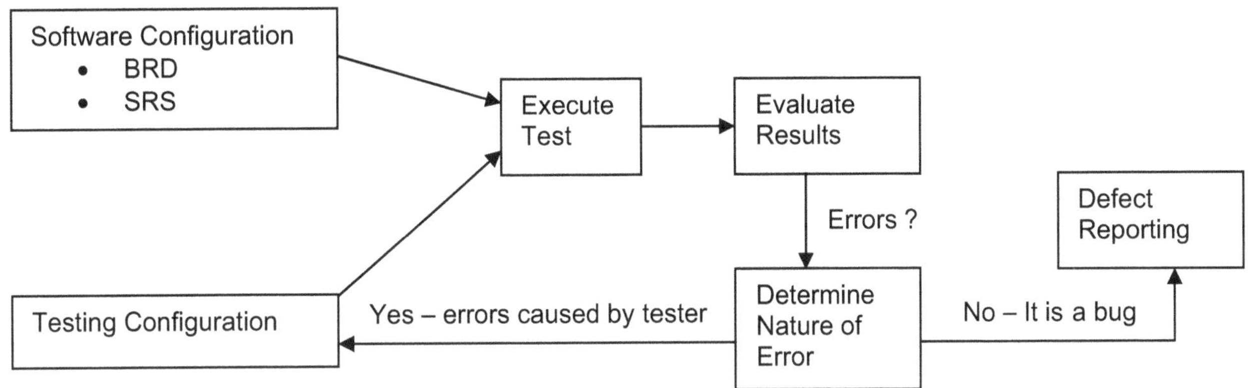

1. Software Configuration

   a. Business Requirement Document (**BRD**)
   b. Software Requirement Specifications (**SRS**) : Module/Program (Detail Technical Design) specs
   c. Technical Design Document (**TDD**)

2. Testing Configuration :

   a. Define a Test Plan
   b. Develop a Hierarchy
   c. Create Test Scenarios
   d. Define Test Cases
   e. Business Rules & Data Validations
   f. Data [Create & Load your : ***test data*** : production data (live, historical, customized, etc …)]
   g. Identify tools (HP QuickTest, IBM Robot, SilkTest, LoadRunner, MS Word, MS Project , etc …)
   h. Write scripts (*record, modify* & *enhance* the *script*)
   i. Identify/Perform the Pre-Conditions and Clean-Up conditions

3. Execute a test :

   a) Manual or Automated
   b) Run the test (Pre-conditions, Actual Test, Clean-up/Reset) in the proper environment

4. Evaluate Results (Determine the *nature* of the error - if any)

   a) Was the ***error*** caused by the tester ?
   b) Didn't perform : *Pre-conditions, Clean-up conditions* , Wrong AUT, Invalid data
   c) Poorly written script (logic does not match with the business rules)

   a) If yes → Correct the problem by returning to test configuration (Step 2B – 2G)
   b) If no → Identify the error as a defect

5. Defect Processing:

   a) Log & Report the bug or *defect* in your system
   b) Assign a priority (typically determined by a team leader, project, leader or project manager)
   c) Sent to the development team for the bug to be *fixed*
   d) Receive the next build (corrected by the Developers)
   e) Re-test the functionality / software until the *error* does not occur !

6. UAT: User Acceptance Testing

Why Use Quick Test Professional (QTP) for Functional Testing ?

If you have ever tested software manually, you are aware of some if its drawbacks.

Manual testing:

1. Time- consuming and tedious
2. Less Reliability
3. Requires a heavy investment in Human Resources
4. Time constraints often make it impossible to manually test every feature thoroughly before the software is released.
5. Inconsistent
6. This leaves you wondering whether serious bugs have gone undetected.

Automated testing with *Quick Test Pro* answers these problems:

1. Quick Test *dramatically* speeds up the testing process.
2. You create test ***scripts*** that check all aspects of your application
3. Then run these tests on each new build
4. As Quick Test runs tests:

   a) It simulates a human user by moving the mouse cursor over the application
   b) Clicking Graphical User Interface (GUI) objects
   c) Entering keyboard input

Benefits of Automated Testing

Fast	Quick Test runs tests significantly faster than human users
Reliable	Tests perform precisely the same operations each time they are run, thereby eliminating human error
Repeatable	You can test how the software reacts under repeated execution of the same operations
Programmable	You can program sophisticated tests that bring out hidden information from the application
Comprehensive	You can build a suite of tests that covers every feature in your application
Reusable	You can reuse tests on different versions of an application, even if the user interface changes

<u>**NOTE**</u>: *Not* all tests can be automated. You begin by using a manual process that could be automated.

QUICK TEST PROFESSIONAL (QTP)

Introduction

1. QTP is HP's automated testing tool for GUI & Web Testing applications
2. Test's are conducted by **recording** operations (*actions*) as you *perform* the business process
3. Quick-Test ***records*** (or captures) each ***step*** as you navigate through your web site or application
4. It generates a ***test*** or *component* and displays it graphically in a table-based **Keyword** view
5. In addition – every **step** in your test includes *automatically* generated documentation
6. An ***Object Repository*** is also created
7. The script creates **steps** in an **KeyWord View** (**icon-based *test tree***)
8. HP QuickTest helps you to ***automate*** the *testing* process, from test development to execution
9. You create adaptable and reusable test scripts which challenge the functionality of your application
10. Prior to a software release, you can execute these tests in a single overnight run— enabling you to detect defects and ensure superior software quality (via TestDirector or Quality Center)
11. Basic Concept: **Record, Enhance (** ***modify script*** **)** & **Playback** → Via **Keyword** Driven Approach

QTP ENVIRONMENT

1. Standard (Default) :
   - Windows Application (GUI or client/server)
   - Web environment (Basic)
   - Supports the following Operating Systems : (Windows NT, ME, 2000, XP, Vista, Windows 7)
2. Optional (add-ins):
   - JAVA
   - .NET
   - RTE (Terminal Emulator)
   - ERP (SAP, Peoplesoft, Oracle Apps, Siebel, Web Services, Terminal Emulator applications, etc …)

MINIMUM CONFIGURATION

1. IBM PC or Compatible with PII 600 Mz
2. Memory : 256Kb RAM
3. Free Disk Space: 500 Mb (before installation)
4. Operating System
   - Window NT 4.0 – SP 6
   - Windows 2000 – SP 2 or SP 3
   - Windows XP – SP 1
5. Browsers
   - Microsoft IE (V5, V6) – Required for completing the registration & downloading any updates
   - Netscape V6 (Optional)

The HP QuickTest Professional Workflow (Testing Process)

Testing with Quick-Test Pro involves four (4) main stages:

1. **Planning**
2. **Creating Tests (or *business components*)**
3. **Running Tests (or *business components*)**
4. **Analyze Results**

5. Log & Report Defects (Assumed)

Planning

1. Determine the required infrastructure :

   a) Facilities (Location)
   b) Hardware
   c) Network
   d) Human Resources (People or *manpower*)
   e) Security (Access to all of the above)

2. Application Under Test (AUT)

   a) Understand the Customers Business (non technical)
   b) Review requirements with Users
   c) What is the impact of the AUT with respect to other Applications within the organization ? :

      - Independent
      - Dependent
      - Replacement
      - Temporary Bridge
      - One-Time

3. Identify & <u>document</u> the functionalities that are to be tested (from the view point of the customer)

   a) Purpose or goal of the test
   b) Input fields
   c) Output fields
   d) Required & dependent fields
   e) Formatting
   f) Business Rules
   g) Data to be used
   h) Checkpoints & Verifications
   i) Actions or Steps to complete the functionality
   j) Reports

   - All of the above should help you in developing & writing test cases and most important in learning or identify the objects required for a test

CREATING TESTS (or business components)

1. Recording a session on your Application (AUT) or Web Site (SUT) *or*
2. Build an *object repository* and add steps manually to the **Keyword View** *or*
3. Modify the test with special testing options and/or with Programming

To Create a test or *component*

1. Add <u>*steps*</u> to your test, in either or both of the following ways:

   a) Record a session:

   - As you *Point & Click* – the action (*step*) is displayed graphically in the **Keyword View**

   b) Build an <u>*object repository*</u> and use those objects to manually add steps:

   - Keyword View
   - Expert View

   c) Create <u>*steps*</u> by selecting items and operations in the Keyword View

   - Keyword View

2. Insert CheckPoints into your <u>test</u>

   a) A Checkpoint <u>*verifies*</u> the behavior of the AUT/SUT

   b) Validate specific *values* or <u>characteristics</u> of a *window, page, object* or *text* strings

   c) Standard or Customized checkpoints

3. Enhance script (Programming)

   a) Customized: *checkpoints, messages, calculations, verifications*, etc …

   b) Loops (To repeat or *iterate*)

   c) Conditional execution statements (**if – else** statements)

   d) Parameterize the test (Data Table parameters)

   e) Output or *save* messages and data from the AUT

   f) Data Driven Test to achieve :

   - Volume Testing
   - Load Testing

To Run a test or *component*

1. After you create a test (or component) you can run it to check the AUT/SUT

   a) Executes from the *first* line of the test to the *last* line

   b) Each line it executes is an action that occurs automatically

   c) The test may also iterate more than once :

   - it may repeat the same lines with different data (parameters)
   - perform different calculations
   - create *output* values (data/information to be used in a subsequent test or component)

   d) Debugging : (Why the test did not execute properly ?)

   - Identify errors to help eliminate defects
   - You can use the ***Step Into***, ***Step Over***, ***Step Out*** commands t
   - Set pre-defined ***breakpoints*** in the script to pause execution
   - You can examine/inspect values in the **DeBug Viewer**

Analyzing the results

1. After you run the test (or component) you can view the results

   a) The output of the test is displayed in the **Test Results** window

   b) A summary of the test is available (the *script*)

   c) By expanding the Keyword (icon tree) and highlighting each *step* you can further inspect detailed information for *that* action

   d) You can view the ***Active Screen*** to visually help you understand the action being performed

   e) Green check mark indicates a successfully *step* execution

   f) Errors are displayed in **red** and indicates that QTP was not able to perform *that* action successfully

   g) Work Backwards from the error (line in the script) to help determine the cause of error

LOG any defects (Assumed)

1. Determine the error was NOT caused by you the ***tester***

   a) Correct the problem and re-run the test

2. Log the Defect (Word, Excel, Access, Database or TestDirector, Quality Center, etc ...)

3. Set the Priority (determined by the QA: *Team Lead, Project Lead, Project Manager*)

4. Report defects to the Development Team

5. Re-test the AUT/SUT in a subsequent build

6. Verify the error does not exist

CREATE (The Basic Test)

1. Configure your **Recording** & **Run-Time Option** settings

2. Create a test by **recording**, programming, or a combination of both.

3. Recording represents the users *actions* executed to complete a business process

Recommendation :

1. Gather Information relevant for your test (business process)

   - Resources : ideally your **Customer / End –User**
   - Subject Matter Experts (SME) *or* Functional Experts
   - Determine the different types of **valid** data & *conditions* to use for that business process
   - Identify the **Business Transactions** or *modules* to test

2. Determine the **functionality** you want to test

3. Identify which **information** you want to validate *during* the test

4. Determine the ***strategies*** to use for testing different *data* and *conditions*

5. Decide *how* you want to organize the **Object Repository**

CREATING THE BASIC TEST

1. Configure your **Recording** & **Run-Time Option** settings

2. Create a test by **recording**, programming, or a combination of both.

3. Recording represents the users *actions* executed to complete a business process

4. While recording a test you can:

   a) Insert synchronization points

   b) Create checkpoints to *validate* or check the response of the application under test (AUT)

5. A Checkpoint checks *specific* values or characteristics of a window, page, object or text string

Recording Mode

1. **Normal** (Default) : captures the ***objects*** in your application and the operations performed as you point & click

2. Analog : records the exact mouse movement and keyboard operations (signatures)

3. Low-Level : Records on any object in your application, whether or not HP QuickTest recognizes the specific object or the specific operation.

   - It can also be used if the exact coordinates are important for your test or *component*

BUSINESS COMPONENTS:

1. Each scenario that the subject matter expert (SME) creates is called a *business process*

2. A *Business Process* is composed of a serial flow of components (modules)

3. Each component performs a *specific* task

4. A component can *pass* data to a subsequent *component*

5. Helps to better understand the organizations business complex rules & validation

6. Always ask yourself :

   a) What is the purpose of the *test* ?
   b) Identify all the *inputs*
   c) Understand the output (expected value)
   d) Define the validations & rules
   e) Perform any calculations (if required)
   f) Understand the process from the organization / department point of view
   g) Compare expected results with actual values

7. Components work with HP's ***Quality Center*** (Originally: TestDirector) product

Understanding Components:

1. Shell

   a) Outer layer
   b) Visible or *available* at the test level (parent)
   c) Helps to understand the flow of action to complete the business process
   d) The script does NOT have to be created / available

2. Implementation

   a) Inner layer
   b) Includes the actual script (actions to be performed)
   c) Visible only in the *inner* layer (script)
   d) Created as a QTP (VB Script)

Example:

1. Generate and Print an order (*Business Process*)

Component (Business Process)	Description
Login	
Create Order	Save order # 101-300
Open Order	Verify Trans # 101-300
Print Invoice	Print Order # 101-300
Logoff	

TEST CASES (Two Sample Test Cases)

Test Case	**FR-100**
Purpose	**Login to FRS**
Application	**Flight Reservation (FRS)**
Business Process	**Login**
Build	**flight1a.exe**
PreConditions	**N / A**
Test Data	**{agent_name} = { Harold, Kriss, Mary, Robert }**
	{agent_password} = { mercury }

Step	Action	Expected
1	Invoke the Application	Login Window Opens
2	Enter *YourName*	Name Appears
3	Enter "mercury"	Asterisks appear
4	Click OK	Splash Screen→ Main Window
5	Checkpoint	Title Bar: **Enabled** & **Text**

Test Case	**FR-200**
Purpose	**Create a New Order**
Application	**Flight Reservation (FRS)**
Business Process	**Reservations**
Build	**Flight4a.exe**
PreConditions	**Main Window Exists : FR-100**
Test Data	**{Date Of Flight } = { Todays Date + 20 Days }**
	{Fly From} = { London }
	{Fly To} = { Frankfurt }
	{Flight} = 1st available flight
	{Name } = { *Yourname* }

Step	Action	Expected
1	Enter Todays Date + 20 days	Date appears as: **mm/dd/yy**
2	Select "London"	**London** appears
3	Select "Frankfurt"	**Frankfurt** appears
4	Click **Flights**	Flight Button Appears
5	Select First Row → Click **OK**	First available flight
6	Enter *yourname*	Your name appears (**First Last Name**)
7	Click **Insert Order**	Progress Bar
8	Verify the reservation was created	Validate: Insert Done …

ADD-IN MANAGER

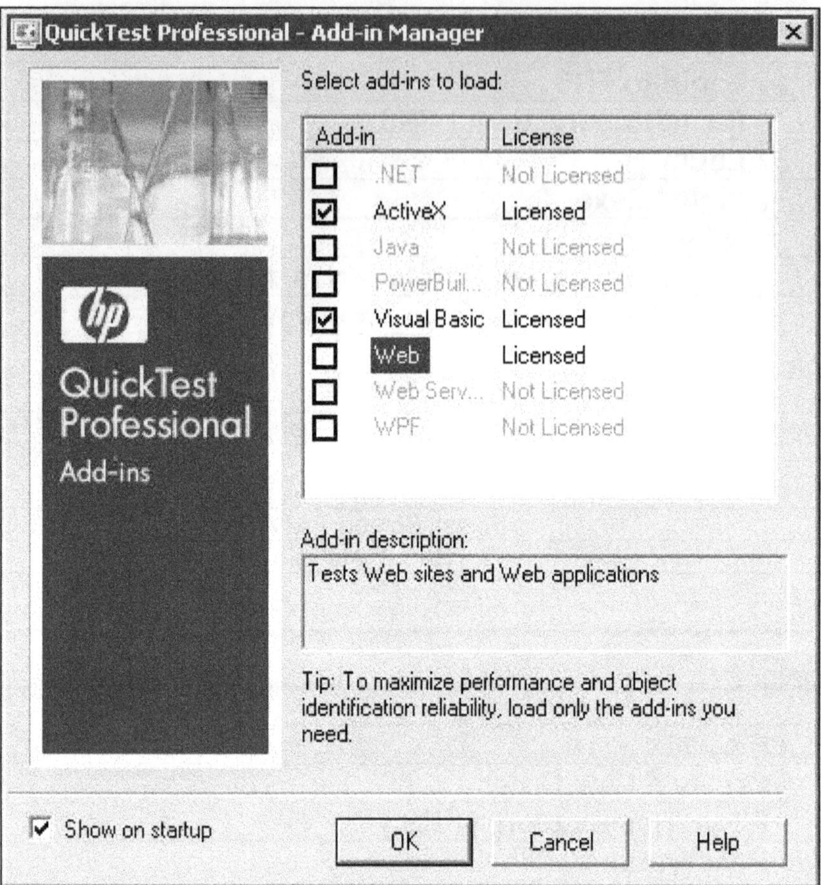

1. Add-In Manager appears after opening Quick-Test

2. Lists all available **environment/technology** for selection

3. **ActiveX** is used for most ***Windows*** (GUI) applications

    - Make one selection
    - Choose other technology (if required)
    - A combination may affect the Object Recognition's ability to learn an object

4. The *Add-ins* make additional available **functions** and **methods**

5. Make only on *expected* and *actual* results

6. The settings can be viewed/modified after Quick Test is opened :

    File (menu bar) → **Settings ...** → **Properties** (tab)

QTP CONFIGURATION SETTINGS

Record and Run-Time Options (Automation → Record and Run Settings…)

- Configure HP QuickTest on how Applications should be ***opened*** before you record (or run) a test :

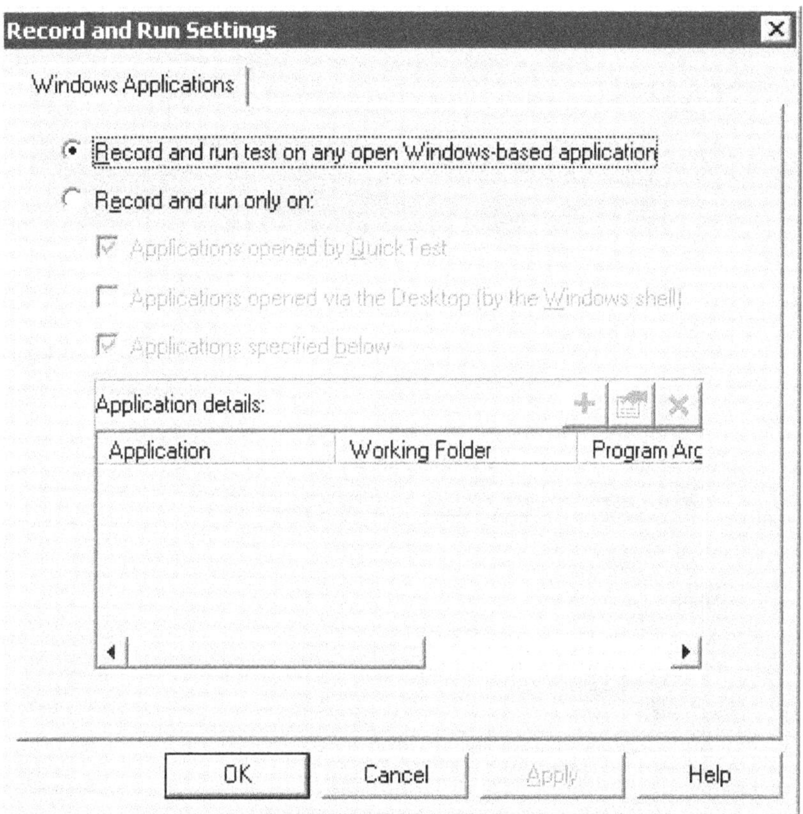

1. **Record and run test on any application** (1st radio button)

   a) Look for and use any open *window*, *screen* or *page* that you *point* and *click* on

   b) If the AUT/SUT is opened – you can begin recording from *that* screen

   c) Typically used when constructing tests that:

      - Use multiple *action* files
      - Build tests using modular tree
      - Creating / Building components

2. **Record and run only on :** (2nd radio button)

   a) Applications Opened by HP QuickTest : (Invoked by QTP as a result of Child Windows within the script)

   b) Applications via Desktop: (Invoked by : **Start** , **Windows Explorer** or by **icons** on the Task Bar)

   c) Applications specified below: (Launch only those listed and they must not be running)

The QTP MAIN WINDOW

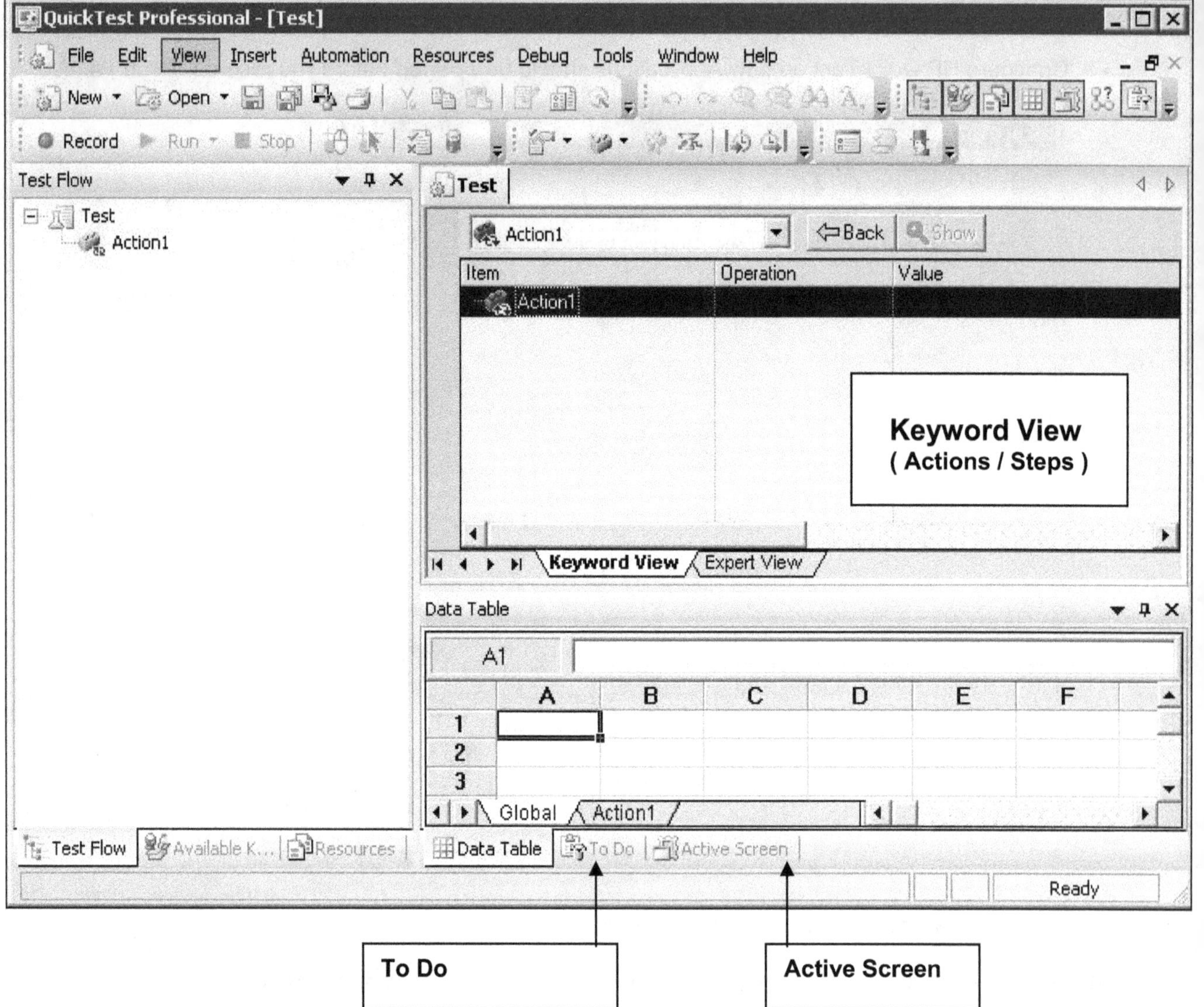

Major Components

1. **Test Flow** (Left Pane)

2. **Action1** (Top Right) : Contains the *steps* or user actions in ***Keyword View*** mode

3. **Data Table :** (Bottom Right) Data created (or used) at **Test & Run Time** time

4. **Active Screen** (Optional : Bottom Right) : Snapshot of the *AUT/SUT* screen or page

5. **To Do** (Optional : A reminder of Tasks)

6. **Available Keywords** : List of available *Local* & *Library Functions* and *Test Objects*

7. **Resources :** A list of the current Resources for this test : Associated libraries, functions & repositories

WHAT IS AN ACTIVE SCREEN ?

1. QTP allows you to *capture* **window/page** screen as you record the *business process*

2. The Active Screen provides a ***snapshot*** of your application as it appeared when you performed a certain step while recording your test

Advantages:

1. If your application is NOT available, you can use the ***snapshot*** saved from any steps to create or refer to objects even if the AUT is not running (in memory)

2. At each step – a picture or snapshot is recorded and saved and is referred to as the **Active Screen**

3. The snapshots contain objects in that **window/page** that the Object Spy can reference

ACTIVE SCREEN SETTINGS (Tools → Options… → Active Screen)

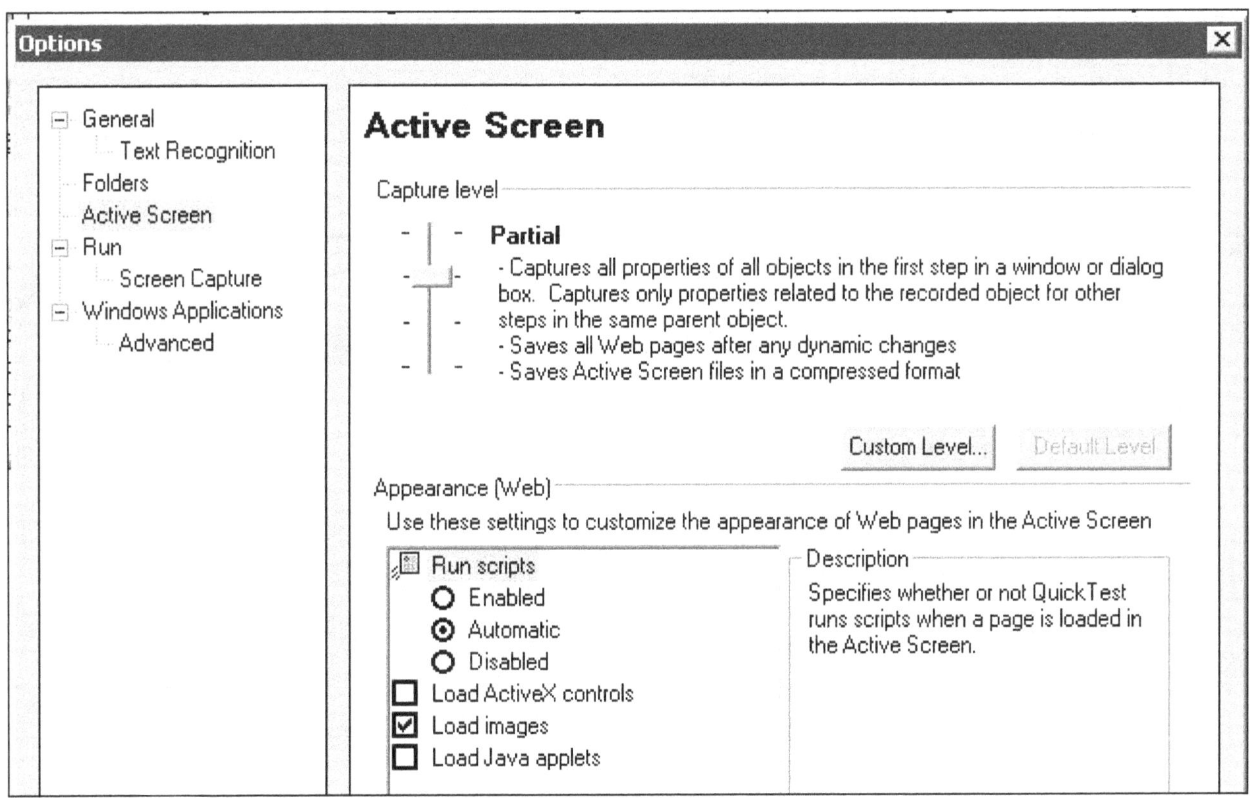

- **Complete** : Captures ***all*** properties of all objects in the AUT's current window/page for each step in a compressed format

- **Partial (Default)** : Captures ***all*** properties of all objects in the AUT's current window/page in the first step , plus all properties of the recorded object in subsequent steps in the same window/page

- **Minimum** : Save properties only for the recorded object and its parent in the Active Screen of each step
- **None** : Disables capture of Active Screen files for Windows applications

TEST STEP & ACTIVE SCREEN

1. Each selected step in the *Keyword View* highlights the captured object in the Active Screen
2. Each Quick Test ***Step*** is represented by an icon

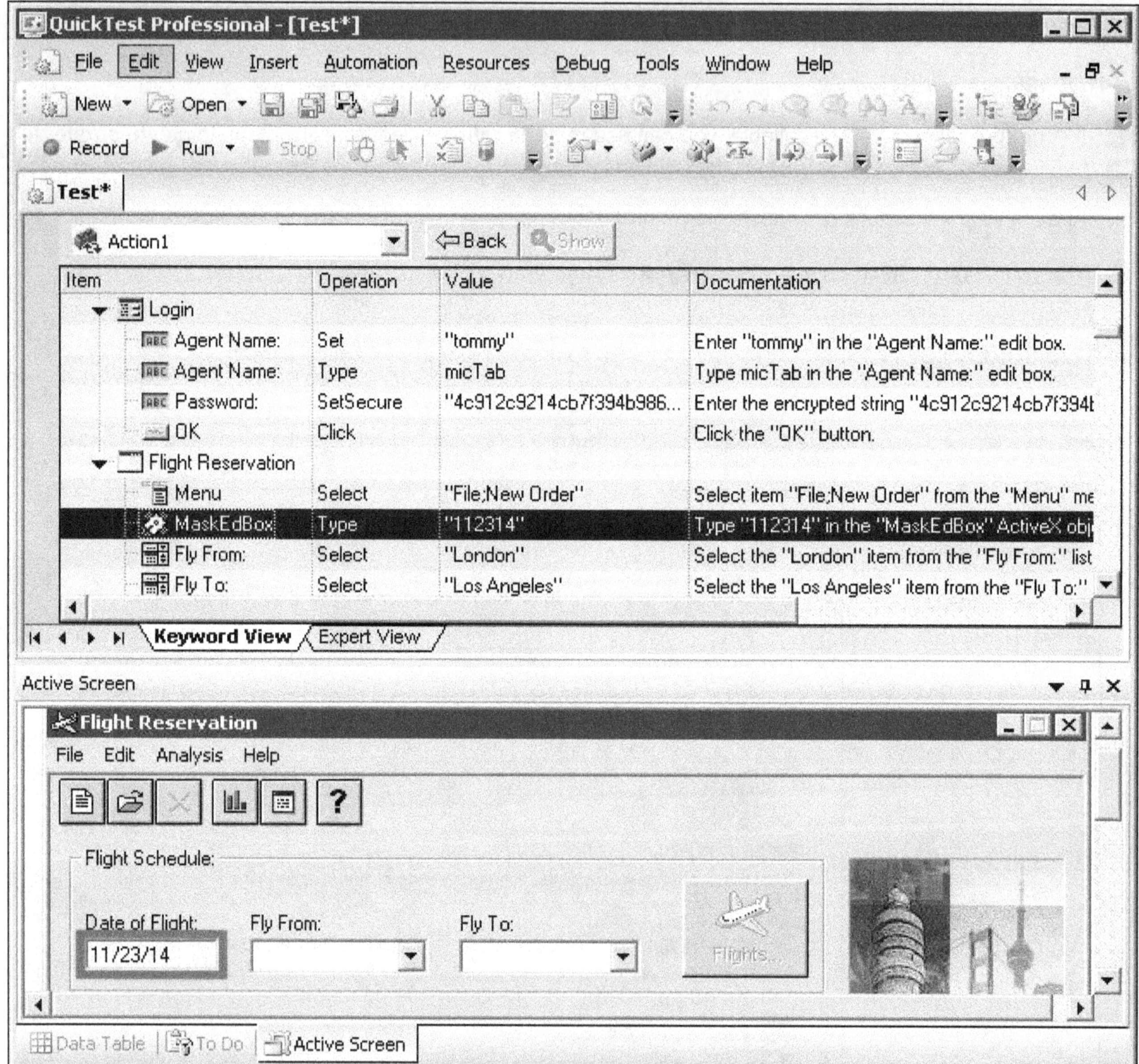

3. A Quick Test ***Step*** is composed of: **ITEM OPERATION DataValue**

Example 2 : In the above Active Screen – QTP captured what the user has entered in Date of Flight: **11/23/14**

ITEM (Object)	OPERATION (Method)	DATA VALUE	DOCUMENTATION
Menu	Select	File; New Order	Select item "File; New Order" from menu
MaskEdBox	Type	112314	Type "112314" in "MaskEdBox" Activex
MaskEdBox	Type	MicTab	Type "MicTab" in "MaskEdBox" Activex
Fly From:	Select	Denver	Select "Denver" item in the "Fly From" list

Note: Method is the user *action* performed on *that* object

SAMPLE QTP Script (Keyword View)

AUT : Notepad Editor **Functionality :** Enter Data and save the file

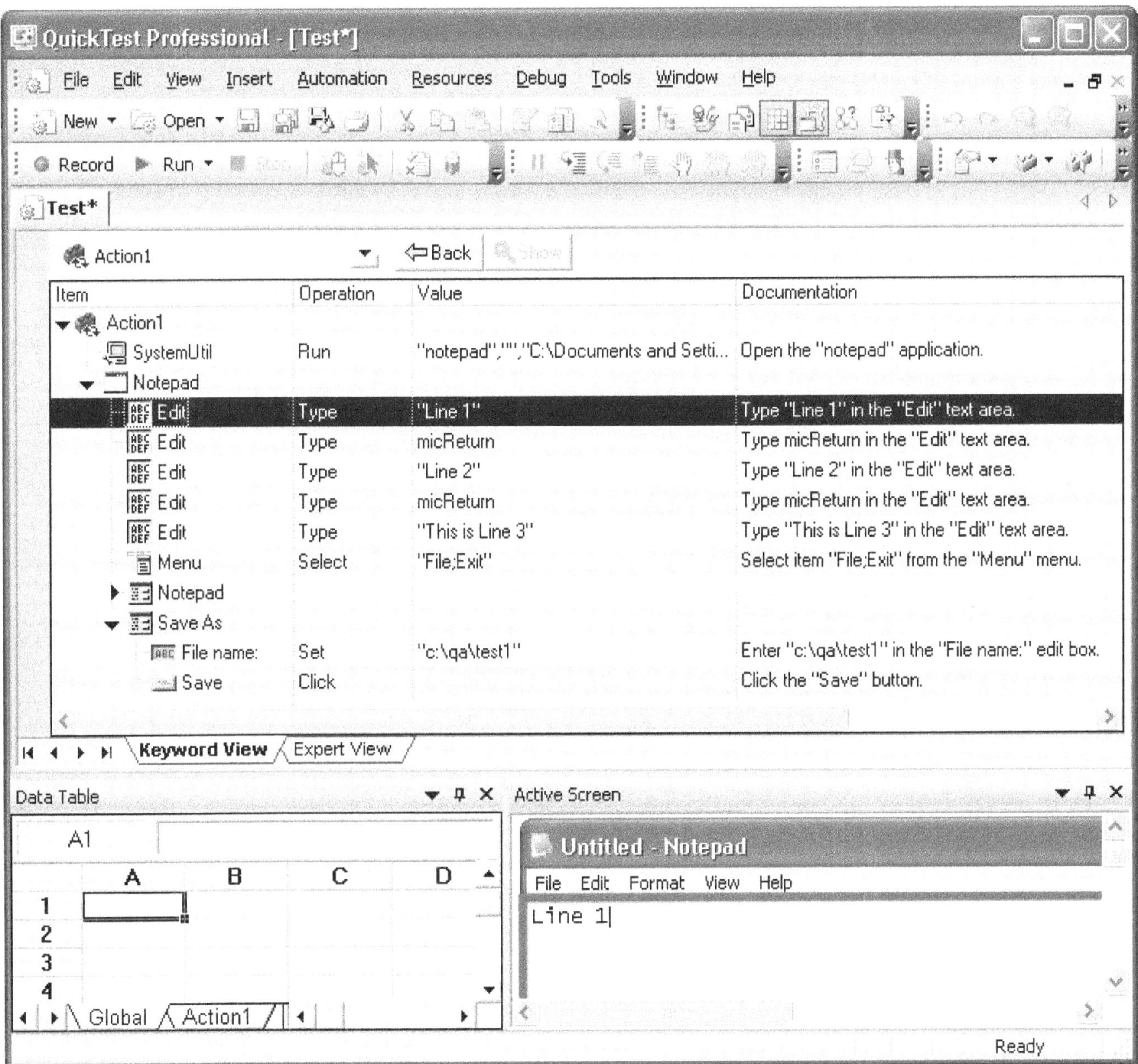

QUESTIONS

1. What statement executes the AUT ?
2. Explain what the highlighted line does (Notepad) ?
3. Explain the steps ?
4. Where is the PATH of the file saved ?
5. What steps must be taken to Playback the test ? Pre-Conditions ????

TESTING FACTORS

Design a Test that does the following :

a) The test can be re-run as many times as required

b) Using the Same Data (not different information)

c) Within exactly the ***same*** environment & conditions

1. Test Case Design

   a) Goal (purpose or objective)
   b) Individual Steps
   c) Expected Results

2. Running the Test

   a) Correct Sequence to match the business rules (policy)
   b) Proper Execution of each step

3. Analysis

   a) Searching & locating errors
   b) Trying to immediately Identify your (QA Tester) mistake or *error* !
   c) Correct & Fix the problems
   d) Isolate or identify true AUT ***bugs*** or problems

AUT / SUT

Five Factors:

1. Execute/Start (run) the AUT
2. Pre-Conditions
3. Feature being tested (functionality)
4. Clean-up / Reset conditions
5. Exit the application normally

Note:

1. Not **all** tests have **Pre-Conditions** or **Clean-Up / Reset** conditions
2. The **goal** or *purpose* of the test typically helps you to make ***that*** decision !

LAB 1 QTP Settings (1 minute)

1. Modify the QTP Environment to be *restored* to the **default** settings

2. If the IDE (Integrated Development Environment) has changed and panes are missing – you can reset it back to the original Layout as when it was first installed

3. The windows & panes can be customized by adjusting & shifting the size & position of each pane & window

4. **Tools → Options...** → The **Options** window opens

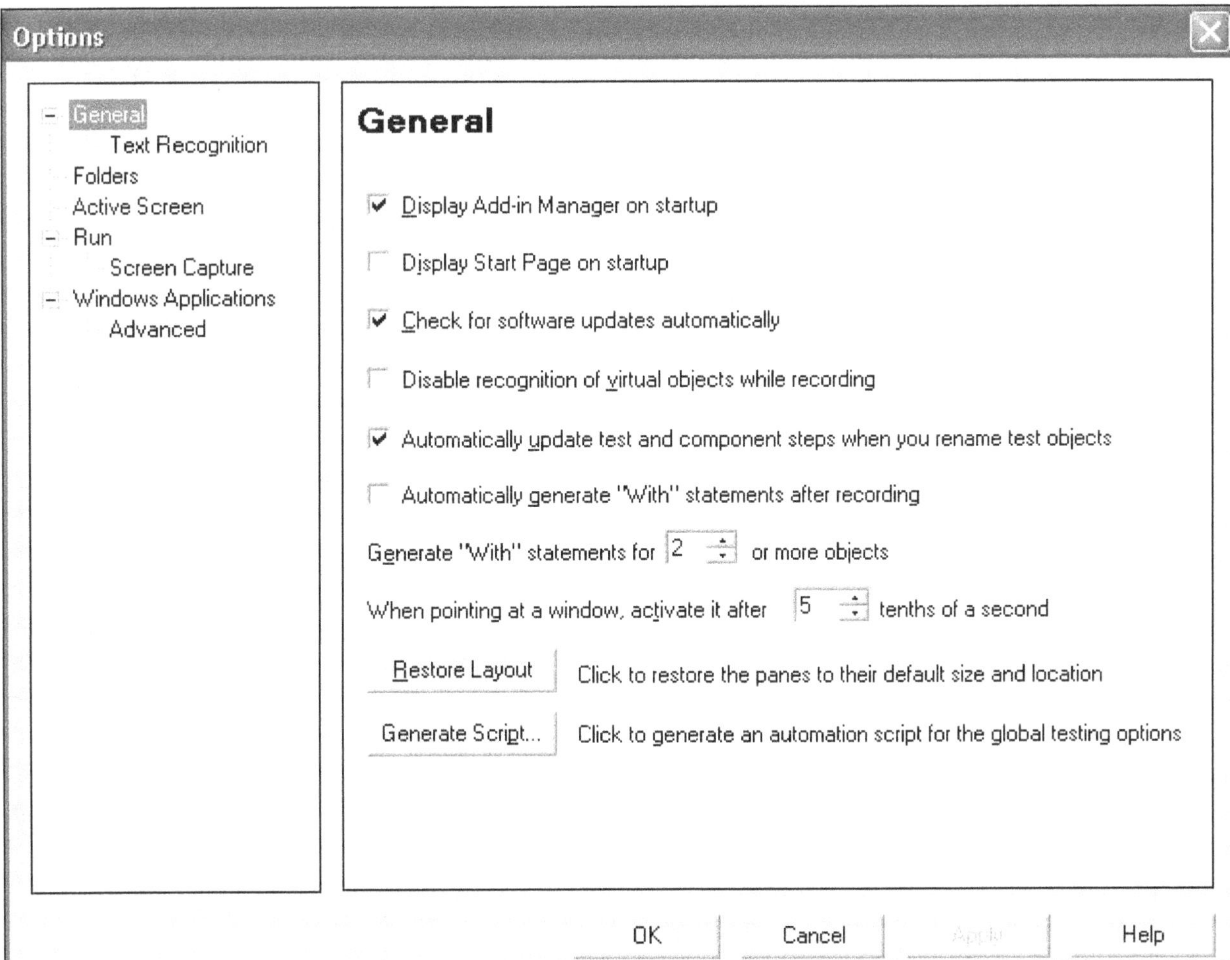

5. Select the following Check Boxes:

    - Display Add-in Manager on startup
    - Check for software updates on startup
    - Automatically update test and component steps when you rename test objects

6. Click **Restore Layout** (Restores the three panes : *Main Window, Data Table & Active Screen*)

7. Click **OK** (to close the **Options** window)

***** END OF QTP Settings *****

LAB 1A – Learn your application *(DO NOT RECORD !)*

Invoke the AUT

1. **Start → Programs → QuickTest Professional → Sample Applications → Flight**

   a) Agent Name: (*yourname*) (Minimum of four characters)

   b) Password: **mercury** → Click **OK**

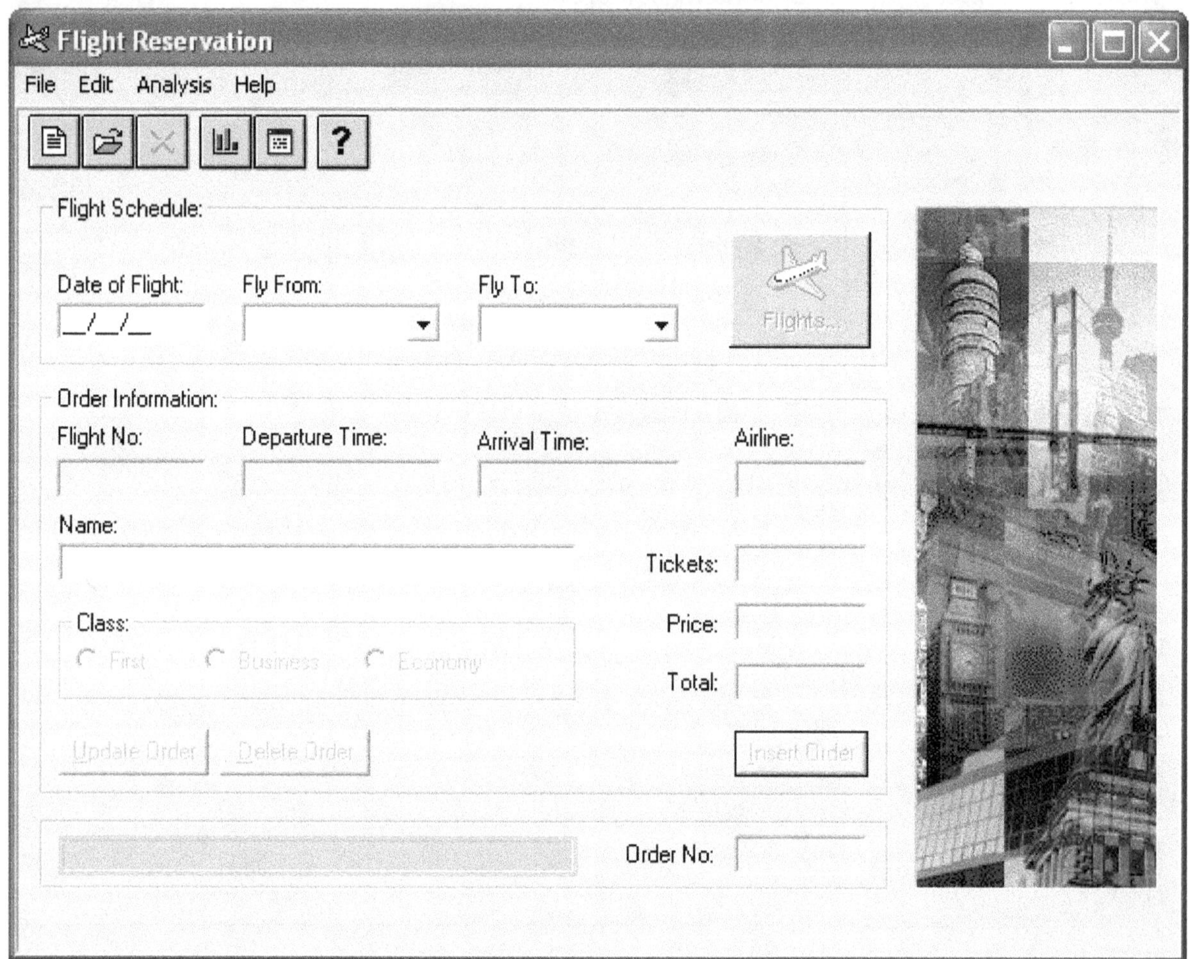

2. FLIGHT Reservation System (Visually review the window & its contents – on your own !):

   a) Flight Date , Flight Schedule (***FLIGHTS*** button, Customer Information , Order No, etc …)

4. Run the application and test the application (Inquiry, Add, Update & Delete, Fax, etc …)

5. Exit the AUT (Flight Reservation)

6. Optionally: Create the FRS Hierarchy (before starting Chapter 2)

***** END OF LAB 1A *****

LAB 1B : (Simple Record and Playback in QTP)

Test Requirements

a) Resize-Size QTP & AUT (side by side)
b) Record : Opening an existing order (reservation)
c) Playback & Verify the test ran successfully
d) Part 1 : Assume AUT is already running *before* recording (i.e: Already logged in)
e) Part 2 : Assume AUT is **NOT** running *before* recording

STEPS (Part 1) :

1. Close all applications (on your own)
2. Create the following folder **C:\YOURNAME\QTP10\LVL1** in your home directory (On your own !)
3. Create the following folder **C:\YOURNAME\QTP10\LVL1\REPO** in your home directory (On your own !)
4. **Start → Programs → QuickTest Professional → QuickTest Professional**
5. **View** (menu bar) → **Data Table** (Hide the ***Data Table***)
6. **View** (menu bar) → **Information** (Hide the ***Information***)
7. **View** (menu bar) → **Active Screen** (Ensure that the ***Active Screen*** is displayed)
8. Verify that the following two panes are displayed (Horizontally):
   a) Top Pane: **Action1** (Under Column labeled: *Item*)
   b) Bottom Pane : The *Data Table* and *Information* is **NOT** visible
   c) Bottom Pane : (Either: **Whats New in Quick Test** or an ***empty*** screen maybe visible)
9. **File** (menu bar) → **New** → **Test ...**
10. **Automation** (menu bar) → **Record and Run Settings ...**
11. Select **Record and run test on any open Window-based application→** Click **OK**

Start the AUT (the *application* you want to test)

12. **Start → Programs → QuickTest Professional → Sample Applications → Flight**
13. Re-Size QTP and AUT (Flight 4A) to be side-by-side
14. Login with ***yourname*** and use the password ***mercury*** → Click **OK**
15. Swap to HP QuickTest
16. **Automation** (menu bar) → **Record ... – F3** (***Record*** message is flashing in red)
17. Click **OK** (for the **Record and Run Settings** window if displayed) → Swap to **AUT**)
18. **File → Open Order ...** → Check **Order No.** (check box) and type **3** → Click **OK**
19. **File → Exit** (Close the AUT – ***not*** HP QuickTest !)

Stop the recording in *Quick Test*

20. **Automation** (menu bar) → **Stop** - **F4**
21. Review the script - it may be *similar* but not exactly as below :

SAMPLE QT Script (Keyword)

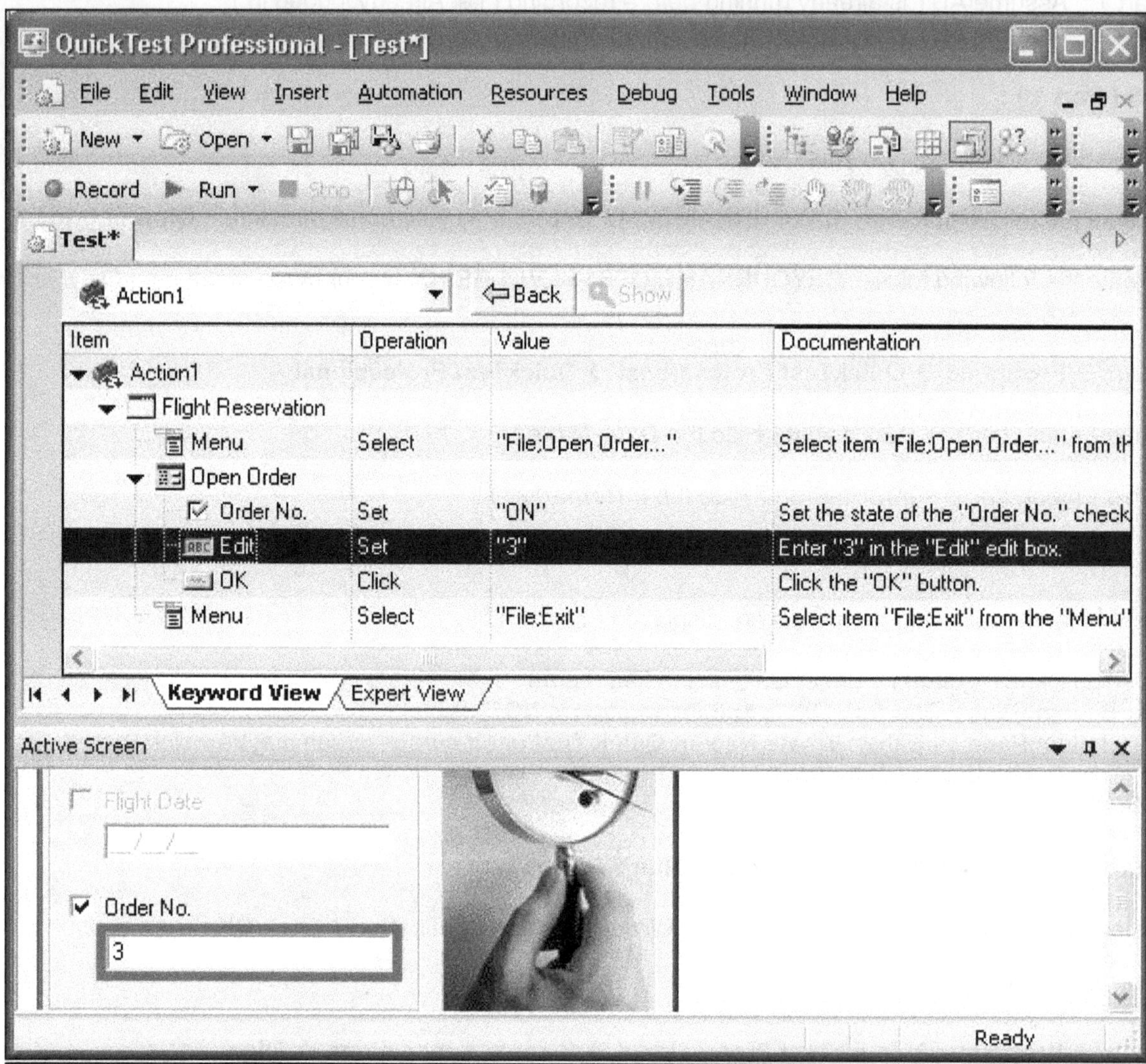

QUESTION : What are the Pre-Conditions for this test ? If you don't know - how can you find out ?

Save QT Script

22. **File** (menu bar) → **Save As ...** → **C:\YOURNAME\QTP10\LVL1\LAB1B_P1**

23. Highlight **Flight Reservation** → What do you observe in the Active Window ?

____

24. Highlight **Menu** → What do you observe in the Active Window ? _______________________

25. Highlight **Open Order** → What do you observe in the Active Window ? _______________________

Playback (Perform the Pre-Conditions !)

26. **Start → Programs → QuickTest Professional → Sample Applications → Flight**
27. Login with ***yourname*** and use the password ***mercury*** → Click **OK**
28. Swap to **HP Quicktest**

29. **Automation** (menu bar) → **Run F5**
30. Check radio button : **Temporary Run Results folder (overwrites any existing temporary results)**
31. Click **OK** → Wait for the *test* to execute

Analyze the Results

32. The **Test Results** window opens for **Lab1b_p1 [TempResults]**

33. Review the Results of your run – it may be similar to the following

34. **View** (menu bar) → **Expand All**

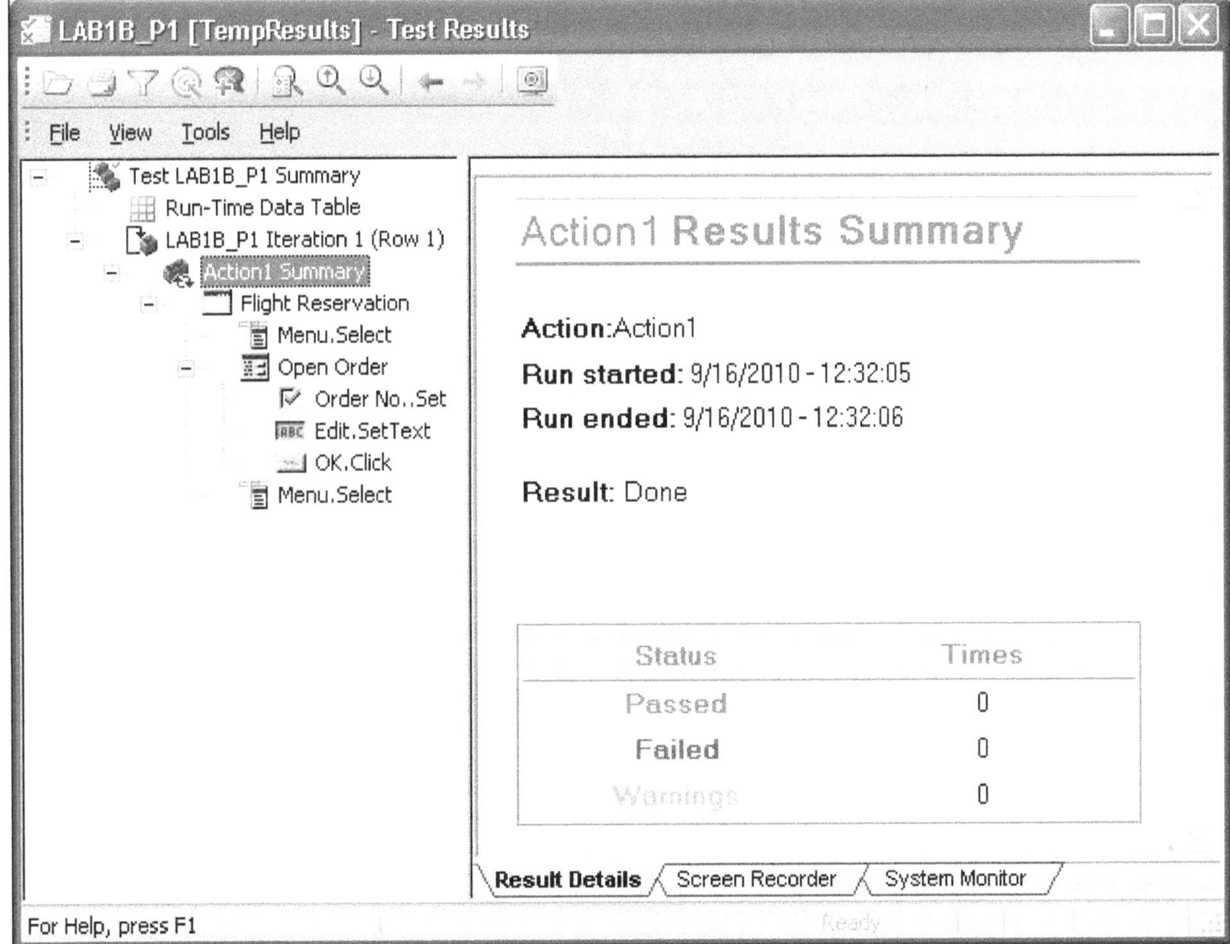

35. Highlight each *step* → Examine the results in the right pane

36. When finished – close the **Test Results** window : **File → Exit**

****** END OF LAB 1B : Part 1 *****

LAB 1B

STEPS (Part 2) :

1. **File** (menu bar) → **New** → **Test ...**
2. ReSize **QTP** window (Make it smaller if required)
3. **Automation** (menu bar) → **Record ...** – **F3** (***Record*** message is flashing in red)
4. Click **OK** (for the **Record and run test on any Windows-based application**)
5. **Start** → **Programs** → **QuickTest Professional** → **Sample Applications** → **Flight**
6. Login with *yourname* and **mercury** → Click **OK** → Wait for the splash screen
7. **File** → **Open Order ...** → Check **Order No.** (check box) and type **3** → Click **OK**
8. Verify **Order No: 3** is displayed on the screen
9. **File** → **Exit** (Close the AUT – ***not*** HP QuickTest !)
10. **Automation** (menu bar) → **Stop ... F4**
11. Review the script & Highlight the following and observe the **Active Screen** :

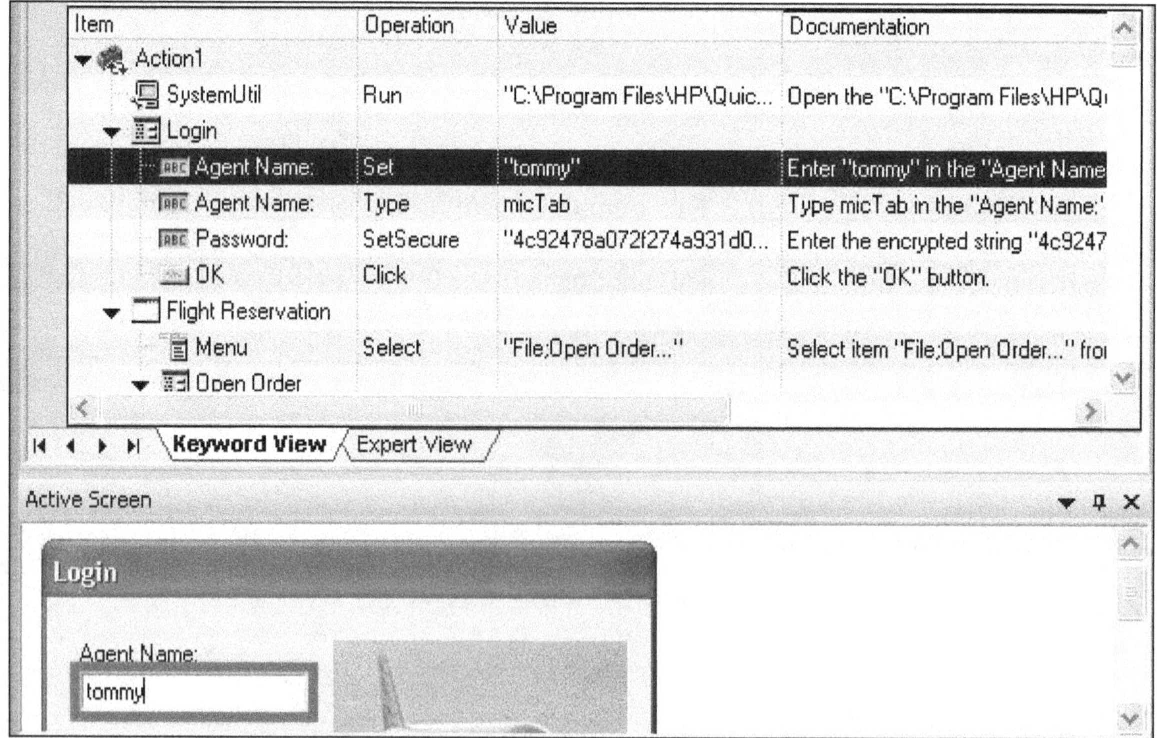

a) Login ________________________________

b) Flight Reservation ________________________

c) Menu _________________________________

Save & Playback the Test

12. **File** → **Save** → → **C:\YOURNAME\QTP10\LVL1\LAB1B_P2**

13. **Automation** (menu bar) → **Run** **F5**

14. Check radio button : **Temporary Run Results folder (overwrites any existing temporary results)**

15. Click **OK** → Wait for the *test* to execute

Analyze the Results

16. The **Test Results** window opens for **Test [TempResults]**

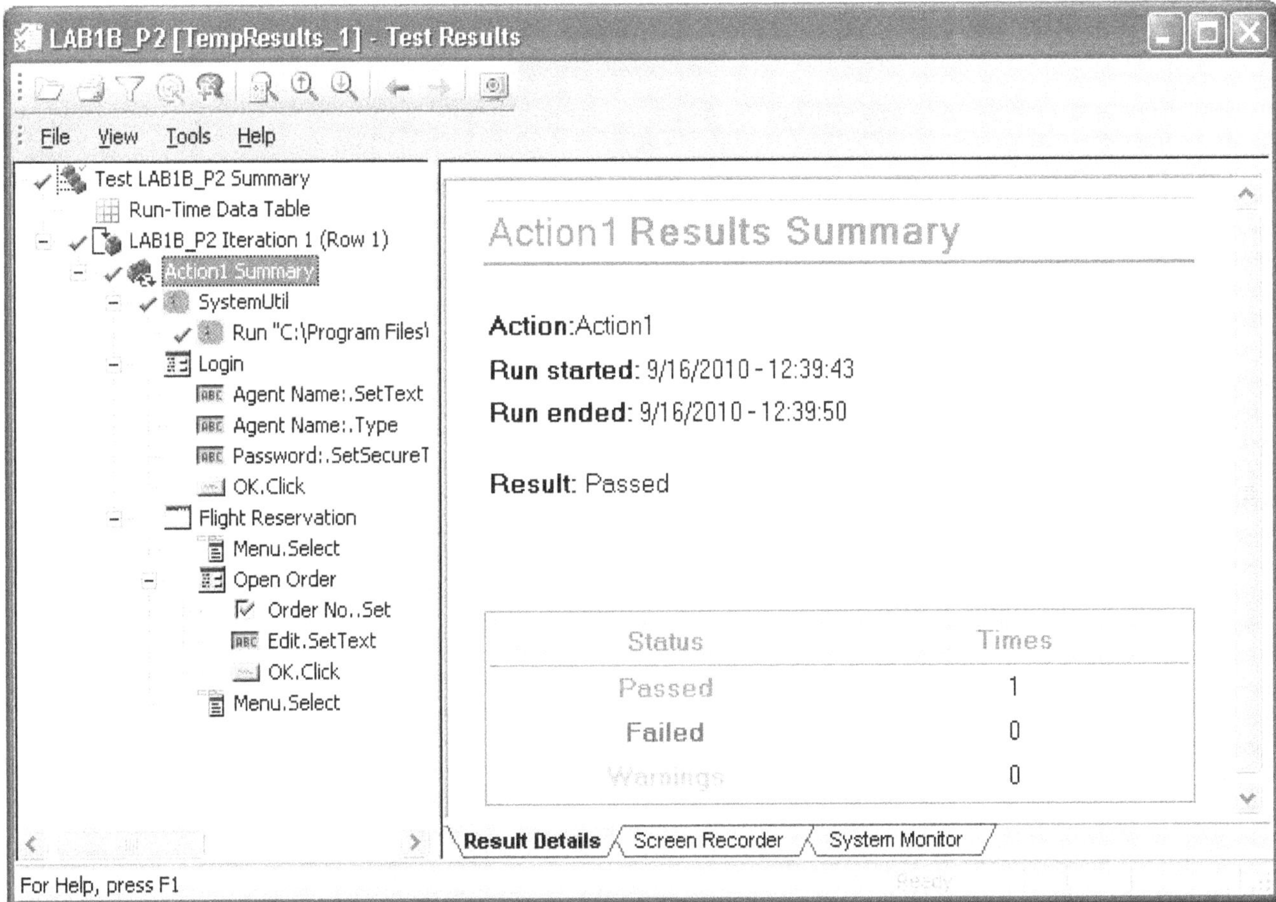

37. Highlight each *step* → Examine the results in the right pane

38. When finished – close the **Test Results** window : **File** → **Exit**

****** END OF LAB 1B : Part 2*****

LAB 1C : Creating a *New* Reservation

Test Requirements

1. Assume the AUT is ***NOT*** running and describe the location of the AUT to be executed
2. **Scenario**: Login → Add an *New* Order (Reservation) → Delete the order → Exit AUT
3. Review the script → Playback & Verify the test ran successfully
4. Locate the **flight4a.exe** (on your own) :

   ____

STEPS:

1. Close all applications (except for *HP QuickTest*)
2. Locate the file **flight4a.exe** → Note the *Drive, Folder* and *sub-folders*
3. **File** (menu bar) → **New** → **Test**
4. **Automation** (menu bar) → **Record F3** (***Record*** message is flashing in red)
5. Select **Record and run only on** radio button → Select the following:

   a) Check : Applications opened by Quick Test
   b) UnCheck : Applications opened via the DeskTop (by the Windows shell)
   c) Check : Applications specified below

6. Click **Add** (Green Plus Sign) → Point to the Path / Location (from Step 2)

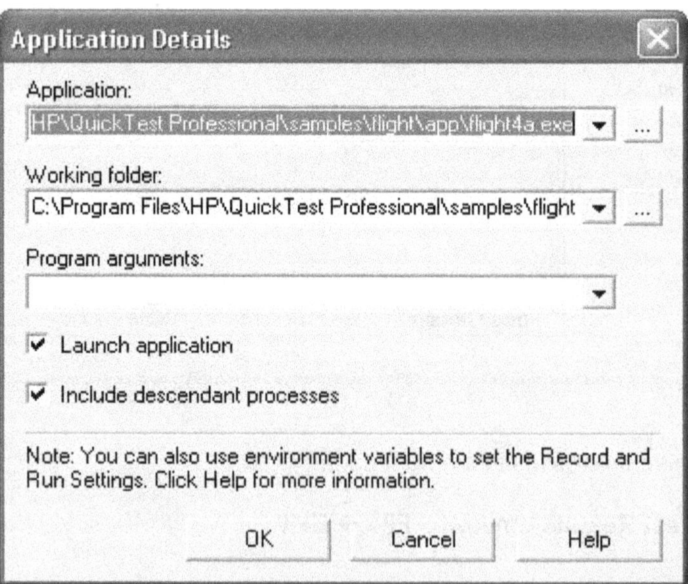

7. Verify you are running **flight4a.exe** → Click **OK** (Close ***Application Details*** window)
8. Click **OK** (Close the ***Record and Run Settings*** window)
9. Login with ***yourname*** and use the password ***mercury*** → Click **OK**

10. **File → New Order …** and Enter the following

    a) mm/dd/yy (enter a *date* greater than today's date)
    b) Select the following: Fly From: Frankfurt Fly To: Los Angeles

11. Click Large Button: **Flights** and highlight & *select* first row → Click **OK**

12. Type *yourname* in the Name: field → Click **Insert Order** button

13. Click **Delete Order** button → Click **Yes** (to *Confirm* the Delete)

14. **File → Exit** (Close the AUT)

15. **Automation** (menu bar) **→ Stop – F4** (Stop the recording in QTP) → Review the QT Script

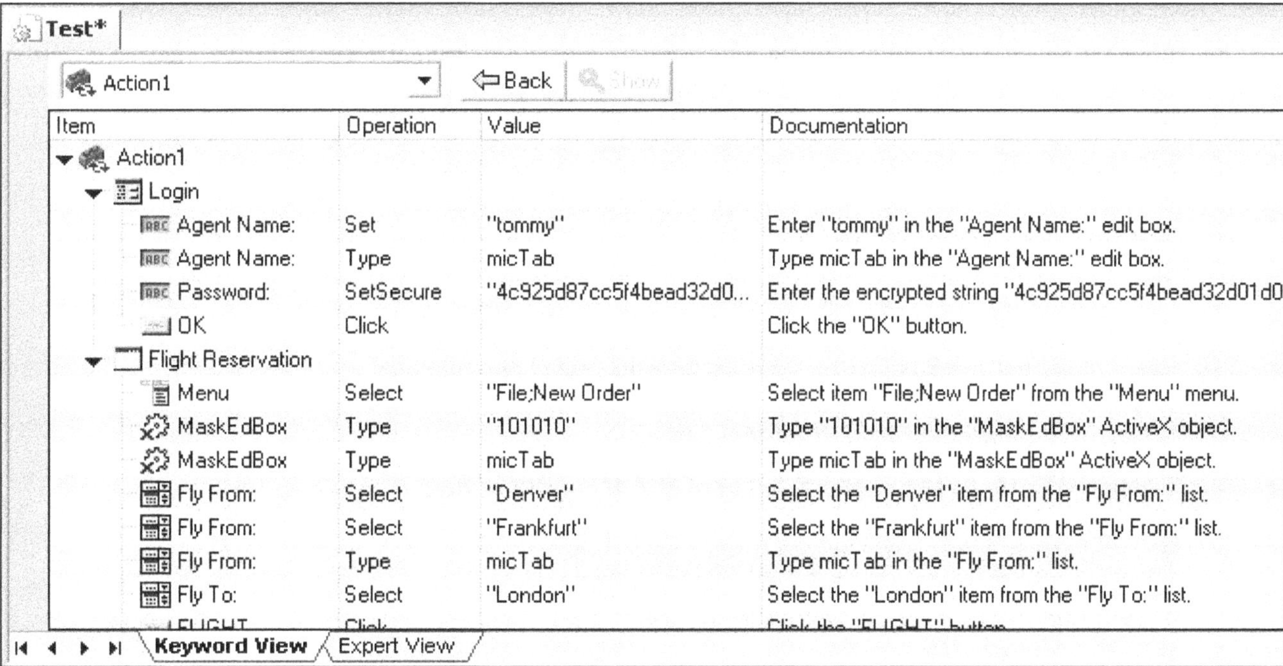

Save & Playback the Automated Script

16. **File** (menu bar) → **Save → C:\YOURNAME\QTP10\LVL1\LAB1C**

17. Review the Script & Active Screen
18. **Automation** (menu bar) **→ Run …. F5**
19. Check radio button : **Temporary Run Results folder (overwrites any existing temporary results)**
20. Click **OK** - Observe the Playback

Review the result

21. Examine the results of the run → **View** (menu bar) → **Expand All**
22. Highlight each step → Examine the results in the right pane
23. When finished – close the **Test Results** window : (**File → Exit**)

**** **END OF LAB 1C** ****

LAB 1D : Normal & Analog Recording

Test Requirements:

1. Send a Fax to an existing Customer (Identify the order id) – Search the database on your own
2. Create a new script and save it as **LAB1D**
3. Switch to Analog for signature → Return to Default Recoding Mode
4. Save & Playback the script

STEPS:

1. Identify the ***Order Id*** for an existing customer : Example - **Order No. 2**
2. Close all applications (except for ***HP QuickTest***)
3. **File** (menu bar) → **New** → **Test ...**
4. **Automation** (menu bar) → **Record F3** (***Record*** message is flashing in red)
5. Select **Record and run test on any open Windows-based application→** Click **OK**
6. **Start** → **Programs** → **QuickTest Professional** → **Sample Applications** → **Flight**
7. Login with ***yourname*** and use the password ***mercury*** → Click **OK**
8. **File** (menu bar) → **Open Order ...**
9. Check **Order Number** (radio button) → Type **2** → Click **OK**
10. **File** (menu bar) → **Fax Order** ... → Enter a 10 digit phone number
11. Check: ***Send Signature with order*** box
12. Swap to QTP → **Automation** (menu bar) → **Analog Recording**

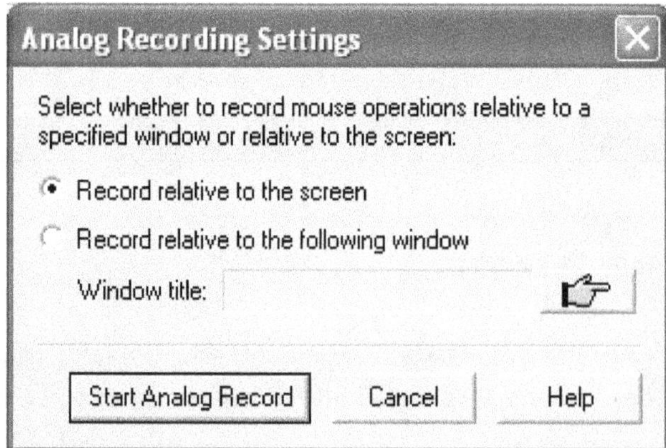

13. Click **Start Analog Record**
14. Swap to AUT (Flight 4A)
15. Move the *mouse* inside the signature area and the cursor changes to a Pen → Sign *your* name
16. Swap to QTP → **Automation** (menu bar) → **Analog Recording** (Stop Analog)
17. Click **Send** (wait 20 seconds) → Returns to the main window

18. **File → Exit** (Close the AUT)

19. **Automation** (menu bar) → **STOP - F4** (Stop Recording)

20. Review & Save the script as **C:\YOURNAME\QTP10\LVL1\LAB1D**

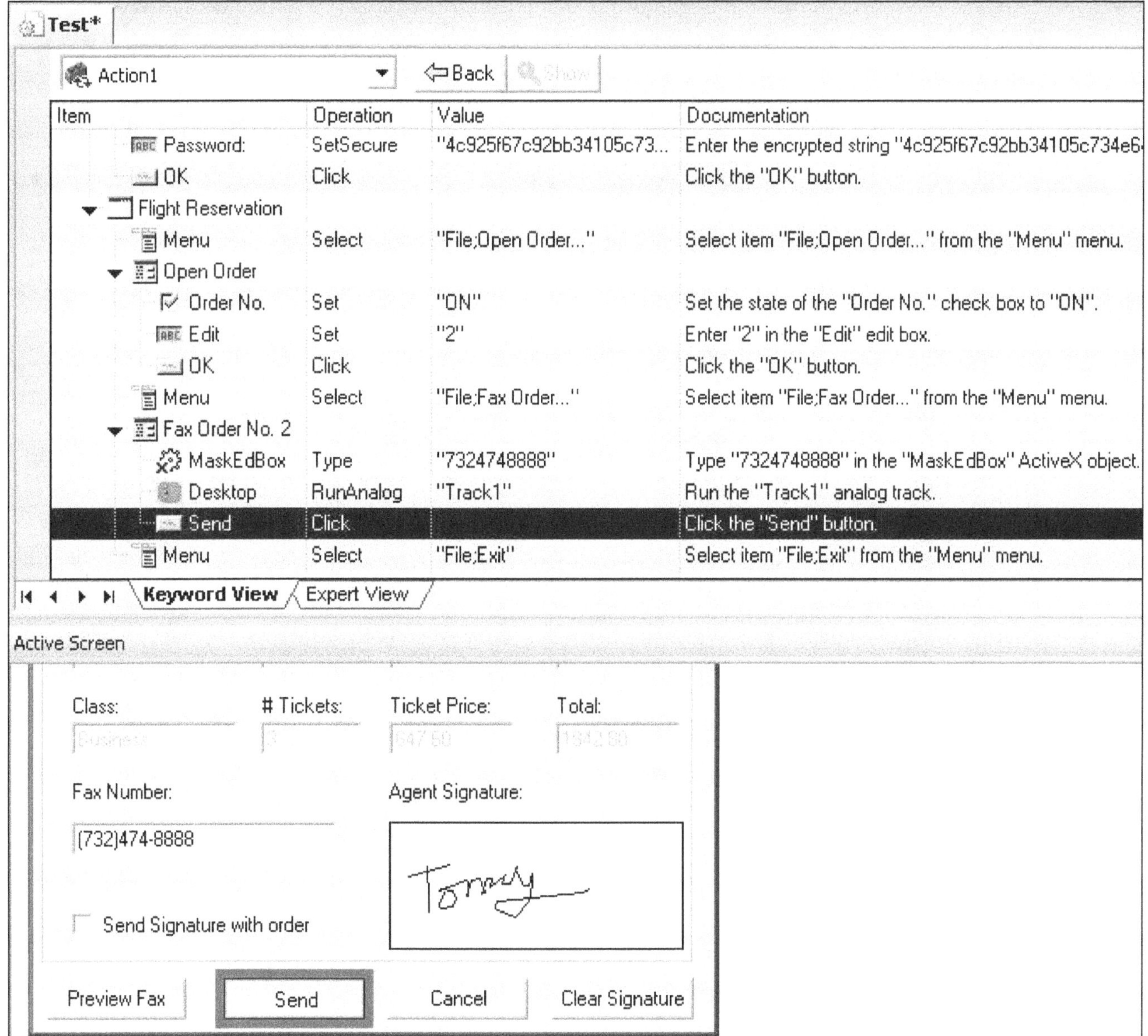

NOTE: Observe the **Desktop** icon and the step **RunAnalog "Track 1"**

The script may not work if run from another machine. (Due to O/S, Position of the window)

21. Playback & Review the results → Close the **Test Results** window

*** **END OF LAB 1D** ***

DIRECTORIES CREATED

1. HP QuickTest creates a ***folder*** for each saved *script* (test)
2. The following is a sample directory for a HP QuickTest *script* file called : **MYTEST**

MYTEST				Folder
	Action0			Folder
		Snapshots		Folder
			*.Z , *.HTML, *.INF, *.PNG	Internal file
		Resource.mtr		Internal file
		ObjectRepository.bdb		Object Repository
		Script.mts		Source
	Action1			Folder
		Snapshots		Folder
			*.Z , *.HTML, *.INF, *.PNG	Internal Zipped Files
		Resource.mtr		Internal file
		ObjectRepository.bdb		Object Repository
		Script.mts		File : Source
	Res1			Folder - Results
		Report		Folder
			Sub-Folders & files	Run-time internal files
	Mytest.usr			File: Virtual User Script
	Default.xls			File : Data Table
	Default.cfg			File : Config Settings
	Lock.lck			Internal File
	Parameters.mtr			Internal File
	Other Files		*.DAT, *.TSP, *.USP	Internal Files

3. The **Action0** folder controls all of the other ***actions*** in the test folder

   Note: This is an internal file that should not be modified or altered !

4. The **Action1** folder contains a log of user *performed* actions on **object's** (during recording)
5. The **script.mts** is the ***script*** file that contains all the user actions
6. For every **Action*x*** folder – a **Snapshots** folder will be generated
7. Each **Action*x*** folder will contain at least two (3) files:

   a) **resource.mtr**
   b) **script.mts**
   c) **objectrepository.bdb** (Local Repository)

8. If you create multiple **Action** files - you will see additional folders & files under the parent folder

NOTE:

1. Save your *script* often or frequently
2. The above is for your information only – there is nothing you need do while testing
3. Copy the entire *script* folder to another location – not just sub-folders

HOW HP QUICKTEST IDENTIFIES OBJECTS

1. As you point and click on objects – HP QuickTest :

   a) Identifies the objects *parent* window

2. Identifies the object type or by the ***class*** it belongs to

   a) Native Class or (MFC – Microsoft Foundation Class)
   b) HP QuickTest Assigned Class

Object	Property	HP Quicktest Class
Edit Field	Nativeclass : Edit	WinEdit
Radio Button	Nativeclass : Button	WinRadioButton
Drop Down List	Nativeclass : Combo Box	WinComboBox

3. Attempts to learn the objects properties (example):

Object Property	Value
Focused	False
Height	32
Width	24
Attached Text	"First Name"
Visible	True
Enabled	False
Text	"James"
State	ON

- Object Property: A set of characteristics that define the *objects* identity
- Each object type will have its own unique characteristics

4. Assigns a logical name to the *learned* object

5. The logical name is the *name* referred to in the script

OBJECT MODEL

1. HP QuickTest Pro records your test in **VBScript** , which is Microsoft's Web scripting language

2. Your test script is a combination of standard **VBScript** statements and statements that use a tests:

   Objects, **Methods** and **Properties**

3. A *test object* is an object that HP QuickTest Pro uses to represent an object in your application

4. An operation *performed* on an object is a ***method***

5. Each test object has one or more methods with which it is associated

6. Each test object also has a number of *properties* that describe the object

7. The *object* is stored in the Object Repository

OBJECT REPOSITORY

1. An object's property is *stored* in the **Object Repository** with a file extension of **.BDB**

2. Each saved test has an **Object Repository** that is saved in the tests *sub-folder* as an internal file

3. **Test Object** (Design Time)

   a) Captured during recording (while performing an *action*)
   b) Each object has its own unique properties & typically the value of those properties will NOT change

4. **Run-time Object** (Execution)

   a) References all the objects in the AUT (*memory*)
   b) Typically – the ***test*** and ***run-time*** object properties & values are identical
   c) It is possible in the *next* build – the AUT physical properties may change or be different
   d) Based on input values for a particular business rules – the objects *property*, *state* and *values* may change

5. **Object Repository** (Resources → Object Repository)

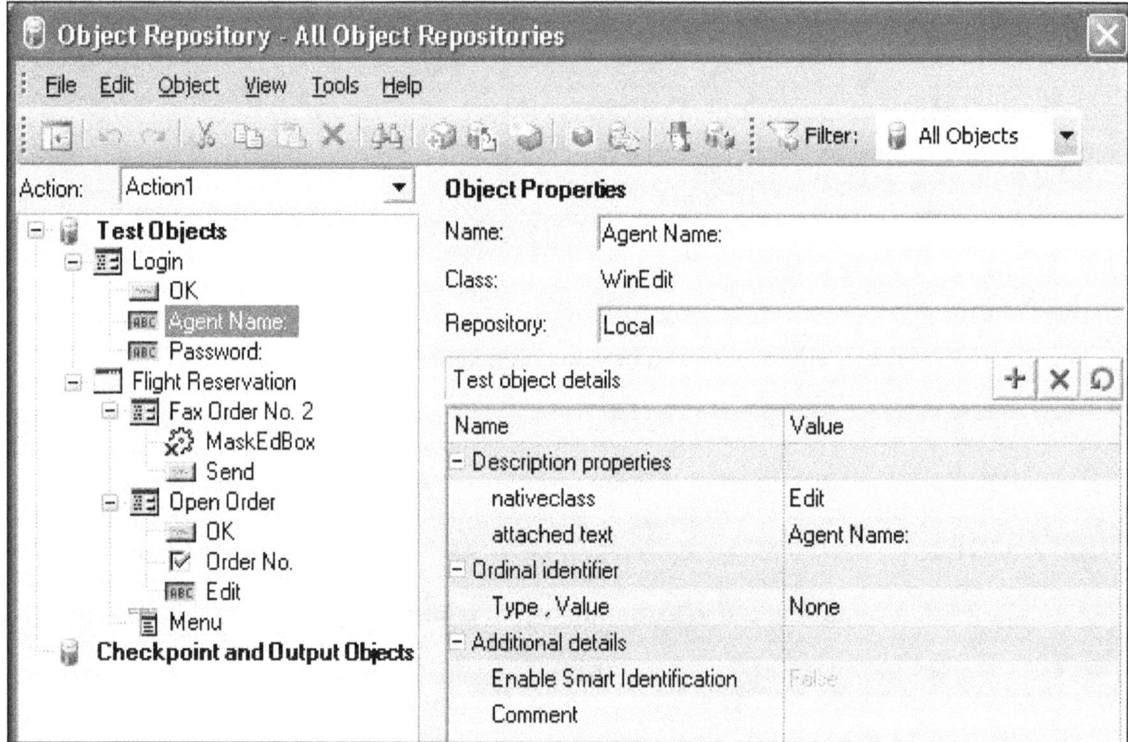

   a) Expand each *window* to see all objects within the parent window
   b) Highlight each *object* and review its properties & value (right pane)

   - Name : **Agent Name:** (Logical Name referenced in your script)
   - Class : **WinEdit** (HP QuickTest Assigned Class)
   - Repository : **Local** (Location of the Object Repository on your hard disk)

     Where is it located : *Folder / Sub-Folder* ? ______________________________

OBJECT REPOSITORY MODE

Action Object Repository (*Default Setting*)

1. Each ***Action*** folder has its own **Object Repository** located in the folder where the file is saved (*Action1*)
2. It is configured or *assumed* to be ***Local*** Repository
3. You can have a separate ***action*** repository for *each* Action file
4. Within each action repository – are **kept/stored** the objects associated with your test

Shared Object Repository

1. You can store all the objects in your ***test*** in a common (shared) object repository file
2. Multiple **tests** can reference the shared repository and it is configured as ***Shared***
3. The repository files are stored as: ***testname*.tsr**

Why use a Shared Object Repository ?

1. Add Objects to an existing Repository

   a) Learn ***new*** objects from an application
   b) During recording only the **objects** you touch are stored in the repository. Not all objects on a window

2. Create new objects that do not yet exist and later update their properties
3. Ensure the logical names (in the repository) are easy for other QA Testers to follow & understand
4. **Copy** or ***Move*** objects from one repository to another
5. Merge and Manage objects thru **Object Repository Manager** :

   a) in a local repository to a common or shared repository
   b) Merge two or more repositories into a single repository

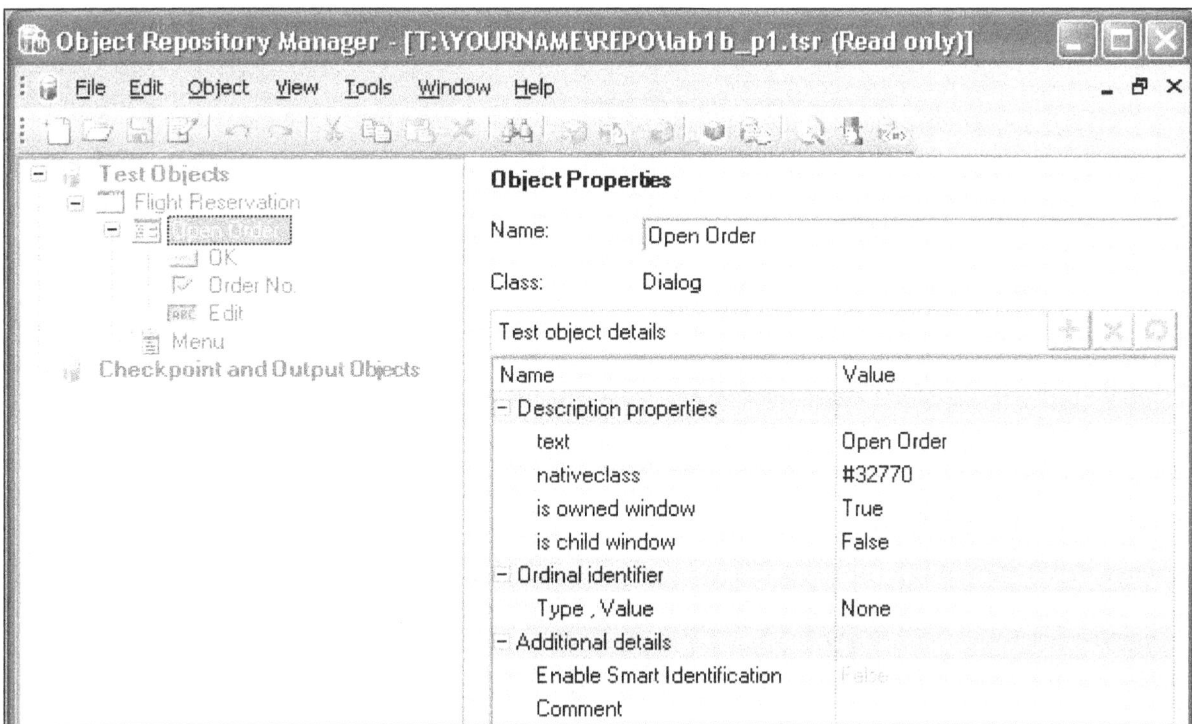

What is the file name of *this* repository ? ____________________

OBJECT SPY

1. A tool to determine & Inspect the *properties* of an **object (AUT or Active Screen)** by pointing to *that* object
   a) All Known Properties recognized by QTP
   b) Values
   c) Description
2. View the available ***methods*** (functions) associated with both :
   a) Test Object
   b) Run-time : available for : *ActiveX , DOTNET (.NET) , Java, Web*
3. Displays an individual Object (pointed to by the hand)

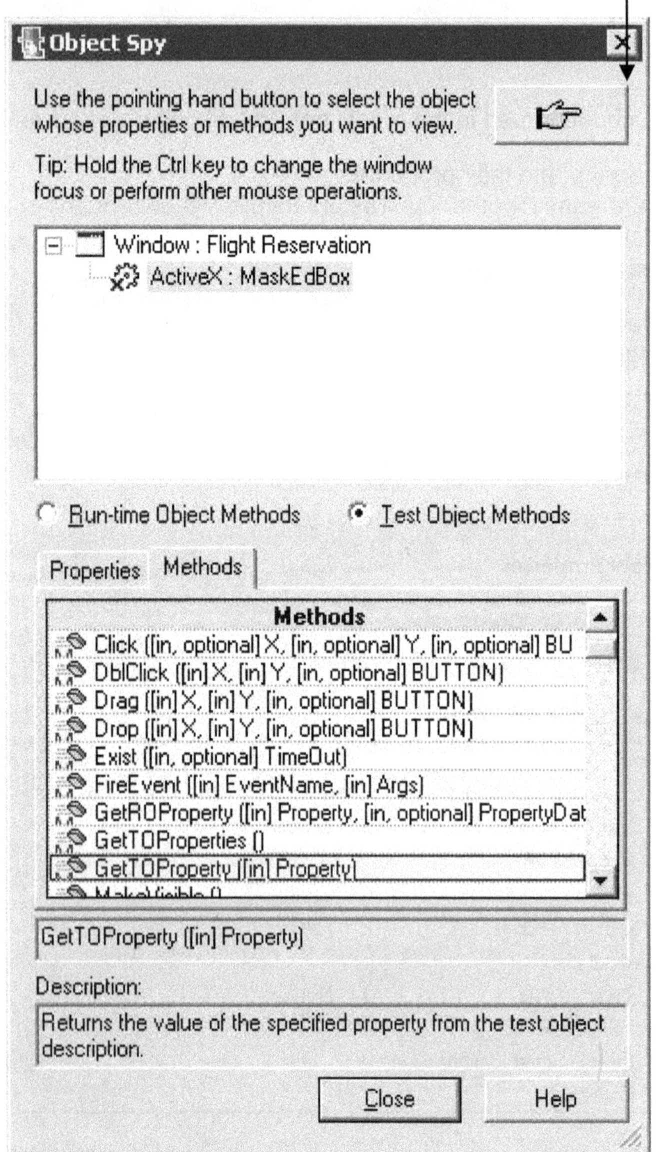

Two (2) Tabs:

<u>Properties</u>: Attributes or Characteristics

<u>Methods</u> : Functions recognized by QTP and available to use in your script

4. What <u>method</u> is highlighted ?

LAB 2A : Object Repository

Requirements:

1. Reviewing an objects *properties* in the Object Repository
2. Delete an object from the Object Repository
3. Add an **object** to the Object Repository thru the Repository feature

PART I

Steps:

1. Open **C:\YOURNAME\QTP10\LVL1\LAB1C**
2. Save As : **C:\YOURNAME\QTP10\LVL1\LAB2A**
3. Playback the Script → Review the Results → It Should Pass → Close the **Test Results** window
4. **Resources** (menu bar) → **Object Repository …**
5. Expand **Flight Reservation** (in the left pane) → Highlight **Insert Order**
6. Right Click → **Delete …** → Click **Yes** (to confirm "***Insert Order*** "Test Object)
7. **File** → **Close** (Close the Object Repository Window)
8. **File** → **Save** (Save the QTP Script)
9. Playback the Script → It Should Fail - you will see the following window → Click **Stop**

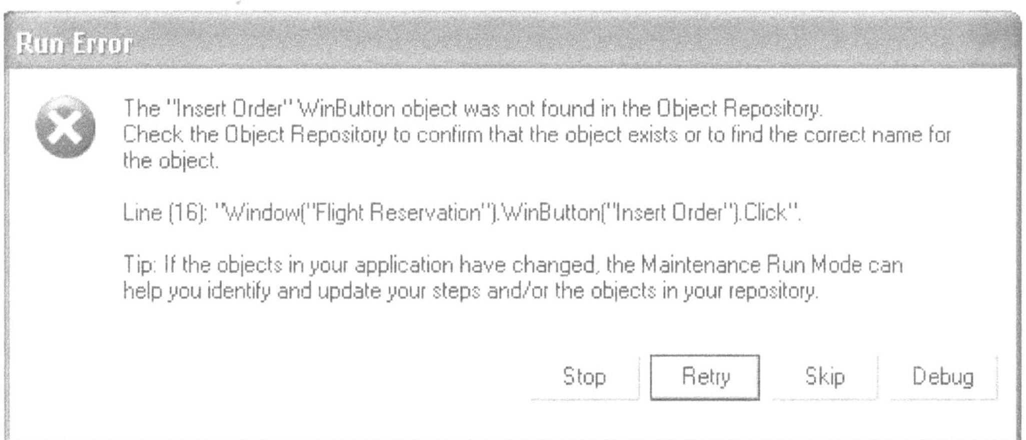

10. In the **Test Results** window → **View** (menu bar) → **Expand All**
11. Identify the ***error*** and line number → Close the **Test Results** window → Leave **AUT** open
12. Swap to HP QuickTest → **Resources** (menu bar) → **Object Repository …**
13. **Object** (menu bar) → **Add Objects to Local** → cursor changes to a hand → Point & Click **Insert Order** button
14. Click **OK** (To Accept ***WinButton : Insert Order*** in **Object Selection – Add to Repository** window)
15. **File** → **Close** (Close the *Object Repository* Window)

16. Close the **AUT**
17. Save **LAB2A**
18. Playback the Test → Review the *Test Results*
19. Close the ***Test Results*** window

***** END OF LAB 2A - PART I *****

LAB 2A - PART II (ACTIVE SCREEN)

Requirements

1. Use the existing ***LAB1C*** (not LAB 2A)
2. Delete two different objects from the Object Repository
3. Add Objects to the Repository thru the *Active Screen* (not from the AUT)
4. Test will Fail !

STEPS:

1. **File** → **Open** → **LAB1C**
2. **File** → **Save As ...** → ***LAB2A_P2***
3. **Resources** (menu bar) → **Object Repository** ...
4. Expand **Flight Reservation** under ***Action1*** (left pane)
5. Highlight **Fly To**
6. Right Click → **Delete ...** → Click **Yes** (to confirm "Delete the Test Object")
7. Expand **Login** window (left pane) → Highlight **OK** button
8. Right Click → **Delete ...** → Click **Yes** (to confirm "Delete the Test Object")
9. **File** (menu bar) → **Close** (Close the ***Object Repository*** Window)
10. **File** → **Save** (Save the *test*)

11. Playback → It should fail with the following **Run Error** message

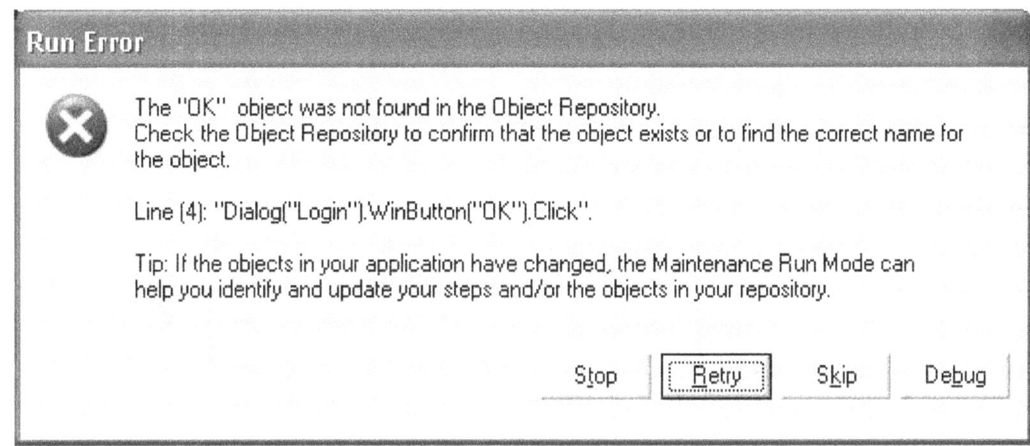

12. Click **Stop** (Stop the Playback)
13. Review the **Test Results** → Identify the error
14. Close the **Test Results** window
15. Close the AUT
16. Highlight **Login** (in Keyword View) → The Active Screen should display the **Login** window
17. In the **Active Screen** scroll & move the mouse over the **OK** button
18. Right Click → Select **View/Add Object …**
19. Verify the object is the **OK** button (**WinButton: OK**) → Click **OK**
20. Click **Add to Repository** (Note: Button label changes to **View in Repository**)
21. Click **OK** (Close the **Object Properties** window)
22. **File** → **Save** (Save the *test*)
23. Playback → It should fail with the following **Run Error** message :

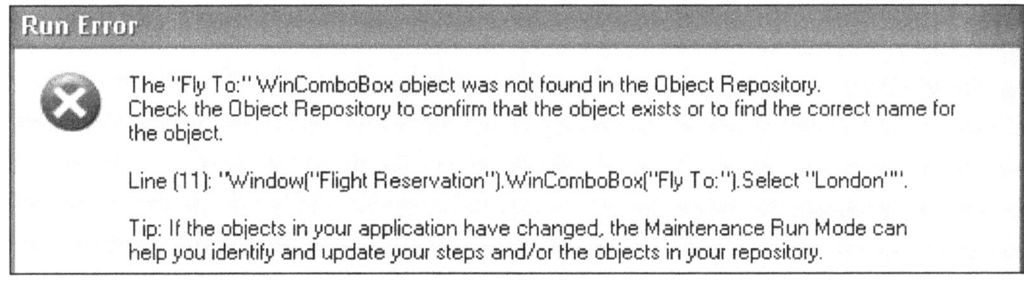

24. Click **Stop** (Stop the Playback) → Close the AUT
25. Review the **Test Results** → Identify the error
26. Close the **Test Results** window

27. Expand ***Flight Reservation*** → Highlight **? Fly To** → The Active Screen displays the **Main** window

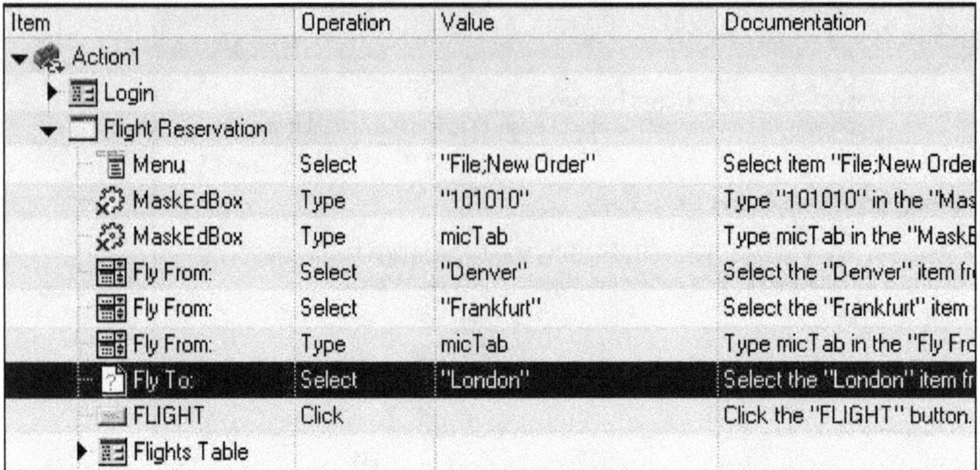

28. In the **Active Screen** move the mouse over the **Fly To** button → Right Click → **View/Add Object …**

29. Verify the object is the **ComboBox** button (**WinComboBox: Fly To:**) → Click **OK**

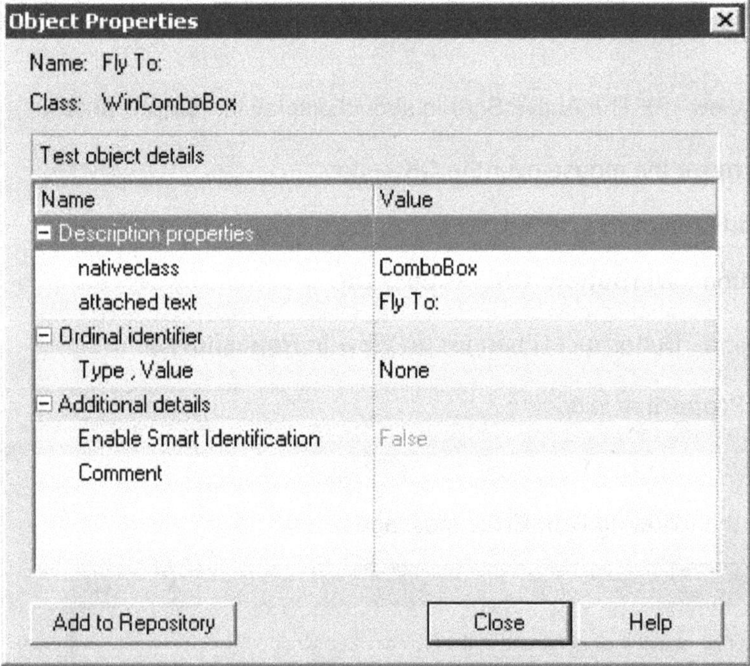

30. Click **Add to Repository** (Note: Button label changes to **View in Repository**)

31. Click **OK** (Close the **Object Properties** window)

32. To Refresh the ***Fly To*** object: Highlight any other object and it will refresh

33. Save **LAB2A_P2** → Close the **Login** window (AUT) if it is opened !

34. Playback, Verify the results → It should Pass ! → Close the **Test Results** window when finished

*** **END OF LAB 2A** ***

LAB 2B : LEARNING OBJECTS

Requirements

1. Learn **<u>All</u>** *objects* in the parent window by ***Adding*** to the Local Object Repository
2. Two Methods : (1) AUT (2) Active Screen : **WinMenu** may <u>not</u> be available in *Active Screen*

Steps (Part 1) : AUT

1. Save ***LAB2A*** as **LAB2B_P1**
2. **Resources** (menu bar) → **Object Repository…**
3. Expand **Flight Reservation** (left pane) → Highlight **Flight Reservation**
4. Right-Click → **Delete …** → Click **Yes** (to confirm deletion of ***Flight Reservation*** object)
5. **File** (menu bar) → **Close** (close **Object Repository** window)
6. **File** (menu bar) → **Save** (save the script *LAB2B_P1*)
7. **File** (menu bar) → **Open** → **Test** → **LAB2B_P1**
8. Click **Yes** (to revert to the saved script)
9. Expand **Login** → Expand **Flight Reservation** → Your script maybe similar to :

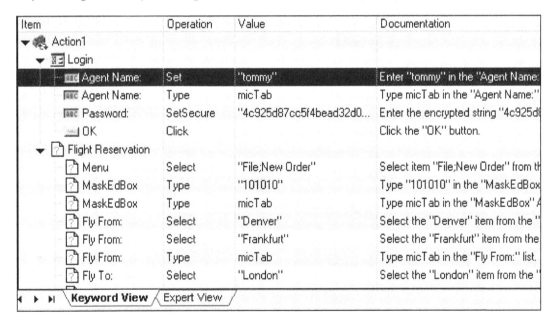

Item	Operation	Value	Documentation
Action1			
Login			
Agent Name:	Set	"tommy"	Enter "tommy" in the "Agent Name:
Agent Name:	Type	micTab	Type micTab in the "Agent Name:"
Password:	SetSecure	"4c925d87cc5f4bead32d0...	Enter the encrypted string "4c925d
OK	Click		Click the "OK" button.
Flight Reservation			
Menu	Select	"File;New Order"	Select item "File;New Order" from th
MaskEdBox	Type	"101010"	Type "101010" in the "MaskEdBox
MaskEdBox	Type	micTab	Type micTab in the "MaskEdBox" A
Fly From:	Select	"Denver"	Select the "Denver" item from the "
Fly From:	Select	"Frankfurt"	Select the "Frankfurt" item from the
Fly From:	Type	micTab	Type micTab in the "Fly From:" list.
Fly To:	Select	"London"	Select the "London" item from the "

Keyword View | Expert View

a) Why does the Active Screen **NOT** display the ***Flight Reservation*** window ?

b) Why does the **KeyWord View** show a Question Mark (**?**)

10. **Start** → **Programs** → **QuickTest Professional** → **Sample Applications** → **Flight**
11. Login with ***yourname*** and use the password ***mercury*** → Click **OK**

12. Wait for the **Flight Reservation** window

13. Swap to *HP Quicktest* → **Resources** (menu bar) → **Object Repository**

14. **Object** (menu bar) → **Add Objects to Local ...**

15. Point & Click on the title bar (in the AUT) on **Flight Reservation**

16. Verify the object is the ***Window*** (**Window: Flight Reservation:**) → Click **OK**

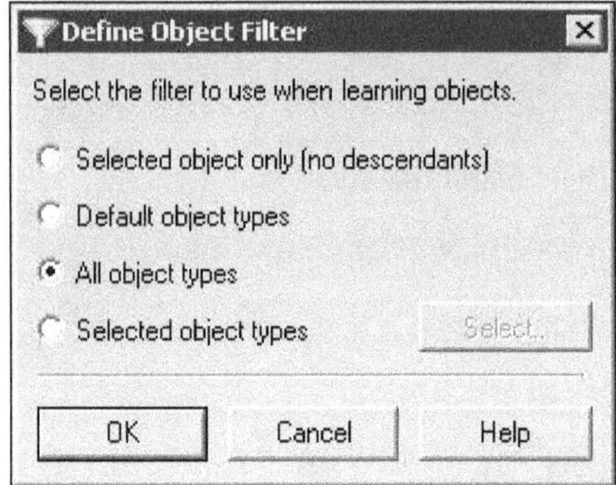

17. Check **All object types** (radio button) → Click **OK**

18. Review the Objects in the Repository → All objects are stored

19. **File** → **Close** (Close the **Object Repository** window)

20. Click **Back** button → Double-Click ***Action1***

21. **File** → **Save** (save **LAB2B_P1)**

22. Close the AUT

23. Playback → It will Fail (***Flights Table*** window was not found) → Click **Stop**

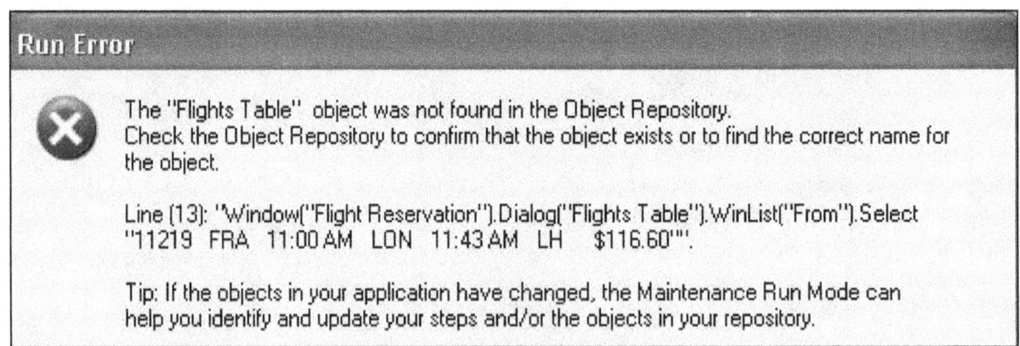

24. Review the **Test Results** → Close the *Test Results* window

25. Verify the AUT is running with The **Flights Table** window open
26. Swap to HP Quicktest → **Resources** (menu bar) → **Object Repository** ...
27. **Object** (menu bar) → **Add Objects to Local ...**
28. Point & Click on the title bar (in the AUT) on **Flights Table**
29. Verify the object is the **Window: Flight Reservation / Dialog Flight Tables:** → Click **OK**
30. Check **All object types** (radio button) → Click **OK**
31. Review the Objects in the Repository → All objects are stored
32. **File → Close** (Close the **Object Repository** window)
33. **File → Save** (save **LAB2B_P1)**
34. Close the AUT (Click **Cancel** → **File** → **Exit**)
35. Playback → It will Fail (***Flight Reservations*** Dialog window is not found) → Click **Stop**
36. Review the **Test Results** → Close the *Test Results* window
37. Verify the AUT is running with The **Flight Reservations** window open
38. Swap to HP Quicktest → **Resources** (menu bar) → **Object Repository** ...
39. **Object** (menu bar) → **Add Objects to Local ...**
40. Point & Click on the title bar (in the AUT) on **Flights Reservations**
41. Verify the object is the **Window: Flight Reservation / Dialog Flight Reservations:** → Click **OK**
42. Check **All object types** (radio button) → Click **OK**
43. Review the Objects in the Repository → All objects are stored
44. **File → Close** (Close the **Object Repository** window)
45. **File → Save** (save **LAB2B_P1)**
46. Swap to the AUT → Click **Yes** (to confirm delete) → **File** → **Exit**
47. Playback → If should Pass ! → Review the **Test Results** → Close the *Test Results* window

ON YOUR OWN: (Optional)

1. Save the **LAB2B_P1** as **LAB2B_OPT1**
2. Delete the **Login** window from the object repository → Save the script (**LAB2B_OPT1)**
3. Start AUT → Add all the objects from the **Login** window to the local object repository
4. Close the AUT (if running) → Playback & Verify the results

***** END OF LAB 2B : Part 1 *****

LAB 2B : WinMenu *Object* Problem (If it is not recorded in the script) OPTIONAL

Steps (Part 2) : Active Screen

1. Save ***LAB2A*** as **LAB2B_P2**
2. **Resources** (menu bar) → **Object Repository...**
3. Expand **Flight Reservation** (left pane) → Highlight **Flight Reservation**
4. Right-Click → **Delete ...** → Click **Yes** (to confirm deletion of ***Flight Reservation*** object)
5. **File** (menu bar) → **Close** (close **Object Repository** window)
6. **File** (menu bar) → **Save** (save the script)
7. In the KeyWord View - expand & Highlight **Flight Reservation** → No *window* is shown in Active Screen
8. Highlight **Menu Select "File;New Order"** → No *window* is shown in Active Screen

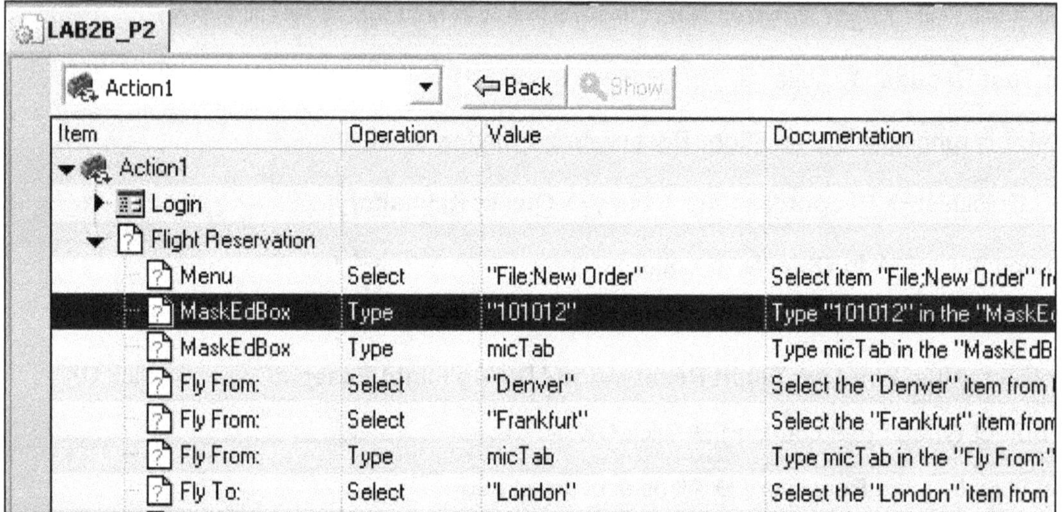

9. Highlight **MaskEdBox** (the first one) → The ***Flight Reservation*** is shown in Active Screen
10. In the Active Screen - move the cursor over the **Flight Reservation** (title bar) → Right –Click
11. Choose **View / Add Object ...** → The **Object Properties** window is displayed
12. Click **Add to Repository** → Click **OK**
13. Click **View Repository** → Verify the object was added
    a) Only the **Window** *Definition* was added to the Repository
    b) No other objects were added
    c) QTP will recognize only a **single object** from the Active Screen
14. **File** → **Close** (Close the *Object Repository* window)

Add the following other objects by performing the following actions for each step:

Note: Ignore any **menu selection** actions

a) Highlight the ***Action*** or Step in the Keyword View (Typically displayed as a question mark (**?**)
b) Move the mouse over the object in the Active Screen
c) Right-Click → **View / Add Objects …**
d) Verify the desired object is visible on the screen → Click **OK**
e) Click **Add to Repository**
f) Click **OK** (to close the Object Repository window)

Logical Name	Window	Description
Fly From :	Flight Reservation	Win Combo Box
Fly To:	Flight Reservation	Win Combo Box
Flight	Flight Reservation	Win Button
Flights Table	Flight Reservation	Dialog Window
From	Flights Table	WinList
OK	Flights Table	Win Button
Name:	Flight Reservation	Win Edit
Insert Order	Flight Reservation	Win Button
Delete Order	Flight Reservation	Win Button
Flight Reservations	Flight Reservation	Dialog
Yes	Flight Reservations	Win Button

15. **File** → **Save** (save the script)

16. Playback (Run) → It Should Fail with the following error !

The "Menu" object was not found in the Object Repository.
Check the Object Repository to confirm that the object exists or to find the correct name for the object.

Line (5): "Window("Flight Reservation").WinMenu("Menu").Select "File;New Order"".

Tip: If the objects in your application have changed, the Maintenance Run Mode can help you identify and update your steps and/or the objects in your repository.

17. Click **Stop** → Review **Test Results** → Close **Test Results** window

18. **Resources** → **Object Repository** → Expand **Flight Reservation**

a) Observe the **Menu** object is missing or not in the *Local Repository*

b) What is the **Window** that the *Menu* object belongs to ? ______________________________

The AUT should be opened or *visible*. DO NOT CLOSE the AUT yet !

Explanation :

c) The **WinMenu** object is not in the Local Repository

d) You must add the single object (***WinMenu***) by referencing the *Parent* window that it belongs to.

19. **Object** (*Object Repository* window) → **Add Objects to Local ...**
20. Point and Click on the **Flight Reservation** (title bar)
21. Click **OK** (to verify object ***Window : Flight Reservation***)
22. Choose **All object types** → Click **OK** :

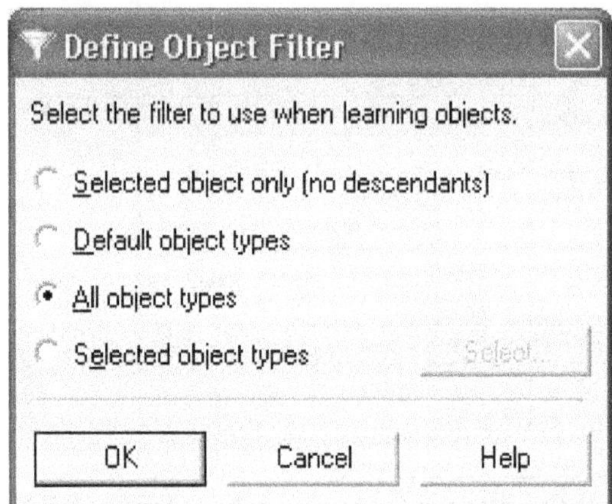

23. **File** → **Close** (close the **Object Repository** window)
24. **File** → **Save** (save the script)
25. Close the AUT (**File** → **Exit**)
26. Playback (Run) → It Should Pass !
27. Review the **Test Results**
28. Close the **Test Results** window

***** END OF LAB 2B : Part 2 (OPTIONAL) *****

LAB 2C : CREATING OBJECTS in the Repository

Requirements: AUT Flight Reservation

1. Prerequisite : **LAB2A** is working with no errors !
2. Create **objects** in the Object Repository **without** recording
3. **WinMenu** may not always be recorded or available
4. This lab (LAB2C) does not have a QTP script !

   Note:

   a) You cannot add **WinMenu** object directly to an object repository using **Add Objects to Local** button
   b) To add a ***WinMenu*** object to the object repository - you can :

   - **Add Objects** or **Add Objects to Local** button to add its parent object
   - Then select to add the parent object together with its descendants or
   - You can record a step on a **WinMenu** object and then delete the recorded step

5. Use the ***new*** Object Repository in another QTP script (it will **fail** !)

Steps: (Part 1) – Create the Object Repository containing all the objects

1. **File → New → Test**
2. **Start → Programs → QuickTest Professional → Sample Applications → Flight →** Do NOT Login
3. Realign or reposition QTP window & AUT side by side
4. **Resources → Object Repository**
5. **Object → Add Objects to Local... →** Point and Click on the **Login** title bar
6. Verify object: **Dialog : Login →** Click **OK**
7. Choose **All object types →** Click **OK**
8. Login into the AUT with *yourname* , *password* → Click **OK**
9. **Object → Add Objects to Local... →** Point and Click on the **Flight Reservation** title bar
10. Verify object: **Window : Flight Reservation →** Click **OK**
11. Choose **All object types →** Click **OK**
12. **File → Close** (close the **Object Repository** window)
13. **File → Exit** (close the AUT)
14. **File → Save → LAB2C_REPO_ONLY** (Script is empty)

Question :

a) What folder is the **Object Repository** kept ? ______________________________

b) What is the file name of the **Object Repository** ? ______________________________

STEPS: (Part 2) – Use the New Object Repository with an existing script

- Modify script ***LAB2A***
- Rename the existing Object Repository
- Copy the new Object Repository to the existing script

1. **Start → Run…** → Type **CMD** → Click **OK**
2. **C: < enter >**
3. **CD \YOURNAME**
4. **CD QTP10\LVL1\LAB2A**
5. **CD ACTION1**
6. **DIR** *** Verify the **ObjectRepository.bdb** file exists
7. **REN ObjectRepository.bdb ObjectRepository_orig.bdb**
8. **DIR**
9. **COPY .. \ .. \LAB2C_REPO_ONLY\ ACTION1 \ *. BDB**
10. **DIR**
11. **EXIT** # Exit or *close* the DOS window
12. **File → Open → Test → C:\YOURNAME\QTP10\LVL1\LAB2A** → Click **Open**
13. Playback the Test → It should Fail with following message :

Run Error

The "Flights Table" Dialog object was not found in the Object Repository.
Check the Object Repository to confirm that the object exists or to find the correct name for the object.

14. Click **Stop** → Review the **Test Results**
15. Close the **Test Results** window
16. Close the **AUT** (Flight Reservation)

Question:

a) Why was QTP not able to find the object **Flights Table** ?

b) Review the **Object Repository** (**Resources → Object Repository**)

c) How would you fix the problem ?

17. **File → Open → Test → C:\YOURNAME\QTP10\LVL1\LAB2C_REPO_ONLY** → Click **Open**

18. Start the AUT → Login → **File** → **New Order …** → Enter data for the following fields (only) :

    a) Date of Flight
    b) Flight From
    c) Flight To

19. Click **Flights** button → Swap to QTP

20. **Resources** → **Object Repository** → **Object** → **Add Objects to Local …** → cursor changes to a hand

21. Point and Click on title bar **Flights Table** → Verify **Dialog : Flights Table** Object → Click **OK**

22. Choose **All object types** → Click **OK**

23. **File** → **Close** → (close the Object Repository window)

24. Highlight any row → Click **OK** (close **Flights Table** window)

25. Enter *Name* → Click **Insert Order** → Wait for confirmation

26. Click **Delete Order** → Swap to QTP

27. **Resources** → **Object Repository** → **Object** → **Add Objects to Local …** → cursor changes to a hand

28. Point & Click on title bar **Flight Reservations** → Verify **Dialog : Flight Reservations** Object → Click **OK**

29. Choose **All object types** → Click **OK**

30. **File** → **Close** → (close the Object Repository window)

31. Click **Yes** (confirm the **Delete**) → **File** → **Exit** (close AUT)

32. **File** → **Save** (***LAB2C_REPO_ONLY***)

33. **Start** → **Run**… → Type **CMD** → Click **OK**

34. **C: < enter >**

35. **CD \YOURNAME**

36. **CD QTP10\LVL1\LAB2A**

37. **CD ACTION1**

38. **COPY .. \ .. \LAB2C_REPO_ONLY\ ACTION1 \ *. BDB /Y**

39. **EXIT**

40. **File** → **Open** → **Test** → **C:\YOURNAME\QTP10\LVL1\LAB2A** → Click **Open**

41. Playback the Test → It should PASS ! → Review the **Test Results** → Close the **Test Results** window

*** **END OF LAB 2C** ***

RECORDING MODE : ANALOG & LOW-LEVEL RECORDING

1. The default recording mode is **Normal**

   a) Makes the maximum use of the Test Object Model
   b) Can easily <u>Recognize</u> object (s) in the AUT regardless of the object's location (x, y coordinates)

2. **ANALOG**

   a) Captures the *<u>exact</u>* mouse movements of the mouse & keyboard operations
   b) It is based in relation to either the screen (location) or the *specific* window
   c) It records every actions movement of the mouse as you drag the mouse through out the screen

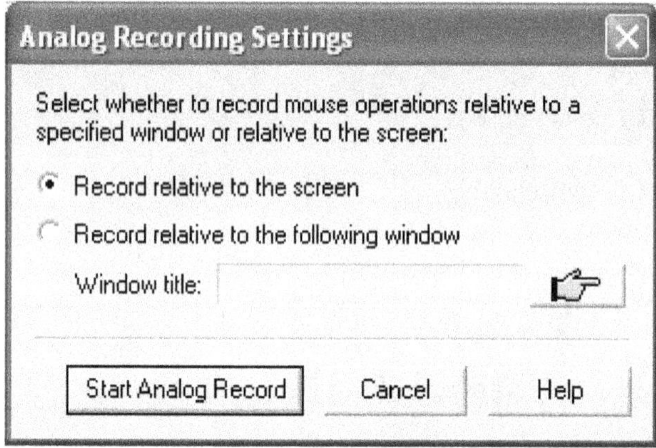

   d) The actions (steps) recorded are saved into a **separate** file in the current **Action*x*** folder

   e) The file name is **AnalogTrackList.dat** & you should ***NOT*** edit or modify the file

   f) Inside the QTP script it will add a **RunAnalog** statement only for that Step (Keyword & Expert)

   g) The **Active Screen** contains the last action (not all the movements)

3. **LOW-LEVEL**

   a) It records the object – whether or not QTP recognizes (class & properties) of an object

   b) It identifies all run-time objects as a **Window** and **WinObject** test object

   c) You can also use low-level recording if the exact coordinates (x, y) are important to the AUT

   d) If the coordinates have changed – the test will FAIL !

   e) The **Low-Level** supports the following methods :

      - Window test objects : **Click, DblClick, Drag, Drop, Type, Activate, Minimize, Restore & Maximize**
      - WinObject objects : **Click, DblClick, Drag, Drop, Type**

   f) Each Actions (or *steps*) are save & captured and can be seen in either the <u>Keyword</u> or <u>Expert</u> mode

<u>Note</u>: In *Analog* recording mode – you **<u>cannot</u>** perform any **Checkpoints** or **Validations !**

LAB 2D : Analog Recording

Test Requirements:

1. Re-Size the QTP Window to a smaller size; Close & Re-Open QTP
2. Run-Time Settings – configure to run **Record and run only on** option
3. Record all events in Analog : *Login → Open an Existing Order → Exit AUT*
4. Create a new script and save it as **LAB2D**
5. Review the script and Playback the script
6. Note : The test may fail due to a synchronization error **or** Unable to execute a step at the specified location

STEPS:

1. Close all applications (except for ***HP QuickTest***)

2. Manually: Identify the ***Order Number*** for an existing customer : Example : **Order No. 3** (on your own)

3. Resize-QTP (make it smaller) , Restart QTP and verify that it is NOT in Full Screen

4. **File** (menu bar) → **New** → **Test …**

5. **Automation** (menu bar) → **Record & Run Settings..**

6. Select **Record and run only on** radio button → Select the following:

   a) Check : Applications opened by Quick Test
   b) UnCheck : Applications opened via the DeskTop (by the Windows shell)
   c) Check : Applications specified below

7. Click **Add** (Green Plus Sign) → Point to the Path / Location (Same as LAB1C)

   Path: **C:\Program Files\HP\ QuickTest Professional\samples\flight\app\flight4a.exe**

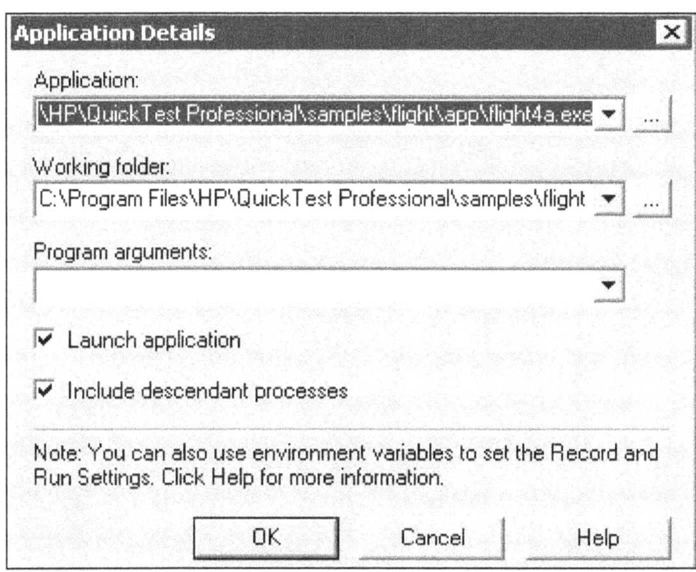

8. Click **OK** → **Automation** (menu bar) → **Record F3** → Recording Mode is *flashing*

9. **Automation** (menu bar) → **Analog Recording**

10. Choose **Record relative to the screen** → Click **Start Analog Recording**
11. Login with ***yourname*** and use the password ***mercury*** → Click **OK**

    Note: QTP will NOT record statements into the script – it is OK !!

12. **File** (menu bar) → **Open Order ...**
13. Check **Order Number** (radio button) → Type **3** → Click **OK**
14. **File** (menu bar) → **Fax Order** ... → Enter a 10 digit phone number
15. Check: ***Send Signature with order*** box
16. Move the mouse inside the signature area and the cursor changes to a Pen → Sign *your* name
17. Click **Send** (wait 20 seconds) → Returns to the main window
18. **File** → **Exit** (Close the AUT) → **Automation** (menu bar) → **STOP - F4** (Stop Recording)
19. It should be similar to :

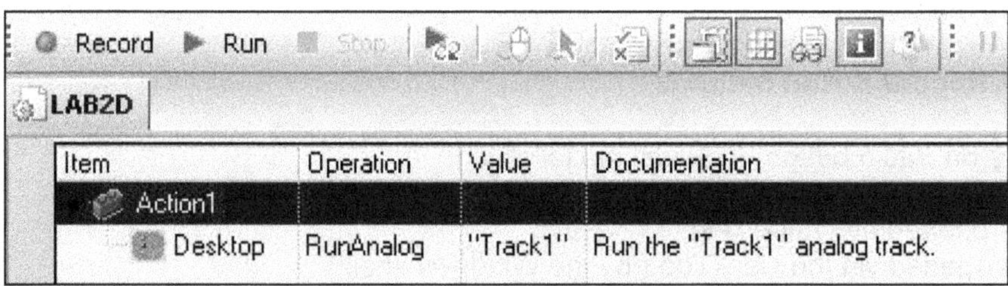

Observe :

a) All of your actions were NOT recorded in the script
b) There is only one (1) step (below *Action1*)
c) The actual movements are saved in a separate file

20. Review & Save the script as **C:\YOURNAME\QTP10\LVL1\LAB2D**
21. **Automation** (menu bar) → **Run - F5** → Check **Temporary run results folder** → Click **OK**

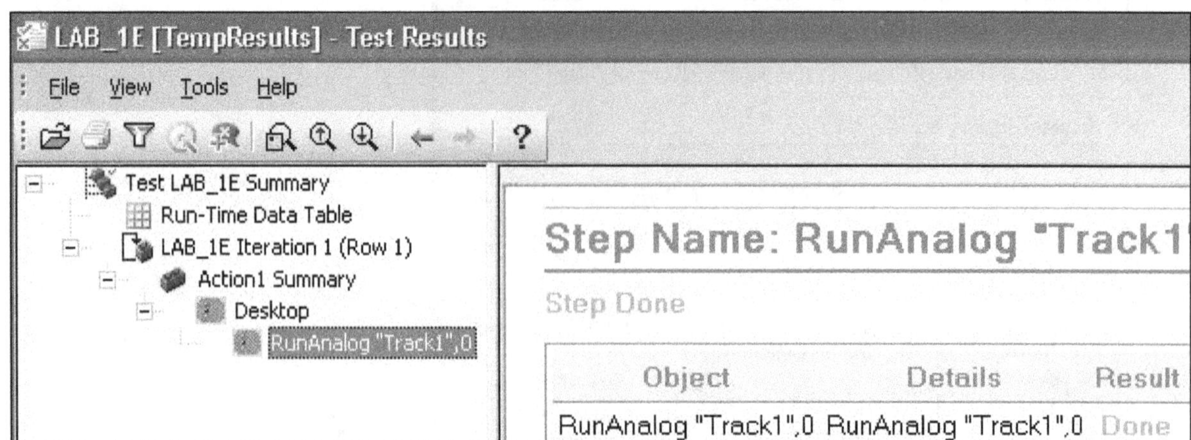

Note : The test may not execute succesfully due to synchronization issues

22. Close the **Test Results** window

23. If the AUT is still open, then :

    a) Close the AUT
    b) Increase the wait time to from 20 to 40 seconds (File → Settings… → Click *Run*)
    c) Rerun the test

24. **Resources → Object Repository** → The Object Repository is empty !!

    Why ? ____

25. **File → Close** (Close the **Object Repository** window)

Verify the **AnalogTrakList.dat** file exists

26. Open the **Windows Explorer**

    Move mouse over **Start** (bottom left of the Task Bar) → Right-Click → select **Explore**

27. Expand folder **C:\YOURNAME\QTP10\LVL1\LAB2D**

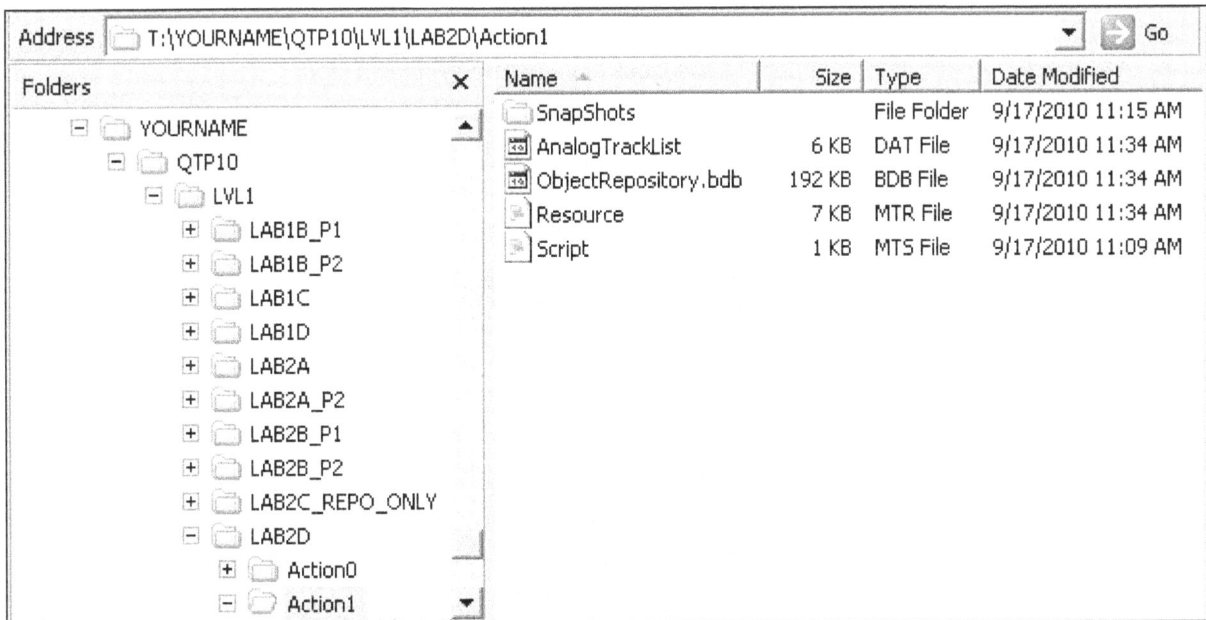

28. Expand **YOURNAME\QTP10\LVL1\ LAB2D** → Highlight **Action1**

29. Verify that **AnalogTrackList.DAT** file exists

30. Close the **Windows Explorer** when finished

*** **END OF LAB 2D** ***

LAB 2E : Low-Level Recording (It should Fail !)

Test Requirements:

1. Run-Time Settings – configure to run **Record and run only on** option
2. Record all actions in both : **Low-Level** & **Analog**
3. Login → Open an Existing Order → Exit AUT
4. Create a new script and save it as **LAB2E**
5. Review , Save & Playback the script
   Note : If in a subsequent recording – it does NOT default to *Normal* mode :
   Close QTP, Open QTP in web setting → Close QTP → Open QTP in *ActiveX* & *Visual Basic*

STEPS:

1. Identify the ***Order Number*** for an existing customer : Example - **Order No. 3**
2. Close all applications (except for ***HP QuickTest***)
3. **File** (menu bar) → **New** → **Test ...**
4. **File** (menu bar) → **Settings** → Click **Run** tab →
5. Change **Object Synchronization timeout:** to **10** seconds → Click **OK**
6. **Automation** (menu bar) → **Record F3** (***Record*** message is flashing in red)
7. Select **Record and run only on** radio button → Select the following:

   d) Check : Applications opened by Quick Test
   e) UnCheck : Applications opened via the DeskTop (by the Windows shell)
   f) Check : Applications specified below

8. Click **Add** (Green Plus Sign) → Point to the Path / Location (Same as LAB 1C)

   Path: **C:\Program Files\HP\QuickTest Professional\samples\flight\app\flight4a.exe**

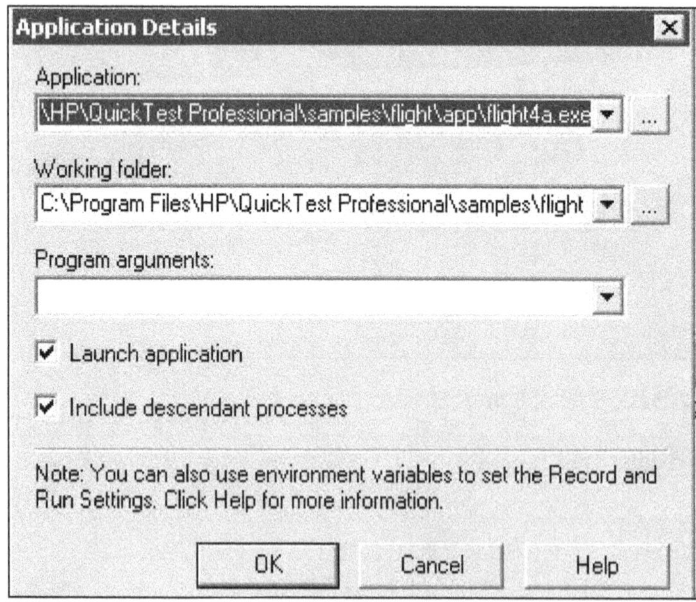

9. Click **OK**
10. **Automation** (menu bar) → **Low-Level Recording**

11. Login with ***yourname*** and use the password ***mercury*** → Click **OK**

    Note: QTP will NOT record statements into the script – it is OK !!

12. **File** (menu bar) → **Open Order ...**
13. Check **Order Number** (radio button) → Type **3** → Click **OK**
14. **File** (menu bar) → **Fax Order** ... → Enter a 10 digit phone number
15. Check: ***Send Signature with order*** box
16. **Automation** (menu bar) → **Analog Recording** → Choose **Record relative to the screen** → Click **OK**
17. Move the *mouse* inside the signature area and the cursor changes to a Pen → Sign *your* name
18. **Automation** (menu bar) → **Low-Level Recording**
19. Click **Send** (wait 20 seconds) → Returns to the main window
20. **File** → **Exit** (Close the AUT) → **Automation** (menu bar) → **STOP - F4** (Stop Recording)
21. It should be similar to the following :

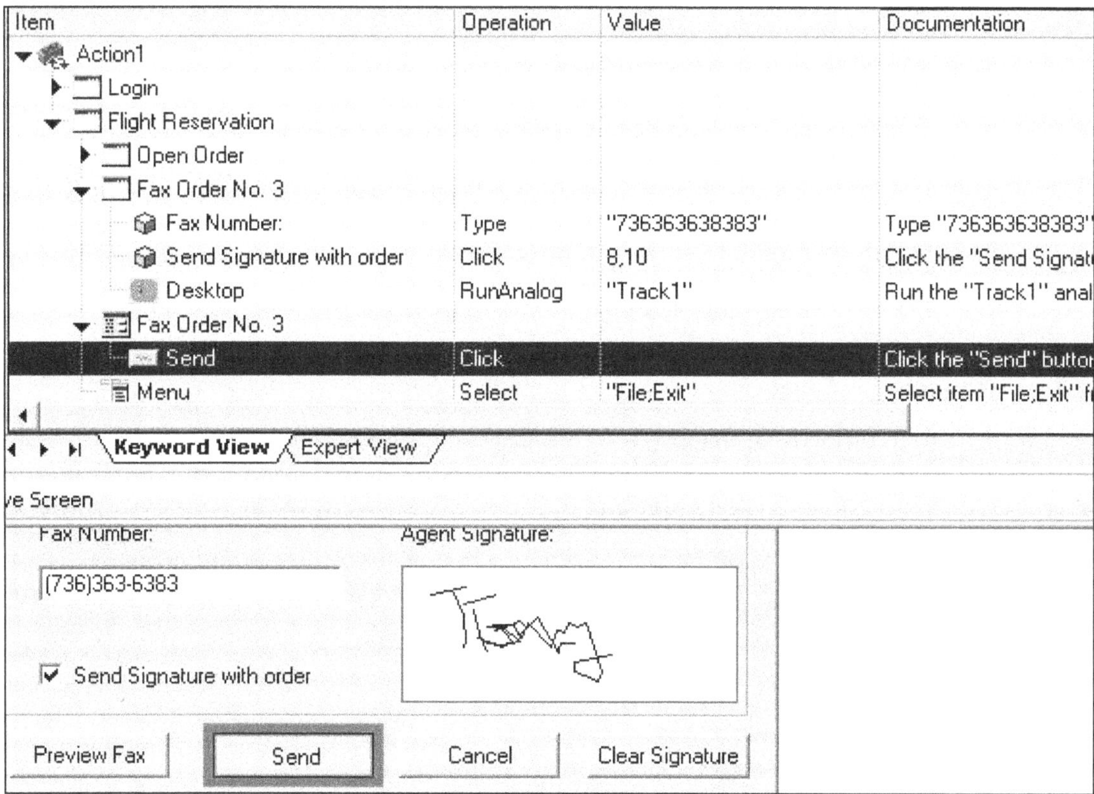

Observe :

a) All steps are recorded
b) Screens are captured in the Active Screen
c) All objects are recorded as : **Window** and **WinObjects** (Either a *logical* or *generic* name)

22. Save script as : **C:\YOURNAME\QTP10\LVL1\LAB2E**

Playback & Review the Test Run

23. **Automation** (menu bar) → **Run - F5** → Check **Temporary run results folder** → Click **OK**

24. It should fail with :

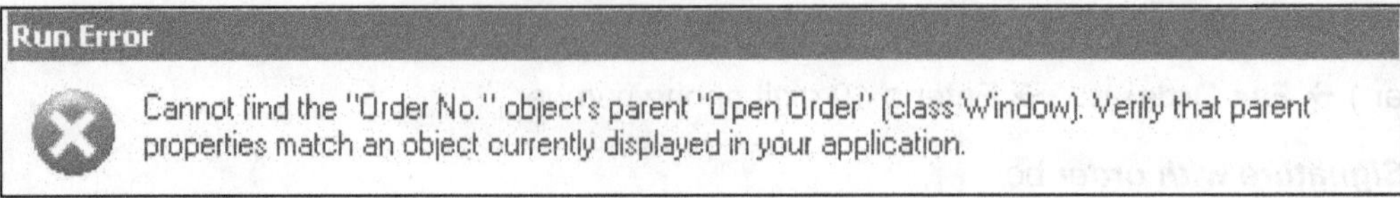
Run Error

Cannot find the "Order No." object's parent "Open Order" (class Window). Verify that parent properties match an object currently displayed in your application.

25. Identify the Line Number : ______ → Click **Stop**

26. Review the **Test Results** window → Analyze the reason as the Test Failure ?

    a) What may have happened ?

    b) What action (step) was QTP trying to execute ?

27. Close the **Test Results** window → Close the **AUT**

Problem :

a) While running the *script* one of the steps was NOT executed !

b) The **menu** action (step) did not get recorded !

c) The **Low – Level** recording mode – captures window objects.

d) Menu actions are so fast quick it cannot recognize the step

28. Start the AUT → Login → Wait for the **Flight Reservation** window

29. Go to QTP → **Resources** → **Object Repository** → **Object** → **Add Objects to local…**

30. Point & Click on the **Flight Reservation** title bar → Verify **Flight Reservation** → Click **OK**

31. Choose **Selected object types** → Click **Select…**

32. Click **Clear All** → Scroll & Choose **Menu** object

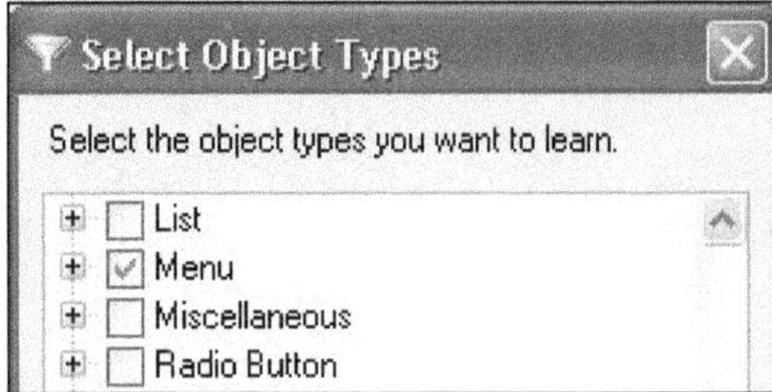

33. Click **OK** (close **Select Object Types** window)

34. Click **OK** (close **Define Object Filter** window)

35. Verify the **Object Repository** contains the **Menu** object

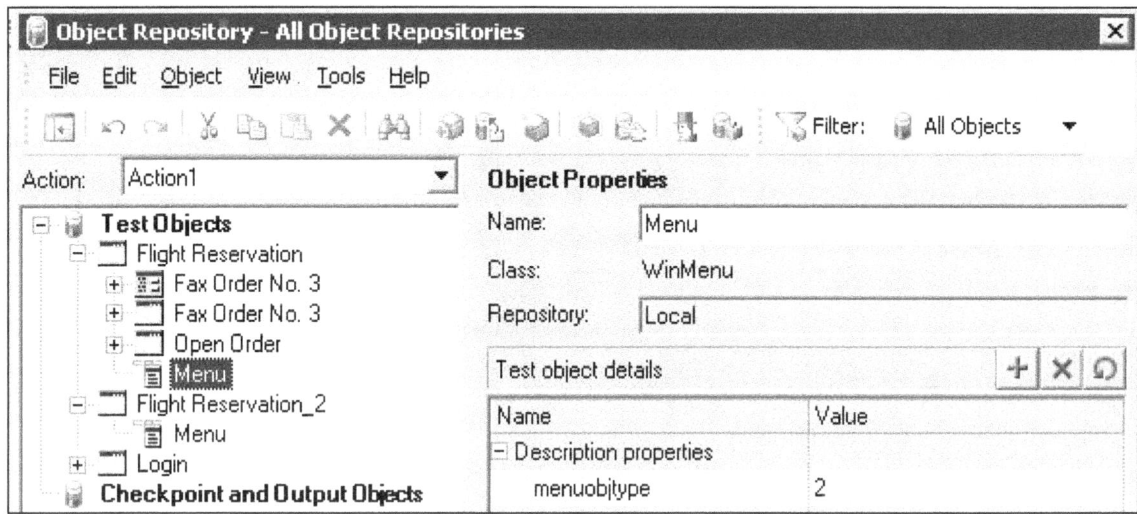

36. **File** → **Close** (close the **Object Repository** window)

37. Close the **AUT**

38. In **Keyword View** → Highlight **Flight Reservation** (in **Item** column)

39. Right-Click → **Insert Step** → **Step Generator…** → Click **Select Object** button

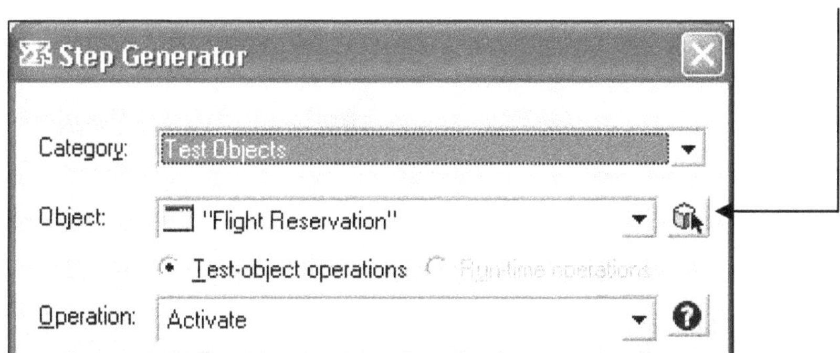

40. Highlight **Menu**

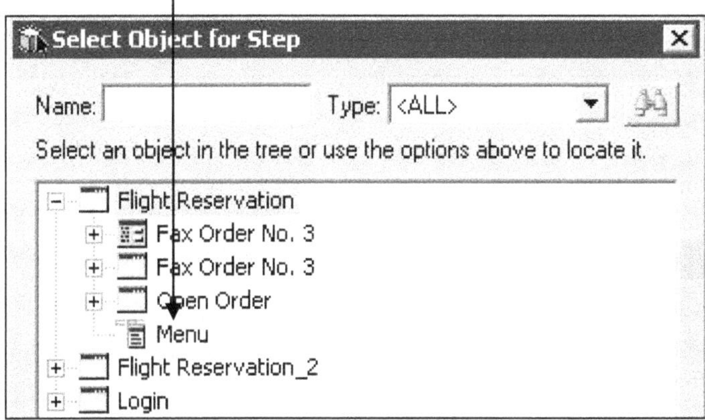

41. Enter the following information in the **Arguments** section under the Value column :

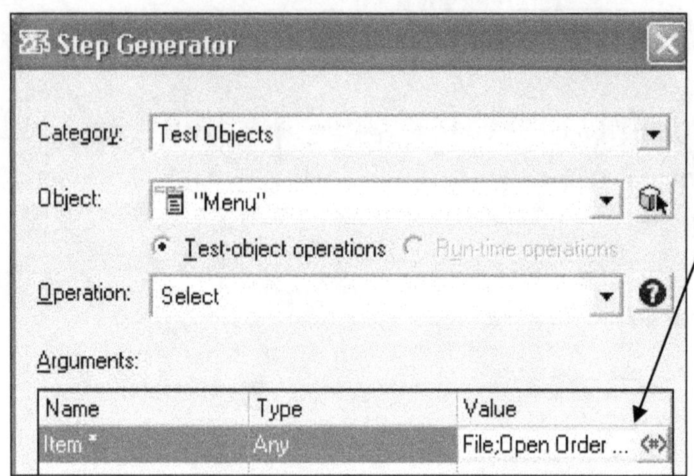

File; Open Order… (There are no spaces after **Open Order** and quotes at all !)

42. Click **OK** (close **Step Generator** window)

43. Rearrange the Steps by Dragging & Dropping the **Menu** below **Flight Reservation** window

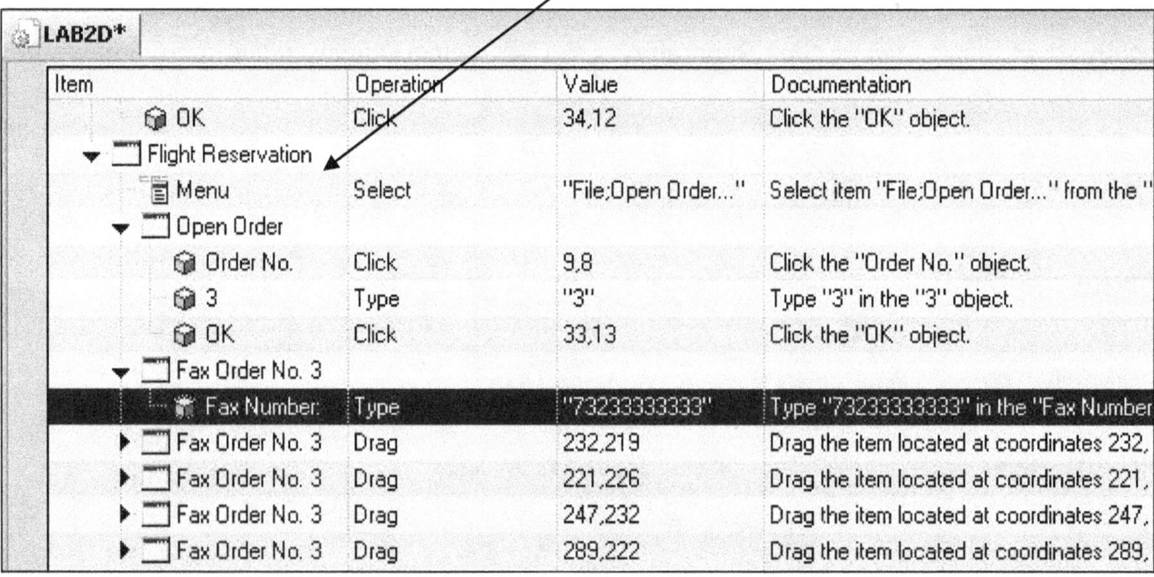

44. Close the **AUT**

45. Save **LAB2E** → Playback & it should fail with the following error :

46. Click **Details >>** → Click **Stop**

47. Review & Close the **Test Results** window

48. Close the **AUT**

How would you fix the Problem ?

a) Problem: Another step is missing ! : **File → Fax Order…**

b) Solution: Insert the missing step

49. Highlight the first **Fax Order No. 3** (under **Item**)

50. Right-Click → **Insert Step → Step Generator…**

51. Select **Menu** (in the Object: field)

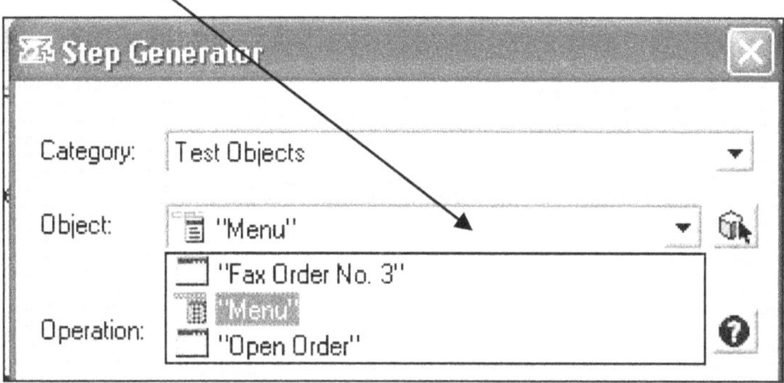

52. Type the following in the *Value* field :

File;Fax Order… (No Quotes & spaces after ***Order***)

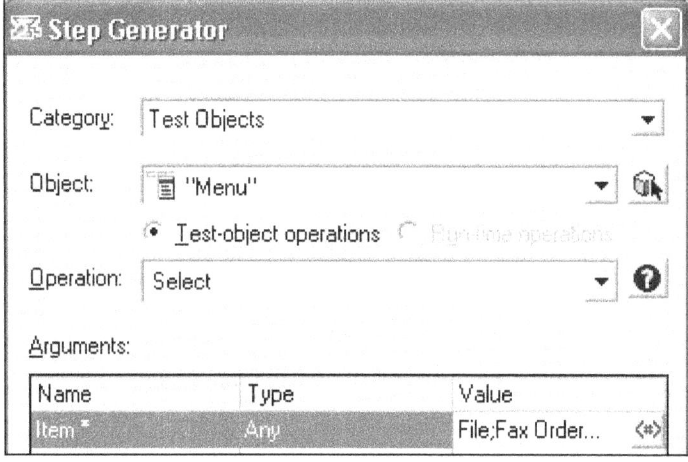

53. Click **OK** (close **Step Generator** window)

54. Rearrange the Steps by Dragging & Dropping the **Menu** above the first **Fax Order No. 3** window

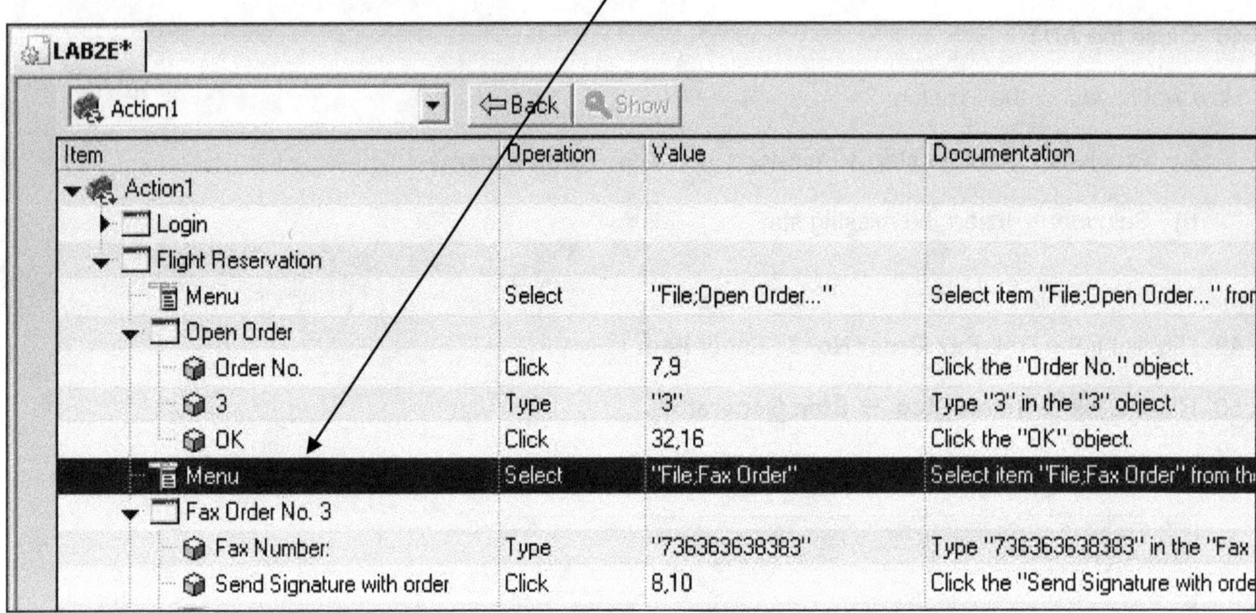

55. Save **LAB2E** → Playback & it should PASS ! Close & review the **Test Results** window

OPTIONAL

- If the AUT is still open , ask yourself : ***Why is the AUT still open*** ? How would you fix it ?

1. Add the Missing Step **File;Exit** after the last line in the script

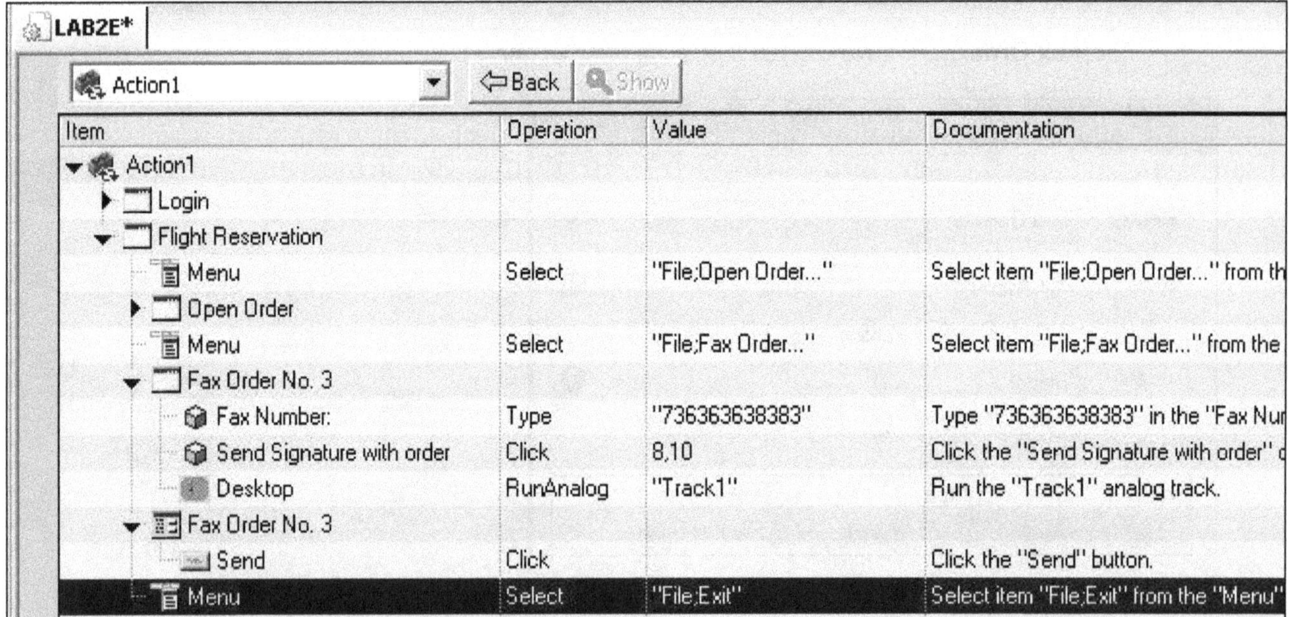

2. Save **LAB2E** → Playback & Verify the Test PASSED !

***** END OF LAB 2E *****

LAB 2F : Object Spy

OPTIONAL

Requirements:

1. Spy on objects learn their property values
2. Identify the methods (functions) of an object

Steps:

1. **Start → Programs → QuickTest Professional → Sample Applications → Flight**
2. Login with ***yourname*** and use the password ***mercury*** → Click **OK**
3. Open an existing order (On your own)
4. **Tools** (menu bar) → **Object Spy ...** → Click on the hand pointer
5. Point to the following **Objects** and record their values → Choose **Test Object Properties** :
   a) Choose and write down any two methods available for the specified object
   b) Skip (writing) any property you cannot find or *locate*

Property / Attribute	**Date of Flight**	**Fly From**	**Name**	**First Class**
class name:				
acx name:				
enabled :				
focused				
text				
attached text				
hwnd (handle)				
nativeclass				
window id				
progid				
Method 1				
Method 2				

6. Click **Close** (close the ***Object Spy*** window)

*** **END OF LAB 2F** ***

TO DO Pane (Reminder)

1. Create and Manage Tasks

   a) Test related reminders to the current script
   b) Manage Tasks thru TO DO pane & Task Editor

2. TO DO Comments

   a) Reminder comments inserted as **comment steps** in either a ***Action*** or ***Function Library***
   b) Accesses via either TO DO Pane or directly from the Testing Document
   c) Enter comments from either ***Keyword*** or ***ExpertView***

3. Tasks and TO DO Comments can be exported to **MS Excel** or **XML**

4. Example :

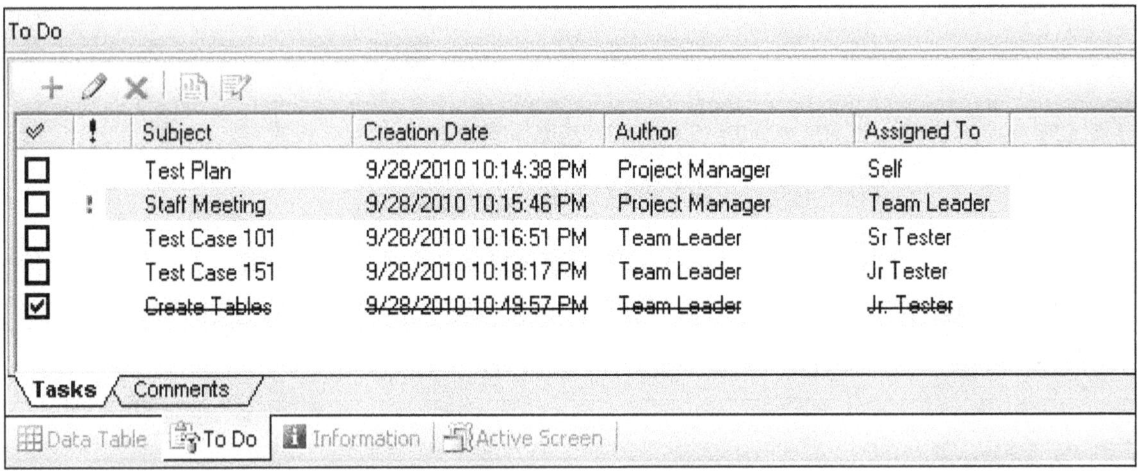

To Do

✓	!	Subject	Creation Date	Author	Assigned To
☐		Test Plan	9/28/2010 10:14:38 PM	Project Manager	Self
☐	!	Staff Meeting	9/28/2010 10:15:46 PM	Project Manager	Team Leader
☐		Test Case 101	9/28/2010 10:16:51 PM	Team Leader	Sr Tester
☐		Test Case 151	9/28/2010 10:18:17 PM	Team Leader	Jr Tester
☑		~~Create Tables~~	~~9/28/2010 10:49:57 PM~~	~~Team Leader~~	~~Jr. Tester~~

Tasks | Comments

Data Table | To Do | Information | Active Screen

   a) Five Tasks (in the above example)

   b) *Staff Meeting* is a High Priority

   c) One Task is Completed (*Create Tables*)

5. Sort the Tasks (Column headings)

6. Duplicate

   a) Creates a copy of an existing task

7. Delete

   a) Permanently removes the task from the Test

8. Toggle *off* or *on*

   a) View → Comments

9. Example (COMMENTS) :

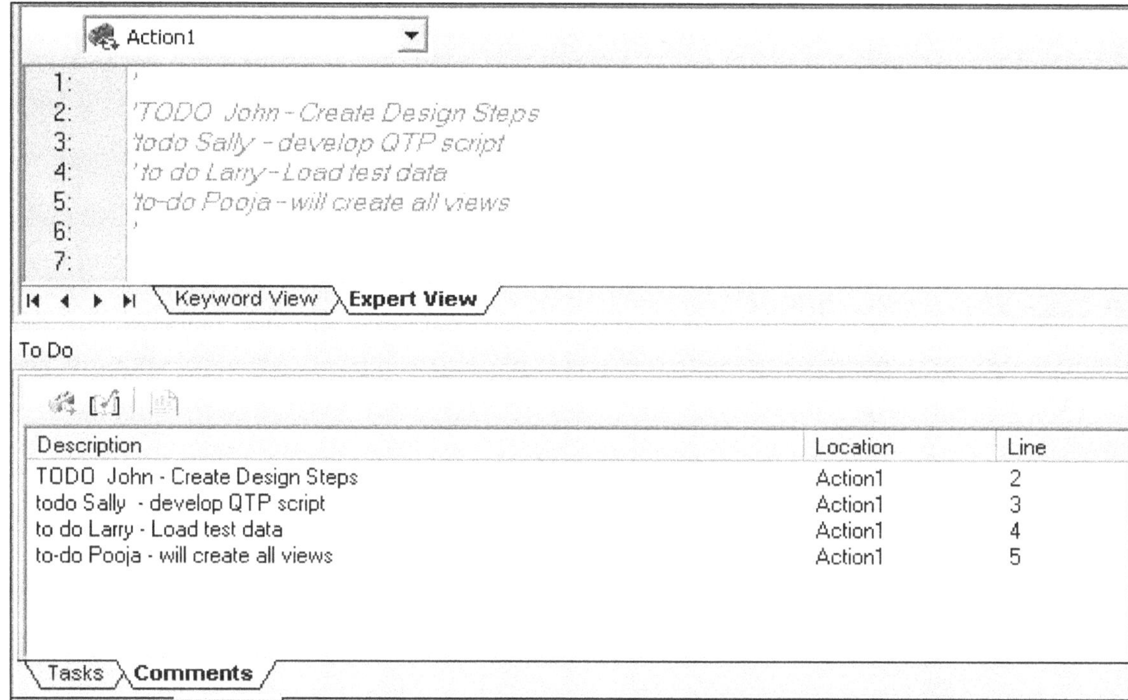

a) Enter the **key word** anywhere in the script as a comment line preceded with a single quote
b) The keyword **TO DO** is NOT case sensitive with a maximum of 260 characters

- TO DO or to do
- TODO or todo
- TO-DO or to-do

c) In ***Keyword*** view the comments are NOT visible
d) By Double-Clicking a comment line, it will position you to *that* line number in the script

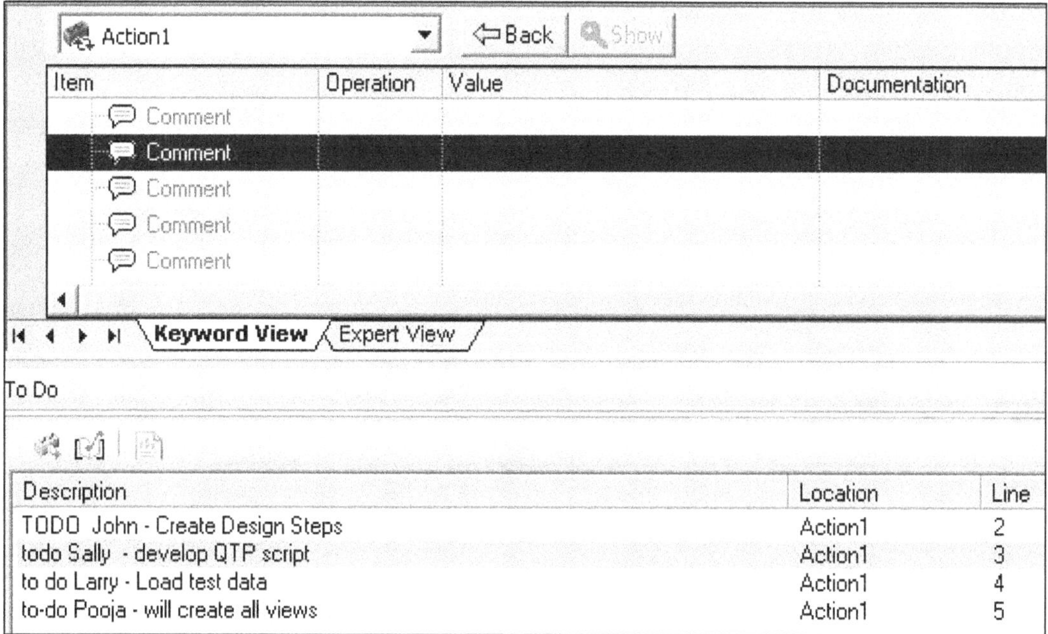

LAB 2G : TO DO LIST

Requirements:

- Create a Reminder of Tasks to complete
- Update existing scripts : *LAB1B_P1, LAB1B_P2, LAB1C*
- Review & Mark Completed tasks

STEPS (Part 1)

1. **File → Open → Test ... C:\YOURNAME\QTP10\LVL1\LAB1B_P1**

2. **View** (menu bar) → **To Do** (Ensure the *TO DO* tab is displayed at the bottom) → Click **TO DO tab**

3. Click the Plus Sign (**+**) [*Add Task*]

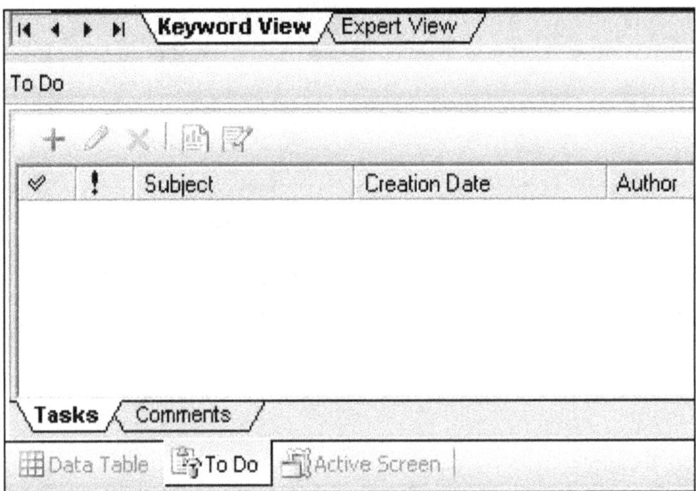

4. The **Task Editor** window opens → Enter the following → Click **OK**

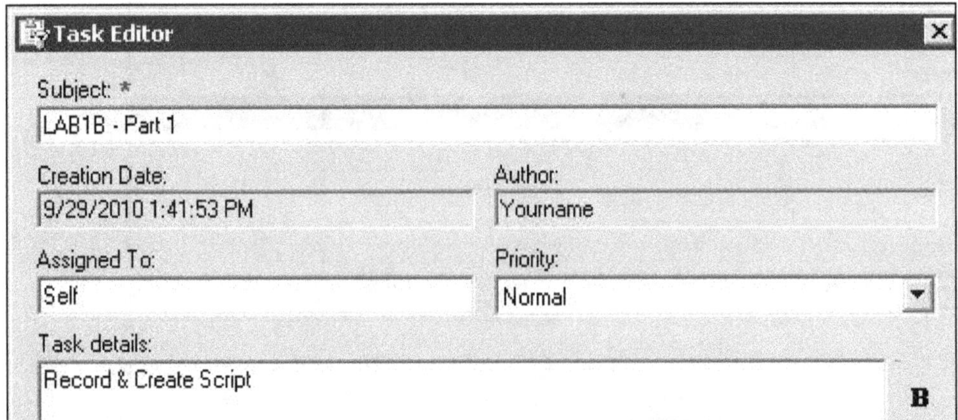

   a) The **Assigned To** is optional and not required if you are responsible

   b) Determine and *set* the **Priority** (Low, Normal, High)

5. Double-Click **LAB1B – Part 1** and <u>edit</u> the Task

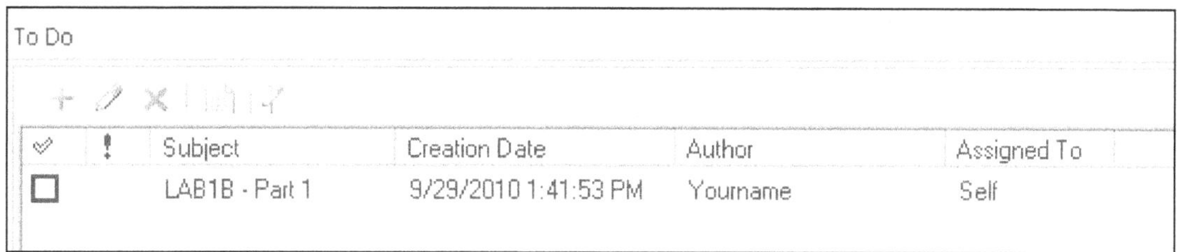

6. Change *LAB1B - Part 1* to **LAB1B - Part 1 : Script** → Click **OK**

7. Click the Plus Sign (**+**) → Enter the Following : → Click **OK**

   - Subject : **LAB1B – Part 1 : Run**
   - Task Details : **Run and Correct the Test**

8. Click the Plus Sign (**+**) → Enter the Following : → Click **OK**

   - Subject : **LAB1B – Part 1 : Analyze**
   - Task Details : **Evaluate the Test Results**

9. It Should be similar to :

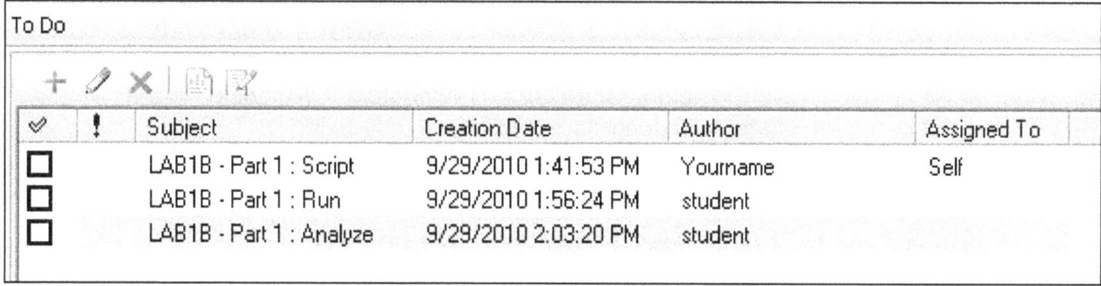

10. Save **LAB1B_P1**

11. Highlight **Action1** → **Insert** (menu bar) → **Co<u>m</u>ment**

12. Type **Launch AUT before running script** → Click **OK**

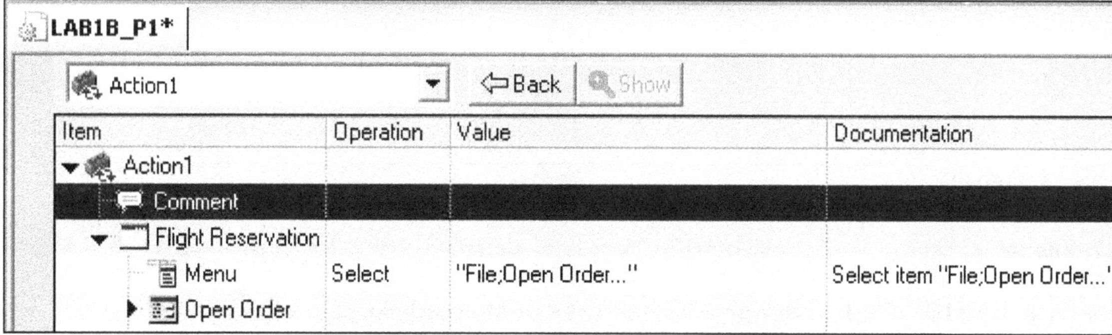

13. Observe : The comments are NOT displayed in the **Keyword View** !

14. **LAB1B_P1**

15. Click **ExpertView** → It should be similar to the following → Examine ***Line 1***

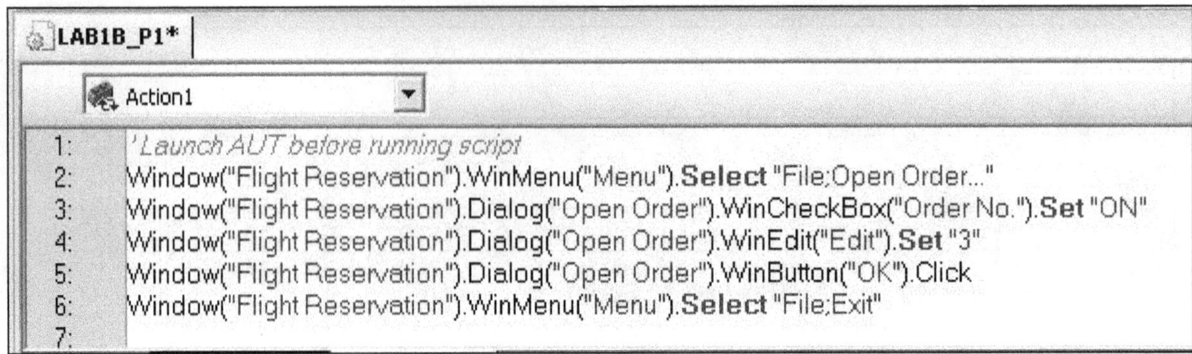

16. Insert a line after Line 1 & enter the following *Task* to be perform: → Start with a single quote

'TO DO Check with Team Leader about increasing the number of transactions

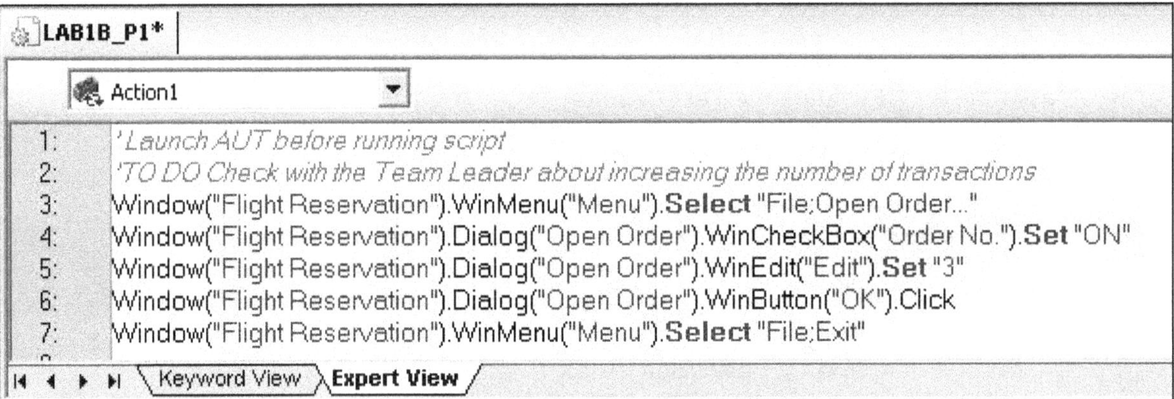

17. Click **Keyword View** → Click the **Comments** tab →Review the *To Do* task

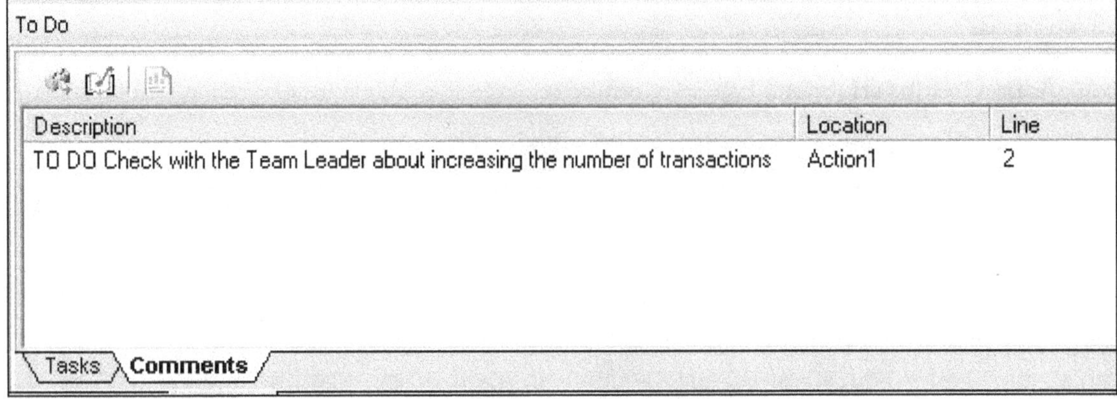

18. Click **Tasks** tab

19. Double-Click **LAB1B – Part 1 : Script** → Check **Task completed** check box → Click **OK**

20. Double-Click **LAB1B – Part 1 : Run** → Check **Task completed** check box → Click **OK**

21. Save **LAB1B_P1**

22. The script should be similar to :

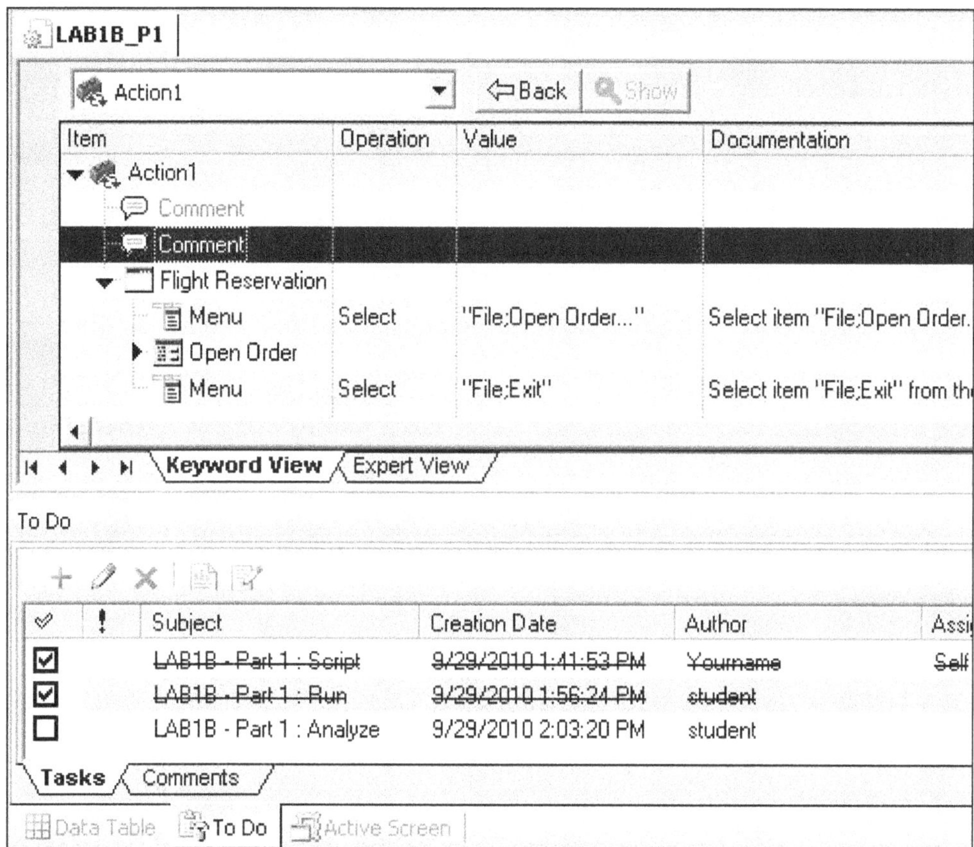

23. Playback **LAB4B_P1** → It Should Pass → Review & Close the *Test Results* window

24. Mark the last Task (*Analyze*) as being completed (on your own)

25. Save **LAB4B_P1**

OPTIONAL

Test	**Subject**	**Task Details**
LAB4B_P2	Review Test Case	Perform Manual Test
	Build Script	Record
	Run Script	Playback the Test
	Analyze	Evaluate the Test Results
LAB1C	Review Test Case	Perform Manual Test
	Build Script	Record
	Run Script	Playback the Test
	Analyze	Evaluate the Test Results
	Meeting	Weekly Status
	UAT	Demo to Customers

- Add Comments
- Mark Items as completed

*** **END OF LAB 2G** ***

LAB 2H : HP QuickTest Help

STEPS:

1. **Help → QuickTest Professional Help**

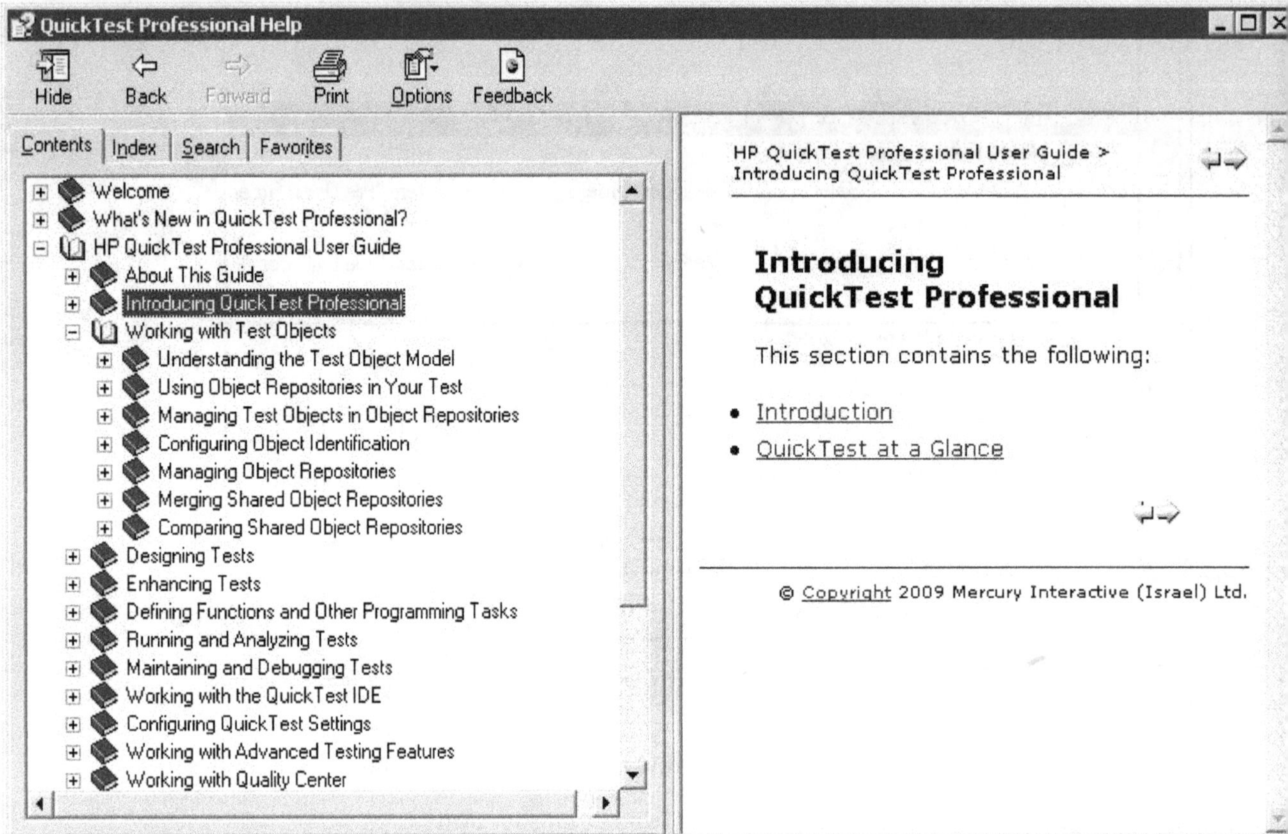

2. **HP QuickTest Professional User Guide**
   - How to use Quick Test
3. **HP QuickTest Professional Add-ins Guide**
   - How to use Add-Ins for support of different environments
4. **HP Quicktest Object Model Reference**
   - Standard Windows, ActiveX, Visual Basic, Web, Multimedia
5. **HP QuickTest Professional Advanced References**
   - Information : *Automation Object Model, Test Results, Object Repository* & *Automation* schema
6. **VBScript Reference**
   - VBScript Language, Script Runtime , Windows Script Host , etc ..

***** END OF LAB 2H *****

NOTES:

CHAPTER 2

Objectives of this section

- AUT (Flight Reservation Design)
- Object Recognition
- LAB 3A : Shared Repository
- Synchronization
- LAB 3B , LAB 3C, Optional LAB 3D
- CheckPoints
- LAB 4A - LAB4C
- Understanding the Different Run Modes : Update & Maintenance
- LAB 4D : Update Mode
- LAB 4E : Maintenance Run Wizard
- LAB 4F : Copy Objects from a Shared Repository
- LAB 4G : Object Repository (Optional)
- LAB 4H : Text Area Checkpoint
- LAB 4I : Zip Files

FLIGHT RESERVATION (AUT)

QUESTIONS:

1) What is the *goal*, *purpose* or use of this application (Chapter 1 : Labs 1 - 4) ?

2) What are all of the features of the application ?

3) When can a passenger make a reservation (for what travel dates) ?

4) While making a new *reservation* - what fields are Mandatory (what button must be pressed or clicked) ?

5) When does the **Insert Order** button get enabled ?

6) What is an order ? What makes it unique ?

7) How would you delete a reservation ?

8) When can you generate a Fax ? (i.e. under what conditions ?)

OBJECT RECOGNITION

1. Native Class : Microsoft Foundation Class (MFC)
2. QTP Class : Sub-set of the MFC defined starting with **WIN*ObjectName***
3. Object : A set of Defined properties
4. Object Repository : (Sample only – Assume the following:)

MFC Native Class	QTP Class	NativeClass Count	Properties QTP Count	Object Repository
Edit	WinEdit	80	60	2
PushButton	WinButton	50	30	1
Radio	WInList	40	10	2

5. Object Repository

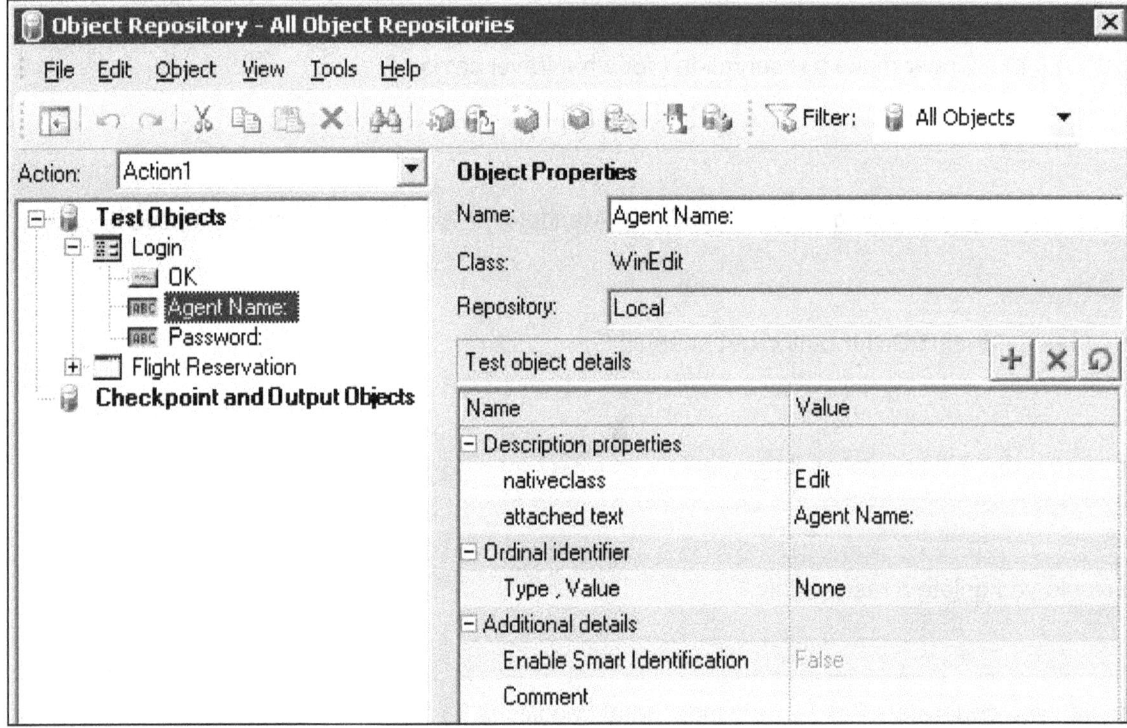

6. How will QTP identify the *object* **Agent Name:** ? By the following two properties :

   a) **Nativeclass Edit and Attached text Agent Name:**

7. Assume you get the error message ***Object Not Found*** - how can you fix it ?

   a) Add more properties to uniquely identify the object
   b) **Nativeclass Edit** , **Attached text Agent Name: , X , Y** coordinates *OR*
   c) **Nativeclass Edit** , **Attached text Agent Name: , Ordinal identifier 1** (position relative to objects of that same <u>class</u> or *type*)

HOW QTP RECOGNIZES OBJECT :

- Microsoft Foundation Class (MFC)
- HP QuickTest Professional Class (QTP Class) : Logical & Physical names (Class)

1. Recording

a) Any ***object*** that you come in contact with is stored in the object repository as a **Test Object**

b) QTP will assign a **Logical** name to the *object*

c) The logical name is *stored* both in the script as well as the **object repository**

d) The Logical name is determined by the : **label, text, value** currently assigned during recording

e) If not – it will look the left , immediately above looking for **Attached Text**

f) If none – it will look in *Memory* for the Developer Assigned Object Name

g) If not available – it will use the QTP class to determine the name

h) For each **Test Object Class** – QTP has a list of mandatory properties it must learn (default properties)

i) Are the default properties enough to **uniquely** identify the object on the window / page ?

j) If **not** – it will insert (add) assistive properties (one by one) until it can uniquely identify the object

k) If NO assistive properties are available or not sufficient (for *that* object) then QTP will add an **ordinal identifier**

l) An ***ordinal identifier*** is the objects location on *that* **window / page** or ***source code*** (if available)

- Assigns a *numerical* value to the test object that indicates the order (position / location) relative to other objects with an identical description in a **clockwise** direction
- Aids in ***uniquely*** identifying the object when the defined set of properties are not sufficient

2. Execution or Playback

a) QTP *reads* one line in the script (the script contains only the **logical** name for a specified object)

b) If you reference an object (via logical name) - QTP searches for the logical name in the **Run-Time** object (in memory) that matches the description in the *Object Repository*

c) It expects to find a **perfect match** for both the mandatory & assistive properties

d) During Run-Time it is possible that the object's physical description may have changed :

- date, data, business rules, access permission, security, etc …

e) QTP will use **Smart Identification** mechanism : a process of elimination to uniquely identify the object

- This feature must be turned **ON**

Sample Application (Jones Shipping)

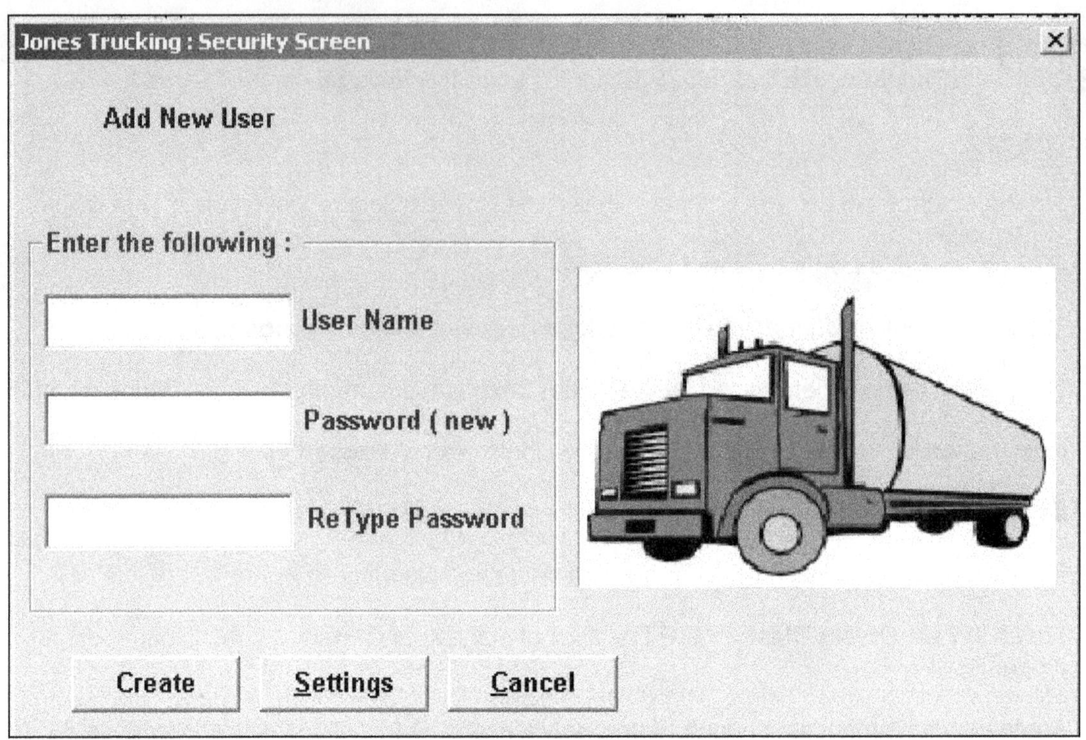

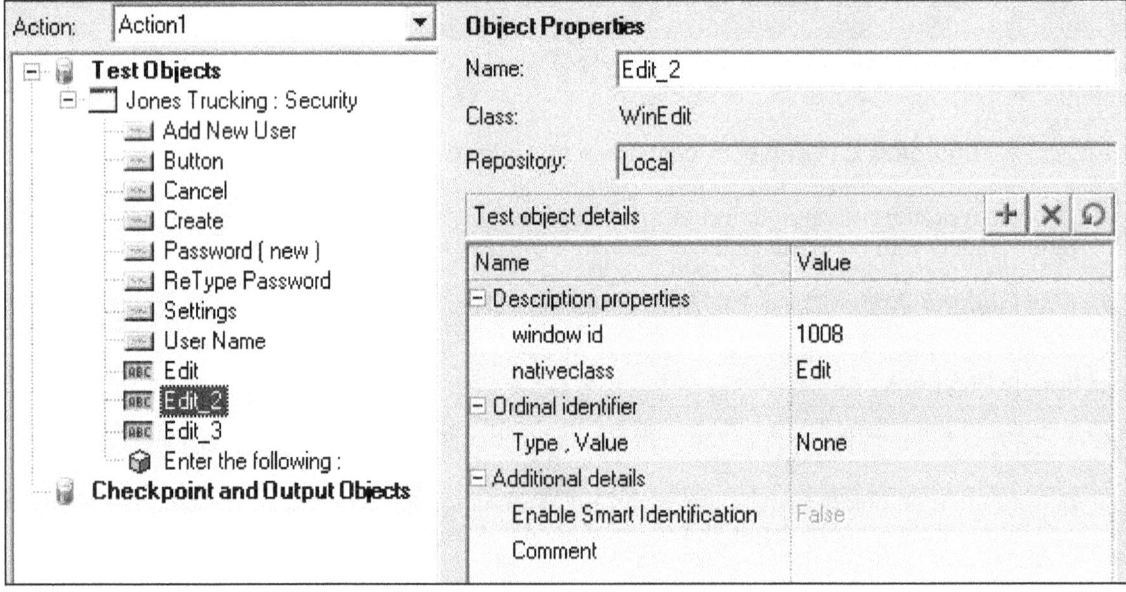

Questions:

1. How many objects in the AUT ? __________ How many objects in the Object Repository ? __________

2. Do the type (classes of objects) match between the AUT & Object Repository ? ______________________

3. What are **Edit** , **Edit_2** & **Edit_3** objects ?

LAB 3A SHARED REPOSITORY

Test Requirements

1. Create All Objects into a single **common** *shared* repository (without *Recording*) :

   - Login
   - Flight Reservation
   - Flights Table
   - Open Order

2. You will not reference the *shared* repository in this script. You will use it later in a subsequent lab.

3. Any script can reference the ***common*** repository

Steps:

1. Create folder : **C:\YOURNAME\QTP10\LVL1\REPO**
2. Start → AUT (Flight4a) → DO NOT Login
3. **File** → **New** → **Test** → ReSize QTP & AUT (side by side)
4. **Resources** → **Object Repository ...** →
5. **Object** (menu bar) → **Add Objects to Local...** → cursor changes to a hand
6. Point & Click on title bar **Login**
7. Verify **Dialog : Login** → Click **OK**
8. Choose **All object types** → Click **OK**
9. Login into AUT (*yourname / password*)
10. **Object** (menu bar) → **Add Objects to Local...** → cursor changes to a hand
11. Point & Click on title bar **Flight Reservation**
12. Verify **Window : Flight Reservation** → Click **OK**
13. Choose **All object types** → Click **OK**
14. **File** → **Open Order ...** →
15. **Object** (menu bar) → **Add Objects to Local...** → cursor changes to a hand
16. Point & Click on title bar **Open Order**
17. Verify **Window : Dialog Open Order** → Click **OK**
18. Choose **All object types** → Click **OK**
19. Choose **Order Number** (check box) → Type **2** → Click **OK**
20. Click **Flights** button → Correct any data validation errors (if any) → Swap to QTP

21. **Object** (menu bar) → **Add Objects to Local…** → cursor changes to a hand →
22. Point & Click on title bar **Flights Table**
23. Verify **Dialog : Flights Table** → Click **OK**
24. Choose **All object types** → Click **OK**
25. **File** → **Close** (close **Object Repository** window)
26. Click **OK** (close **Flight Tables** window)
27. **File** → **Exit** (close FLIGHT4A - AUT)
28. **File** → **Save** → **FRS1_REPO** (QTP Script)
29. **Resources** → **Object Repository …** →
30. **File** (menu bar) → **Export Local Objects …** → cursor changes to a hand
31. Save as C:\YOURNAME\QTP10\LVL1\REPO\ **FRS_REPO.TSR**
32. **File** → **Close** (close **Object Repository** window)

QUESTION

1. What other objects are missing in the repository for **FLIGHT** application ?

ANSWER

1. All dialog windows containing **data** validation & **business** rule errors

2. Settings & other windows requiring data input may generate/open other windows

*** **END OF LAB 3A** ****

Synchronization

Definition : Resolve anticipated timing problems by *including* synchronization statements.

1. You synchronize your test to ensure that your application is ready for HP QuickTest to execute the *next* **step**

2. When you ***run*** tests – the AUT/SUT may not always respond quickly (as expected)

3. It may be waiting for a response from the:

    a) Application
    b) Network
    c) Database
    d) Web server
    e) Application server
    f) etc …

4. The AUT/SUT may respond (*behave*) by displaying a:

    a) Progress Bar to reach 100%
    b) Status or *message* to appear
    c) Button to be enabled
    d) Pop-up window or window to *open*

5. QTP is executing or *running* faster than the AUT – you can instruct / inform QTP to slow down

SYNCHRONIZATION OPTIONS

1. Insert a ***synchronization*** point allowing HP QuickTest to pause the test

    a. HP QuickTest *generates* a **WaitProperty** statement

2. You can insert **Exist** or **Wait** statement instructing Quick Test to ***wait***

    a. Until an object exists
    b. Wait a specified amount of time before continuing the test

3. Increase the object **default** timeout settings in the ***Test Settings*** options

    a. Time is expressed in milli-seconds
    b. 1000 milliseconds = 1 second
    c. Default Timeout is: 20 seconds (20,000 milliseconds) per object
    d. Every Step will wait 20 seconds

WHEN SHOULD I ADD A SYNCHRONIZATION Point ?

If you **do not** want HP QuickTest to perform a ***step*** or ***checkpoint*** until an object in your application achieves a certain status, you should *insert* a *synchronization* point to instruct HP QuickTest to pause the test until the object property achieves the value you specify (or until a specified timeout is exceeded.)

STEPS TO INSERT A SYNCHRONIZATION

1. Valid while performing a **Recording** ! The synchronization is *enabled* in the menu options
2. Display the **screen** or ***page*** in your AUT that contains the object for which you want to insert a synchronization point.
3. In HP QuickTest click **Insert** > **Synchronization Point ...** (cursor changes to a hand pointer)
4. Click on the object in your application for which you want to insert a synchronization point

   Note: It does not matter what property values the object has at the time that you insert the synchronization point.

   If the object's location is associated with more than one object in your application, the **Object Selection - Synchronization Point** dialog box opens.

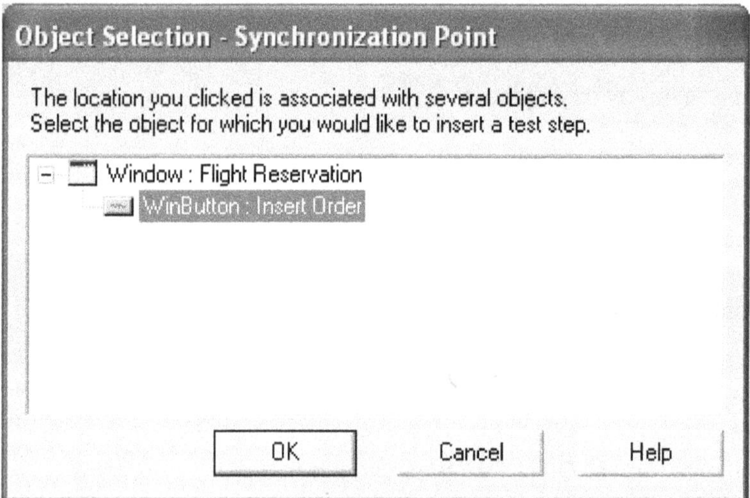

5. Select the object for which you want to insert a synchronization point, and click **OK**

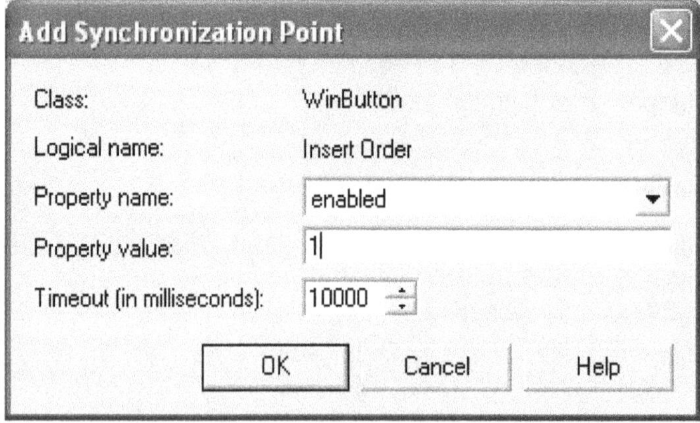

6. The **Property name** list contains the test object properties associated with the object

   a) Select the ***Property name*** you want to use for the synchronization point
   b) Enter the value associated with the **Property name** selected

7. Enter the synchronization point timeout (in milliseconds) – default is 10 seconds

   Note: after which HP QuickTest should continue to the next step in the test, even if the specified property value was not achieved.

8. Click **OK** → (A **WaitProperty** step is added to your test)

Action1			
SystemUtil	Run	"C:\Program Files\HP\QuickTest P...	Open the "C:\Program Files\HP\QuickTest F
Login			
Flight Reservation	WaitProperty	"text","Flight Reservation",10000	Wait until the value of the "text" property of tl
Menu	Select	"File;New Order"	Select item "File;New Order" from the "Menu

   Note: Because the **WaitProperty** step is a ***method*** for the selected object, it is displayed in the Tree View with the icon for the selected object.

9. Click **OK**

Example 1 : Overview: Insert a new order → Click **OK** → Wait for next step

Item	Operation	Value	Documentation
Action1			
SystemUtil	Run	"C:\Program Files\HP\QuickTest P...	Open the "C:\Program Files\HP\QuickTest F
Login			
Flight Reservation			
Menu	Select	"File;New Order"	Select item "File;New Order" from the "Menu
MaskEdBox	Type	"121010"	Type "121010" in the "MaskEdBox" ActiveX
MaskEdBox	Type	micTab	Type micTab in the "MaskEdBox" ActiveX o
Fly From:	Select	"Frankfurt"	Select the "Frankfurt" item from the "Fly From
Fly To:	Select	"Los Angeles"	Select the "Los Angeles" item from the "Fly T
FLIGHT	Click		Click the "FLIGHT" button.
Flights Table			
Name:	Set	"James Stewart"	Enter "James Stewart" in the "Name:" edit bo
Insert Order	WaitProperty	"enabled",1,10000	Wait until the value of the "enabled" property
Insert Order	Click		Click the "Insert Order" button.
Threed Panel Control	WaitProperty	"text","Insert Done...",10000	Wait until the value of the "text" property of tl
Delete Order	Click		Click the "Delete Order" button.
Flight Reservations			
Menu	Select	"File;Exit"	Select item "File;Exit" from the "Menu" menu

Questions :

1. What is the *functionality* of this test ? (the business process)
2. Explain the steps ?
3. What does the **"Insert Order" WaitProperty** do ?
4. What happens after the user *clicks* **" Insert Order "** ?
5. What other synchronization could be added ?

LAB 3B : Synchronization

Overall Scenario :

1. Invoke *Flight4A* → Log-In → Add New Order → Delete Order → Exit the AUT
2. Modify, Enhance & Save the test script → Playback
3. May need to Resize QTP windows (smaller) before recording

PART I : Test Requirements : Business Function: Add New Order (Reservation)

a) Start the AUT by using QTP settings (not inside the script). Locate/Identify the PATH. (On your own)

"C:\Program Files\HP\QuickTest Professional\samples\flight\app\flight4a.exe"

b) Record & ***review*** the script

c) Next, determine where in the script you need to insert synchronization points

✓ **Insert Done ...** Static Text (After Inserting a new order)

✓ **Delete Done ...** Static Text (After Deleting a new order) - *Reset/Clean-up* condition

Steps: (Part I)

24. **File** → **New** → **Test** (Create a new script)

25. **Automation** (menu bar) → **Record... F3**

26. Check **Record and run only on** (radio button) → Choose the following *selections* :

a) Select all check boxes

b) Uncheck **Applications opened via the Desktop (by the windows shell)**

c) Click the **Green Plus sign** → Point to the location of the AUT :

"C:\Program Files\HP\QuickTest Professional\samples\flight\app\flight4a.exe"

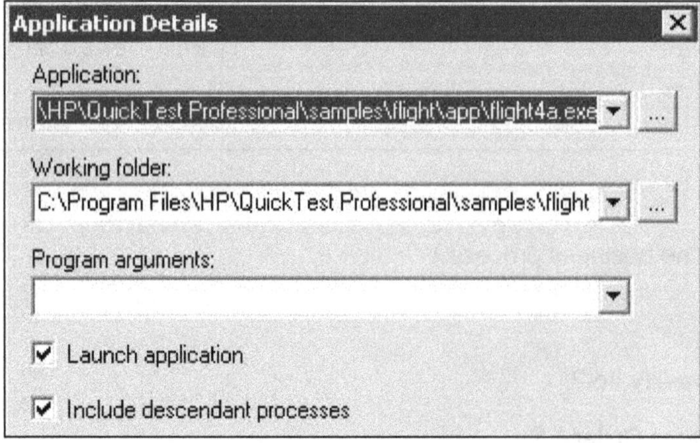

27. Click **OK** (close the **Application Details** window)

28. The settings should be similar to :

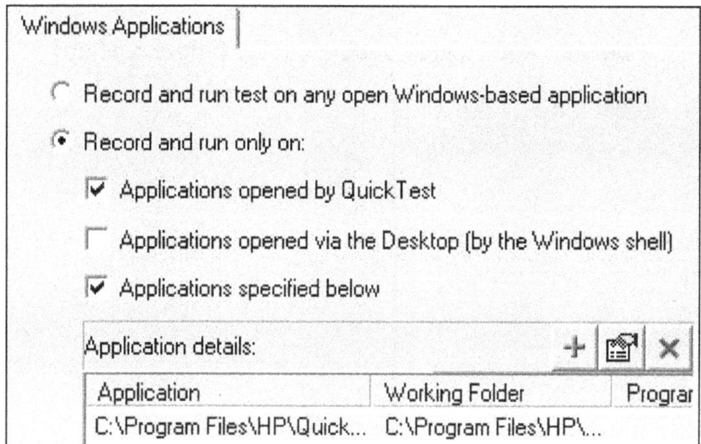

29. Click **OK** → (close the **Record and Run Settings** window) → Wait for the AUT to invoke automatically
30. Login with ***yourname*** and use the password ***mercury*** → Click **OK**
31. **File → New Order** [On your own → *Type* all the information required to add a new order]
32. Click **Insert Order** → Wait for the confirmation message
33. Swap to HP Quicktest → **Insert** (menu bar) → **Synchronization Point …** cursor changes to a hand
34. Point & click on *text* message **Insert Done …**

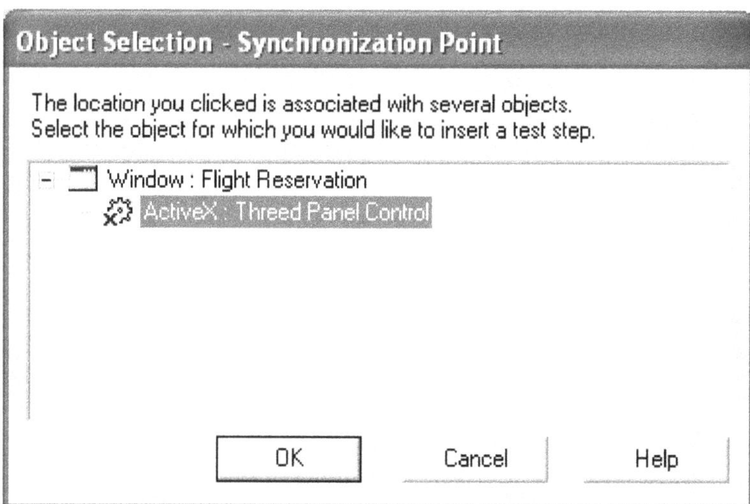

- Verify the object you pointed & clicked is correct

35. Click **OK**

36. Change the following *properties* and Click **OK** :

   a) Property name: **text**
   b) Property value : "Insert Done…" (be sure and include double quotes before & after)

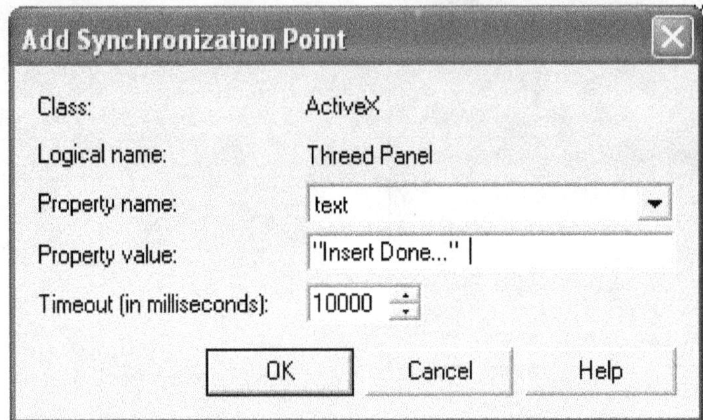

37. Click **OK** (Close the **Add Synchronization Point** window)
38. Click **Delete Order** button → Click **Yes** to confirm the order
39. Swap to HP Quicktest → **Insert** (menu bar) → **Synchronization Point …** cursor changes to a *hand*

40. Point & click on text message **Delete Done …** → Click **OK**

41. Change the following and click **OK** :

   a) Property name: **text**
   b) Property value : "Delete Done…" (be sure to ***include*** double quotes before & after)

42. Click **OK** (Close the **Add Synchronization Point** window)

43. **File** → **Exit** (Close the AUT)

44. Stop Recording (or Press **F4**) → Review the script

45. If there are any errors while typing – you may receive the following message - otherwise go to *next* step :

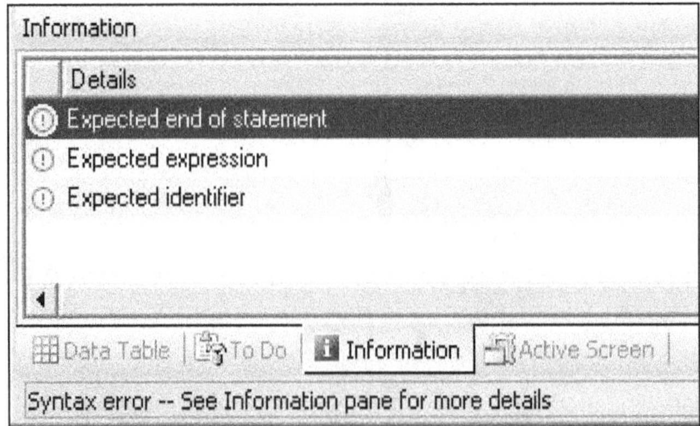

Information Pane:

a) Short Error Description
b) Double-Click the *message* & it will take you to the Line Number ##

Common errors & How to Fix (a – d) :

a) Missing : Commas, left & right parentheses, quotes or parameters
b) Misspelling : *Methods*, *functions* or *property* name
c) Highlight & Double-Click the Line 1 (in the *Information Pane*) → Review & Make Changes !

d) Enclose Double quotes around the text message **"Insert Done…"**

Window("Flight Reservation").ActiveX("Threed Panel").WaitProperty "text", "Insert Done...", 10000

e) Enclose Double quotes around the text message **"Delete Done…"**

Window("Flight Reservation").ActiveX("Threed Panel").WaitProperty "text","Delete Done...", 10000

f) Click **KeyWord View**

46. Review & Correct the Script (it maybe similar to the following) :\

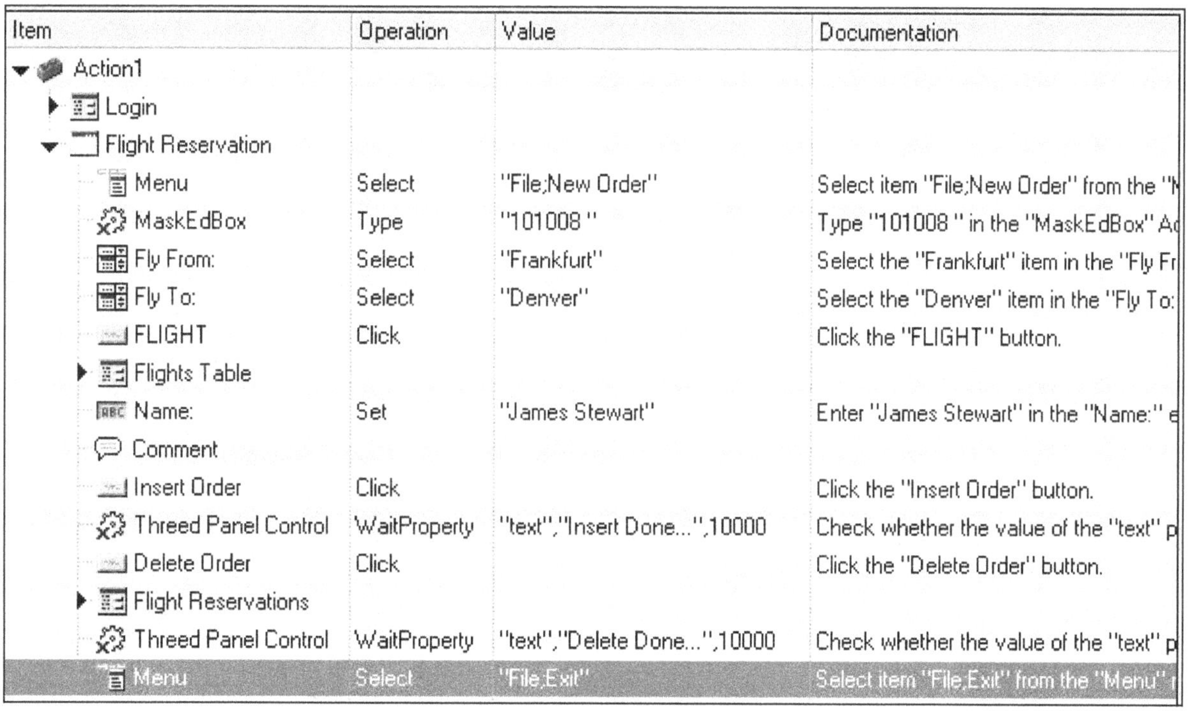

Item	Operation	Value	Documentation
Action1			
Login			
Flight Reservation			
Menu	Select	"File;New Order"	Select item "File;New Order" from the "N
MaskEdBox	Type	"101008 "	Type "101008 " in the "MaskEdBox" Ac
Fly From:	Select	"Frankfurt"	Select the "Frankfurt" item in the "Fly Fr
Fly To:	Select	"Denver"	Select the "Denver" item in the "Fly To:
FLIGHT	Click		Click the "FLIGHT" button.
Flights Table			
Name:	Set	"James Stewart"	Enter "James Stewart" in the "Name:" e
Comment			
Insert Order	Click		Click the "Insert Order" button.
Threed Panel Control	WaitProperty	"text","Insert Done...",10000	Check whether the value of the "text" p
Delete Order	Click		Click the "Delete Order" button.
Flight Reservations			
Threed Panel Control	WaitProperty	"text","Delete Done...",10000	Check whether the value of the "text" p
Menu	Select	"File;Exit"	Select item "File;Exit" from the "Menu" r

<u>Playback</u>:

47. Save the script as : **C:\YOURNAME\QTP10\LVL1\LAB3B_P1**

48. **Automation** (menu bar) → **Run… F5**

49. Check **Temporary Run Results folder (overwrites any existing temporary results)** → Click **OK**

50. Review the **Test Results** → Close the **Test Results** window

<u>Question</u>:

1. What did you observe in the test results on synchronization ?
2. Any warnings related to synchronization may require you to increase the timeout value

**** **END OF LAB3B_P1 - PART I** ****

LAB3B : PART II (Synchronization – *another* method)

Test Requirements :

1. Check the *properties* of the **Update** button instead of verifying the message "**Insert Done…**"
2. Assume that Synchronizations statement have NOT been created in a script
3. After *Adding* a new reservation → the **Update** button should be enabled & **Insert** button is disabled

Steps: (Part II)

1. **File → Save As… → LAB3B_P2**
2. In ***KeyWord View*** mode – Delete the following steps :

   a) Highlight **"Threed Panel" WaitProperty "text" ,"Insert Done…" , 10000**
   b) Right-click → **Delete** → Click **Delete Step** to confirm the delete
   c) Highlight **"Threed Panel" WaitProperty "text" ,"Delete Done…" , 10000**
   d) Right-click → **Delete** → Click **Delete Step** to confirm the delete

3. **Start → Programs → QuickTest Professional → Sample Applications → Flight**
4. Login with ***yourname*** and use the password ***mercury*** → Click **OK**
5. Add a *new* order → Ensure the message **"Insert Done…"** is displayed

6. Highlight "**Insert Order"** (button in *KeyWord View*)

7. **Automation** (menu bar) → **Record… F3** (***Record*** message is flashing in red)

   Note: The **Login** window opens , ignore it and refer to the **Flight Reservation** window

8. **Insert** (menu bar) **→ Synchronization Point … →** Point & Click on the **Update Order** button
9. Verify that the **Winbutton: Update Order** is selected → Click **OK**
10. Change the following and Click **OK** :

    a) Property name: **enabled** b) Property value : **1** c) Timeout : 10,000

11. Verify "**Delete Order**" is Highlighted (in Keyword View)
12. **Insert** (menu bar) **→ Synchronization Point … →** Point & Click on the **Insert Order** button
13. Verify that the **Winbutton: Insert Order** is selected → Click **OK**
14. Change the following and Click **OK** :

    a) Property name: **enabled** b) Property value : **0** c) Timeout : 10,000

15. Stop the recording (or Press **F4**)
16. Close the **AUT** (Close ***Flight Reservation*** window)
17. Close the **Login** window

18. **File → Save** → (**LAB3B_P2**)

19. **Automation** (menu bar) → **Run… F5**

20. Check **Temporary Run Results folder (overwrites any existing temporary results)** → Click **OK**
21. Review the **Test Results** → Expand All
22. When finished → close the **Test Results** window

**** **END OF LAB 3B_P2 - PART II** **

LAB3C : (Modifying Default Timeout)

Test Requirements : Example of a timeout issue : Typically the timeout will not be 1 second

1. Modify the Default **Time-out** Settings to **1** second
2. Login → Open Order → Fax Order → Exit the AUT
3. It should Fail due to a Synchronization problem for a specific object
4. You may need to Re-Size QTP window and AUT before recording

Steps (Part 1) :

1. **File** → **New** → **Test** (Create a new script)
2. **Automation** (menu bar) → **Record... F3**
3. Check **Record and run only on** (radio button) → Choose the following *selections* :
   a) Select all check boxes
   b) Uncheck **Applications opened via the Desktop (by the windows shell)**
   c) Click the **Green Plus sign** → Point to the location of the AUT

   "C:\Program Files\HP\QuickTest Professional\samples\flight\app\flight4a.exe"
4. Click **OK** → (close the **Record and Run Settings** window) → Wait for the AUT to invoke automatically
5. Login with ***yourname*** and use the password ***mercury*** → Click **OK**
6. **File → Open Order ...** → Select **Order No.** check box → Type **2** → Click **OK**
7. **File → Fax Order ...** → Enter any ten (10) digit Phone Number → Click **Send**
8. Wait for the Confirmation message → **File → Exit** (Close the AUT)
9. Click **Press F4** key OR Click **Stop** button → Review the script → Correct any problems (if any)
10. Save the script as **C:\YOURNAME\QTP10\LVL1\LAB3C**
11. **File** → **Settings ...** → Click **Run** tab
12. Change **Object synchronization timeout** to **1** seconds → Click **OK**
13. Click **Run** button *or* press **F5** key → It should fail with the error message → Identify the Line # ___________

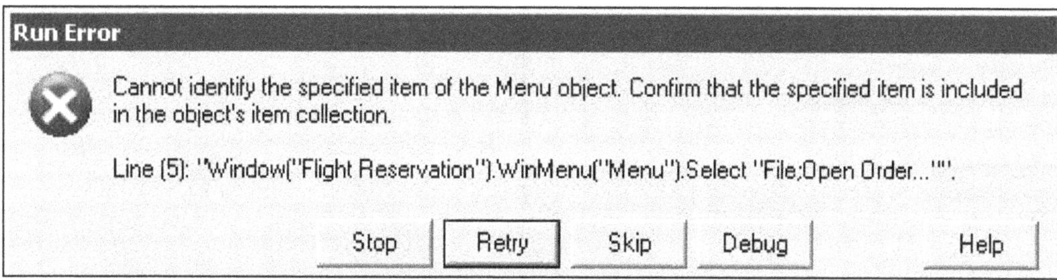

14. Click **Stop** → Review and Close the **Test Results** window
15. Close the **AUT**
16. **File** → **Settings ...** → Click **Run** tab

17. Change **Object synchronization timeout** to **5** seconds (from *1* seconds) → Click **OK**

18. Click **Run** button → it should fail with error : → Identify Line # : ___________

19. Click **Stop** → Review and Close the **Test Results** window

20. Close the **AUT**

21. **File** → **Settings ...** → Click **Run** tab

22. Change **Object synchronization timeout** to **10** seconds (from *5* seconds) → Click **OK**

23. **File** → **Settings ...** → Click **Run** tab → It should Pass

STEPS : Part 2

1. **File** → **Settings ...** → Click **Run** tab

2. Change **Object synchronization timeout** to **5** seconds (from *10* seconds) → Click **OK**

3. Highlight **Flight Reservation** (in Keyword View)

4. **Insert** (menu bar) → **New Step** **F8;Insert**

5. Click inside column **< Select an item >** → List Box appears

6. Choose **Object from repository**

7. Click **Flight Reservation** → Click **OK**

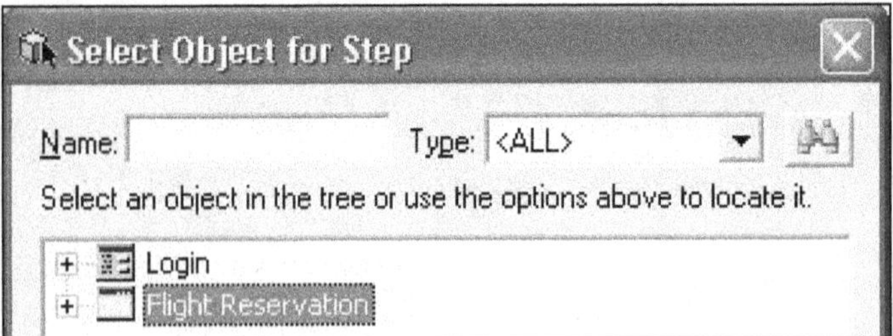

8. Click Inside column **Activate** → Scroll & Select **WaitProperty**

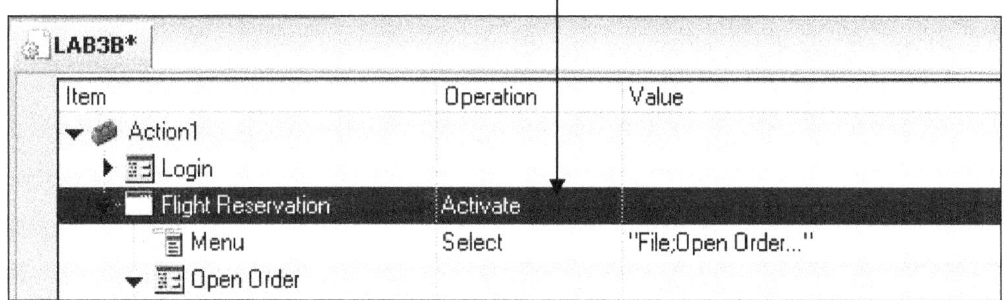

9. Move cursor to **PropertyName**

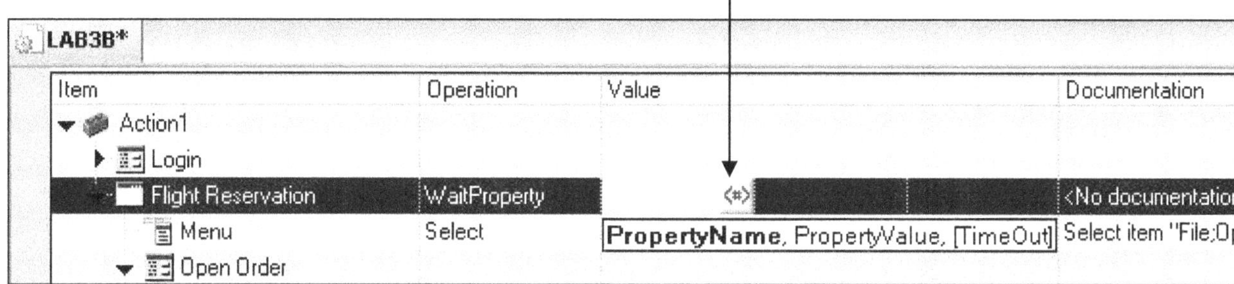

10. Change the **Value** column to specify three (3) parameters :

    "label" "Flight Reservation" 15000

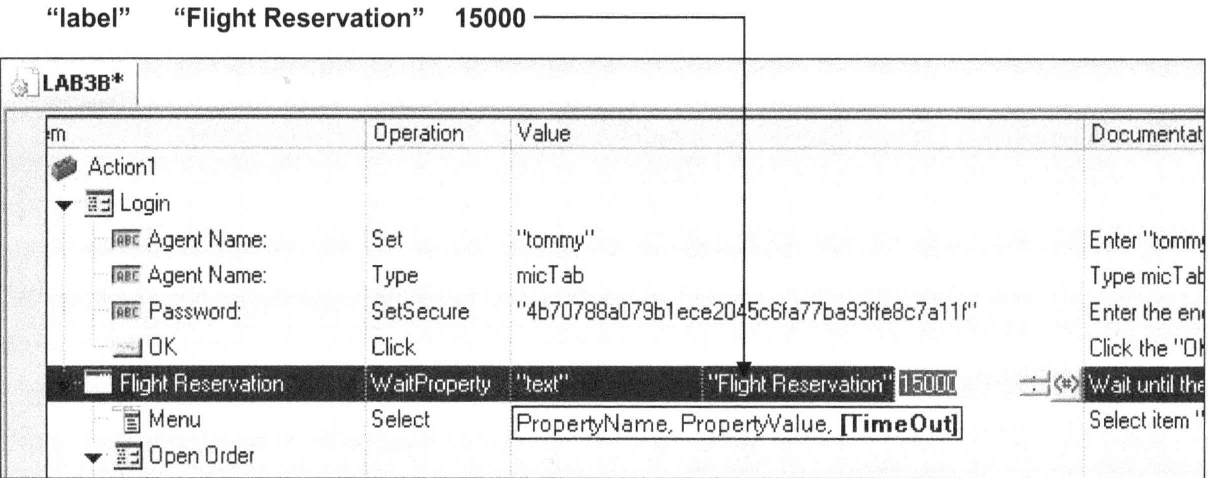

11. Save **LAB3C** → Run the test

    It should fail with : **Cannot identify the specified item of the Menu object ….**

12. Click **Stop** → Review & Close the **Test Results** window

13. Close the **AUT**

14. Highlight **Send Click** step

15. **Insert** (menu bar) → **New Step**

16. Select **Step Generator…**

17. Make the following changes to fields : → Click **OK**

    a) Category : **Functions**

    b) Operation : **Wait**

    c) Seconds * : **10**

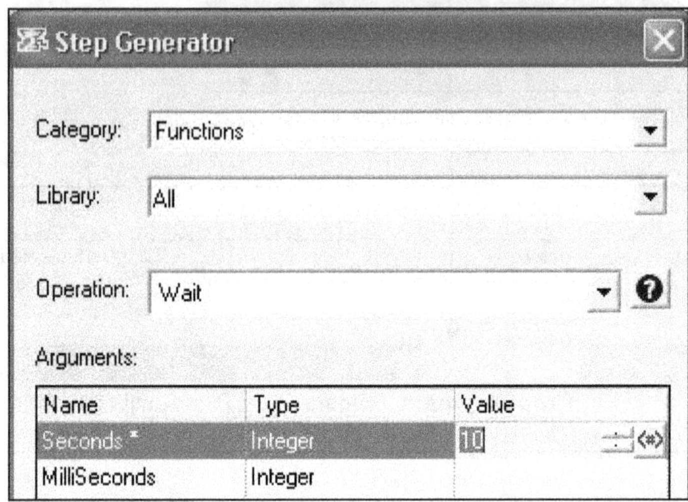

18. It should be similar to :

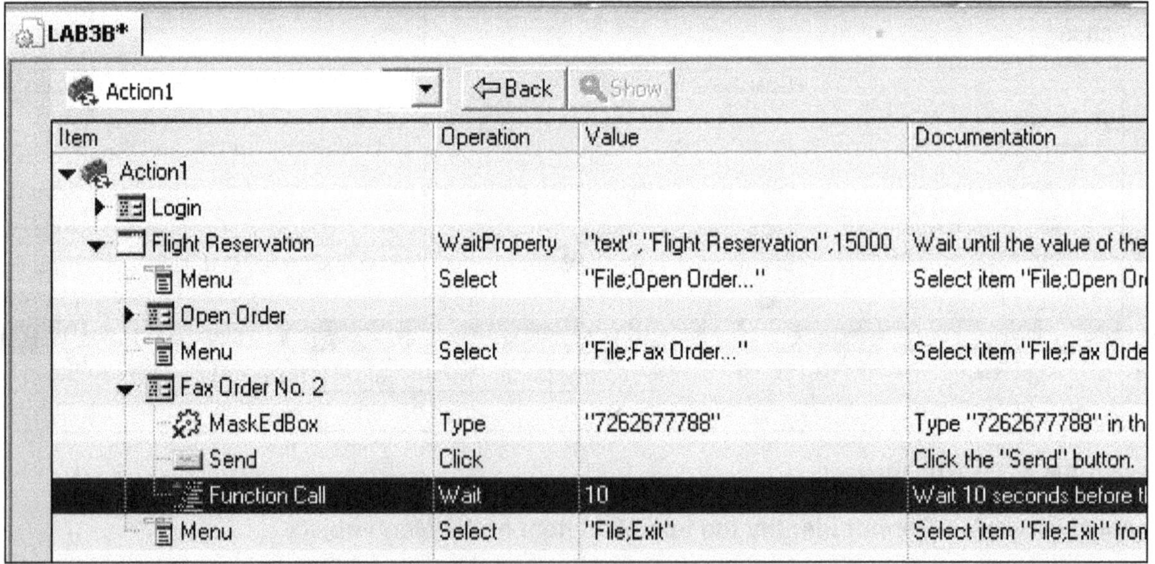

19. Save **LAB3C** → Playback and It should Pass !!

    - If it fails – increase the timeout from 10 to **15** seconds

20. Review & Close the **Test Results** window

*** **END OF LAB 3C** ***

LAB 3D : Synchronization **OPTIONAL**

Part 1 - Test Requirements:

Insert a synchronization at the end of each following functions or *business process* :

1. LAB3D_P1 : Add a new order
2. LAB3D_P2 : Update the order
3. LAB3D_P3 : Fax the order
4. LAB3D_P4 : Delete the order

Other Requirements

1. Create four new scripts
2. Invoke the application from within the script
3. Remember to *Include* a synchronization while Previewing a Fax

RECORDING SETTINGS

1. During a recording (by Pressing **F3**) – you may not get the following window - its ok:

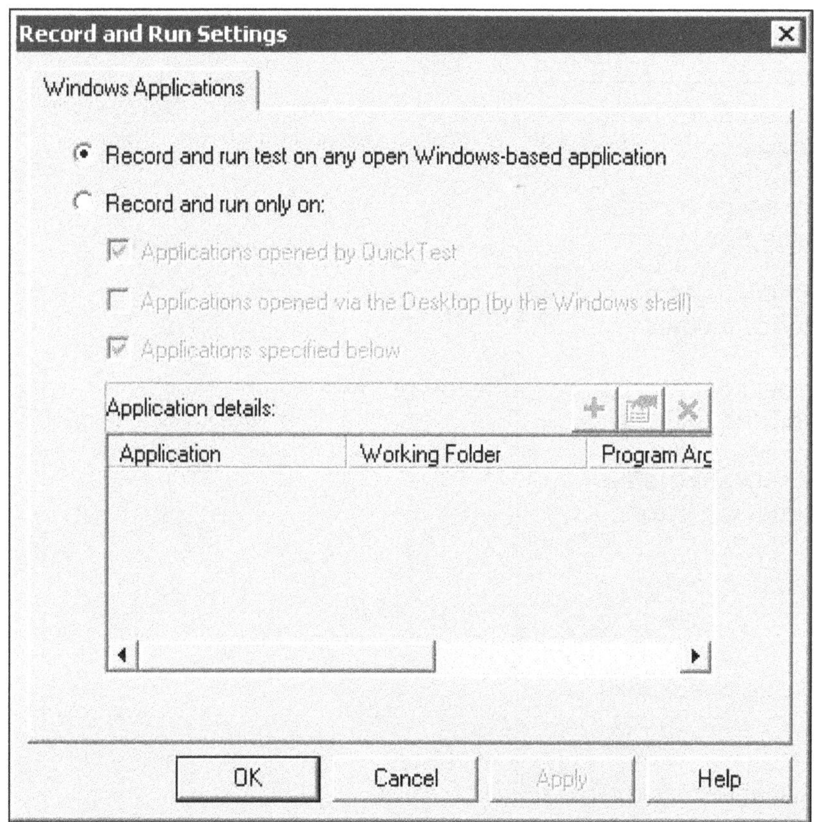

2. After the initial recording – the above window will NOT be displayed

***** END OF LAB 3D *****

CHECKPOINTS

Definition : a ***step*** in your test that compares the values of the specified property (during a test run) with the values stored for the same test object property within the test.

1. Acts as a **verification** point that compares a current value for a specified property with the ***expected*** value

2. You create / insert ***checkpoints*** to verify the behavior of the AUT in accordance with the business rule

3. This enables you to identify whether or not your Web whether or not you AUT/SUT is functioning correctly

4. After creating a *checkpoint* HP Quicktest will add a:

   a) **Checkpoint *icon*** in the KeyWord View mode
   b) **Check Checkpoint** statement in the Expert Mode

5. During the ***execution*** of a test – HP QuickTest

   a) Compares the ***expected*** results with the **actual** results
   b) If the values *match* report the step has **passed**
   c) If the values *DO NOT match* report the test has **failed**
   d) A *boolean* value indicates whether the checkpoint has *failed* or *passed*

6. Sample ICONS of checkpoint on *objects* :

a Window object	a checkpoint on a window
a Dialog object	a checkpoint on a dialog box
a WinObject object	a checkpoint on a Windows object
a WinEdit object	a checkpoint on a Windows edit box
a WinButton object	a checkpoint on a Windows button
a Page object	a page checkpoint
a WebTable object	a table checkpoint
an Image object	an image checkpoint
a WebElement object	a checkpoint on a Web element

DIFFERENT TYPES OF CHECKPOINTS

Standard

1. Checks the *property* **value** of an object in your **Application** or **Web** page
2. The standard checkpoint checks a variety of objects: *edit fields, buttons, radio buttons, combo boxes, etc*
3. Option is *enabled* only during recording

Text (Common)

1. Checks that a **text** string is displayed in the appropriate place in AUT/SUT
2. The checkpoint is taken on the *objects* text value (*Label, Data, Information, String*, etc …)

Bitmap

1. Checks all or a selected area of your AUT / SUT as a bitmap

   Example 1 : Assume you have a Web site that can display a map of a city the user specifies:

   a) The map has control keys for zooming
   b) You can record the new map that is displayed after one click on the control key that zooms in the map
   c) Using the bitmap checkpoint, you can *check* that the map zooms in correctly.

Image

1. Checks the property value of an ***image*** on a Web Page.
   - Properties: *PathName, HTML Tag, HREF, SRC*, etc …
2. HP Quicktest *must* be able to *recognize* the object and all of its properties
3. You create an ***image*** checkpoint by pointing & clicking on the *picture* via a *Standard CheckPoint*
4. Image checkpoints are supported for the ***Web*** environment

Table

1. Checks information (*data*) for an object defined as a table
2. Typically used from a **list box** or objects that allow the user to make a selection

   Example 2: Assume your application or Web site contains a ***table*** listing all available flights

   a) You can add a table checkpoint to check that the ***time*** of the first flight in the table is correct
3. Supported in both **Web** and **ActiveX** environments

Page

1. Checks the characteristics of a **web** page (The total time to load page ***or*** broken links)
2. Created via Standard CheckPoint

Database

1. Checks the contents of a database accessed by your AUT/SUT

   Example 3: You can use a database checkpoint to check the ***contents*** of a database

   a) Requires an ODBC DSN on the local machine

   b) The ***DSN*** must have a valid **UserId** & **Password**

   c) It validates the entire ***Result Set*** from a SQL query

   d) It does **NOT** validate what is displayed on the window/page

XML : Extensible Markup Language

1. Checks the data content of XML documents in XML files or XML documents in Web pages and frames
2. XML is a meta-markup language for ***text*** documents
3. XML makes the complex data structures portable between different computer environments
4. XML primary purpose is to ***share*** data
5. An XML file can be a static data file that is accessed in order to retrieve commonly used data
6. XML files are often intermediary file that retrieves dynamic data
7. Two types of XML checkpoints:

   a) **XML/WebPage Frame** : validates a document within a web page frame
   b) **File** : Checks a specified XML file

SUPPORTED CHECKPOINTS

	WEB	STD	VB	ACTIVEX
Standard	S	S	S	S
Image	S	NS	NS	NS
Table	S	S	S	S
Text	S	S	NS	NS
Bitmap	S	S	S	S
Accessibility	S	NS	NS	NS
XML	S	N/A	N/A	N/A
Page	S	N/A	N/A	N/A
Database	S	S	S	S

S=Supported NS=Not Supported N/A=Not Applicable

- In the HP QTP Workshop Series – Level 1 : We will cover : ***Standard, Text, Bitmap***

VISUAL CUES

1. The behavior of most applications is to ***inform*** the user - the outcome of *that* process
2. Typical responses :
    a) Static *text* message
    b) Flashing message
    c) Dialog Box
    d) Response window
3. A ***visual cue*** is a **measurable** result or *display* that indicates the system is functioning as expected
4. Testers use **visual cues** for manual verification that they ***see*** on the screen (GUI)
5. HP QuickTest can check parts of an application which are not *visible*
6. A ***CheckPoint*** is a specialized step in a script that compares two values and reports the outcome
7. The values can come from one of the objects properties or values calculated from the AUT
8. If the two values match → then the step ***passes***
9. Otherwise the step ***fails***

TIPS (What Property Do I use to *define* a checkpoint ?)

1. Do NOT automatically select the pre-defined list of properties to *check*
2. Scroll down the property list to find property (or *properties*) that best ***meets*** the requirement

    Example 6: Here are just a few :

    a) **Button** : enabled, focused, label

    b) **List** : number of items *or* the selected item

    c) **Edit** : name, enabled, focused, text, formattedtext
3. Text Checkpoint

    a) Purpose: To ***Verify*** the *values* or **information** displayed on the screen

    b) **Standard** : Select the available properties from the selected object

    c) Text Checkpoint : The **text** property from the object (**Label, Value, String , etc ...**)

HINT:

1. Remember to match the test requirement as ***close*** to the objects property
2. Reduce the QTP window such that the **AUT** is adjacent or side by side to QTP

CHECKPOINTS

1. Sample Script :

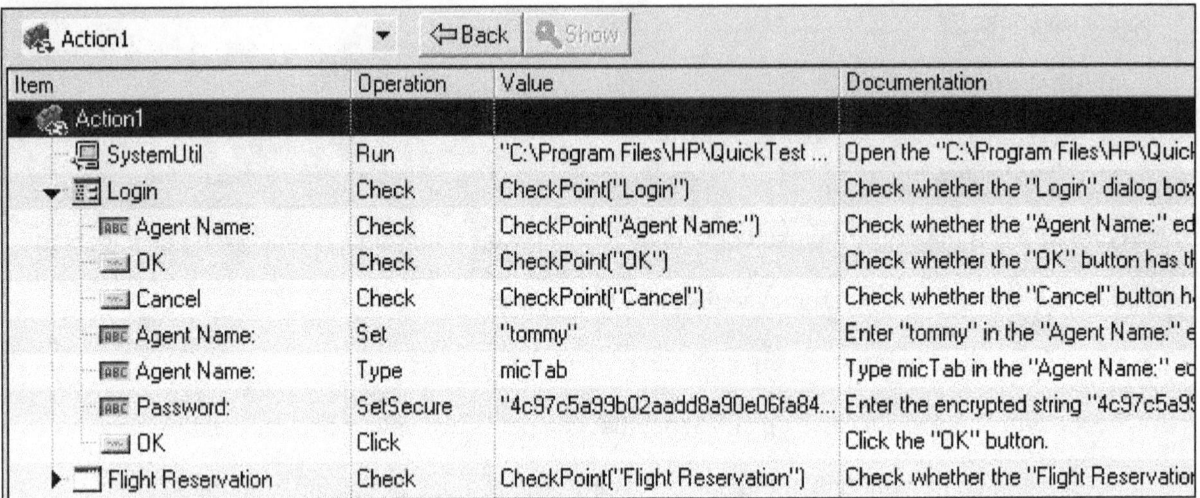

Item	Operation	Value	Documentation
Action1			
SystemUtil	Run	"C:\Program Files\HP\QuickTest ...	Open the "C:\Program Files\HP\Quicl
Login	Check	CheckPoint("Login")	Check whether the "Login" dialog box
Agent Name:	Check	CheckPoint("Agent Name:")	Check whether the "Agent Name:" ed
OK	Check	CheckPoint("OK")	Check whether the "OK" button has tl
Cancel	Check	CheckPoint("Cancel")	Check whether the "Cancel" button h
Agent Name:	Set	"tommy"	Enter "tommy" in the "Agent Name:" e
Agent Name:	Type	micTab	Type micTab in the "Agent Name:" ed
Password:	SetSecure	"4c97c5a99b02aadd8a90e06fa84...	Enter the encrypted string "4c97c5a9!
OK	Click		Click the "OK" button.
Flight Reservation	Check	CheckPoint("Flight Reservation")	Check whether the "Flight Reservatio

2. Checkpoints are typically based on the Business Rules :

   a) Individual Object Checkpoints (*Title Bar, Agent Name, Buttons, Static Text*, etc ...)

   b) Verifying the Property of the object

   c) Specify more than One (1)

   d) Stored in the Object Repository

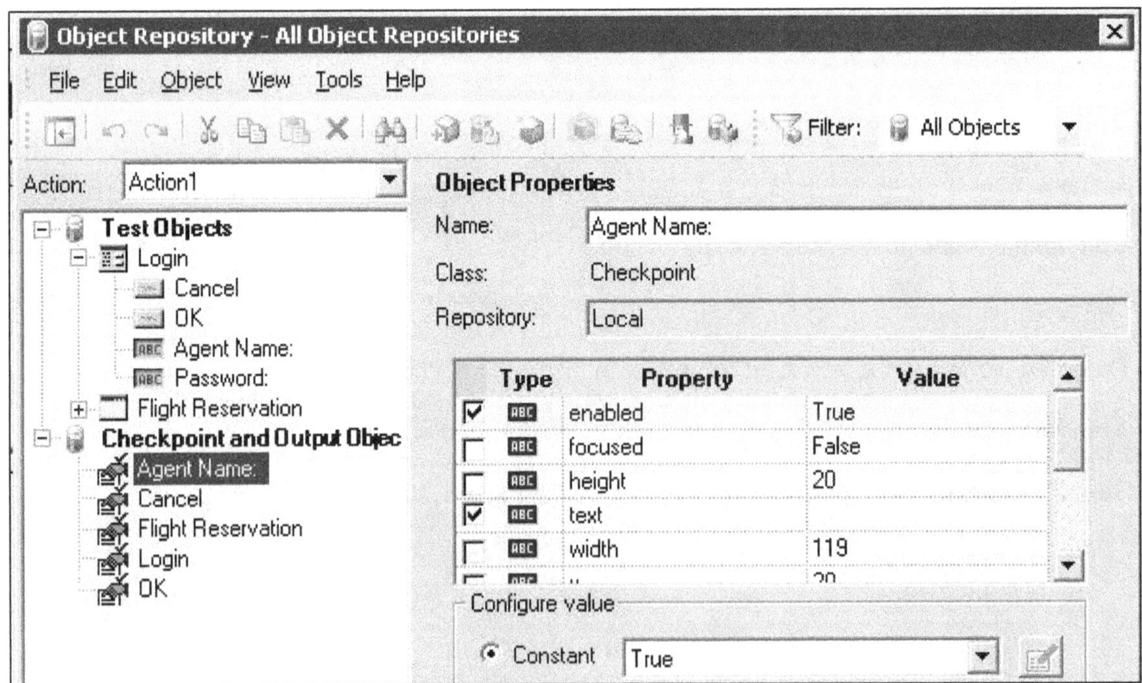

3. Modify an Existing Checkpoint :

   a) Highlight the existing *Checkpoint Object* ***Agent Name:*** in the Keyword View

   b) Insert (menu bar) → Checkpoint → Existing Checkpoint :

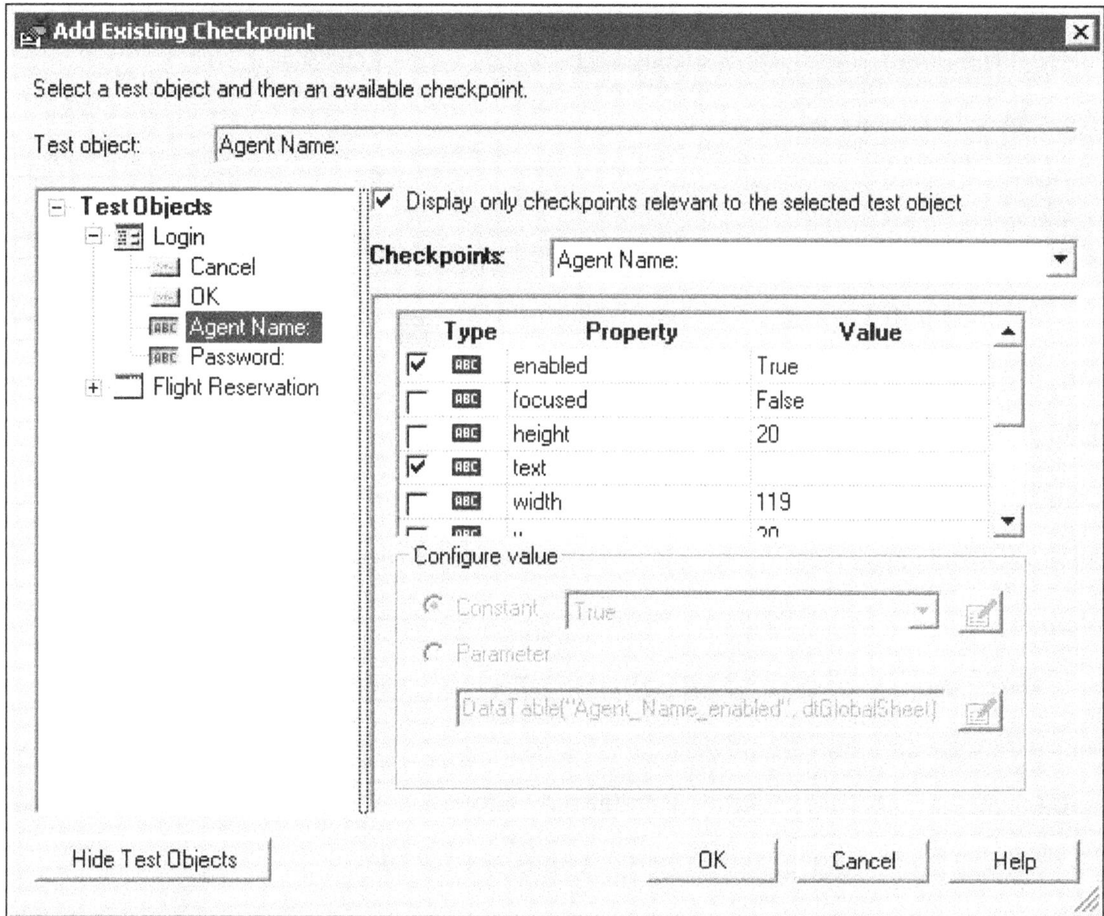

   c) Scroll, Select & enhance the CheckPoint Property Values :

      - Choose to add more properties
      - Change the existing values
      - Remove properties

LAB 4A : Standard CheckPoints (Part I, II, III)

Overall Scenario :

- Part 1 : Invoke Flight4A → Log-In → Exit the AUT
- Part 2 : Invoke Flight4A → Log-In → Add New Order → Delete Order → Exit the AUT

PART I : Test Requirements : Issue a ***separate*** checkpoint for the ***LOGIN*** window

a) The **Login** window displays the text **Login** in the *title bar*
b) The ***Agent Name*** input field is *enabled* and the *text* value is empty
c) The **OK** button is enabled (*on your own*)
d) The **Cancel** button is enabled (*on your own)*
e) The **Flight Reservation** is enabled and the title bar displays the *text* **Flight Reservation**

Steps: (Part I) : Verify the Login Process (*Do not Add Orders ...*) **(Test Will Fail !)**

1. Close all applications (except HP QuickTest)

2. **File → New → Test ...** (create a *new* script)

3. **Automation** (menu bar) → **Record and Run Settings ...**

4. Check **Record and run test on any open Window-based application** radio button → Click **OK**

5. **Automation** (menu bar) → **Record... F3** (***Record*** message is flashing in red)

6. **Start → Programs → QuickTest Professional → Sample Applications → Flight**

7. Click on the title bar **Login** (Activate the Window)

8. **Insert** (menu bar) → **Checkpoint → Standard Checkpoint... F12 →** cursor changes to a hand

9. Click on the title bar **Login** → Verify **Dialog : Login** → Click **OK**

10. The **Checkpoint Properties** window opens

    a) Uncheck all the properties
    b) Check **text** → Under **Configure Value** : Click **Constant** →verify **Login**

11. Click **OK** (Close **CheckPoint Properties** window)

12. **Insert** (menu bar) → **Checkpoint → Standard Checkpoint... F12 →** cursor changes to a hand

13. Click on the object **Agent Name:** (input field) → Verify **WinEdit : Agent Name:** → Click **OK**

14. The **Checkpoint Properties** window opens (Designed to *Fail*)

    a) Uncheck all the properties
    b) Check **enabled →** Under **Configure Value** : Click **Constant** → Select **False**
    c) Check **text** : (It should be empty)

15. Click **OK** (Close **CheckPoint Properties** window)

16. **Insert** (menu bar) → **Checkpoint → Standard Checkpoint... F12 →** cursor changes to a hand

17. Click on the object **OK** button → Verify **WinButton : OK** → Click **OK**

18. The **Checkpoint Properties** window opens

    a) Uncheck all the properties
    b) Check **enabled** → Under **Configure Value** : Click **Constant** → check **True**
    c) Check **text** → Under **Configure Value** : Click **Constant** → verify **OK**

19. Click **OK** (Close **CheckPoint Properties** window)

20. **Insert** (menu bar) → **Checkpoint** → **Standard Checkpoint... F12** → cursor changes to a hand

21. Click on the object **Cancel** button → Verify **WinButton : Cancel** → Click **OK**

22. The **Checkpoint Properties** window opens

    a) Uncheck all the properties
    b) Check **enabled** → Under **Configure Value** : Click **Constant** → Select **True**

23. Click **OK** (Close **CheckPoint Properties** window)

24. Login with ***yourname*** and use the password ***mercury*** → Click **OK** → Wait for the **Flight Reservation**
25. **Insert** (menu bar) → **Checkpoint** → **Standard Checkpoint...** → cursor changes to a hand

26. Click on the **Flight Reservation** (title bar) → Verify **Window : Flight Reservation**

27. The **Checkpoint Properties** window opens

    a) Uncheck all the properties
    b) Check **enabled** → Under **Configure Value** : Click **Constant** → Select **True**

28. Click **OK** (Close **CheckPoint Properties** window)

29. **File → Exit** (Close the AUT)

30. **Automation** (menu bar) → **Stop F4**

31. Review the *script* : Verify the five (5) *checkpoints* are created in the script :

Item	Operation	Value	Documentation
Action1			
SystemUtil	Run	"C:\Program Files\HP\QuickTest ...	Open the "C:\Program Files\HP\Quick
Login	Check	CheckPoint("Login")	Check whether the "Login" dialog box
Agent Name:	Check	CheckPoint("Agent Name:")	Check whether the "Agent Name:" edit
OK	Check	CheckPoint("OK")	Check whether the "OK" button has th
Cancel	Check	CheckPoint("Cancel")	Check whether the "Cancel" button ha
Agent Name:	Set	"tommy"	Enter "tommy" in the "Agent Name:" ed
Agent Name:	Type	micTab	Type micTab in the "Agent Name:" edi
Password:	SetSecure	"4c97c5a99b02aadd8a90e06fa84...	Enter the encrypted string "4c97c5a99
OK	Click		Click the "OK" button.
Flight Reservation	Check	CheckPoint("Flight Reservation")	Check whether the "Flight Reservation"
Menu	Select	"File;Exit"	Select item "File;Exit" from the "Menu"

32. Save the script as : **C:\YOURNAME\QTP10\LVL1\LAB4A_P1**

33. **Automation** → **Run ... F5** → Check **Temporary run results folder** → Click **OK**

34. The Test Should ***Fail*** ! → Review the **Test Results** window → **View** (menu bar) → **Expand All**

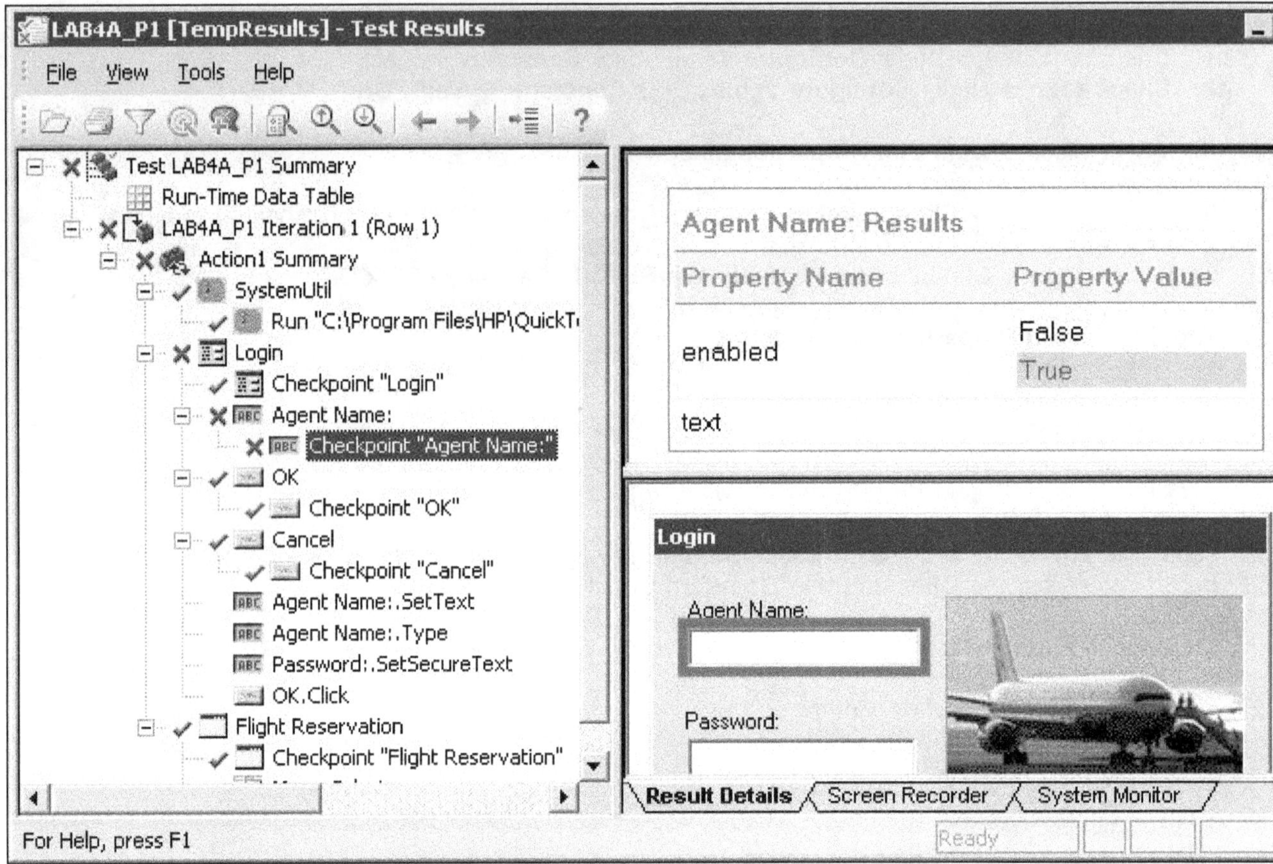

35. Locate & Highlight the *red* ***Checkpoint "Agent Name"*** → Scroll in the top right pane for messages

36. Determine the cause of the error → Close the **Test Results** window

37. In the ***Keyword View*** mode → Expand & Highlight **Checkpoint "Agent Name"** (in ***Value*** column)

38. Click **Blue** Check mark → Under **Configure Value** : Click **Constant** → Change ***enabled*** to **True** → Click **OK**

39. **File → Save C:\YOURNAME\QTP10\LVL1\LAB4A_P1** (Save the Script)

40. **Automation → Run ... F5** → Check **Temporary run results folder** → Click **OK**

41. Review the **Test Results** → Test should PASS → Close the ***Test Results*** window

QUESTION

1. What was the nature of the problem ?

2. How would you correct the error ?

******* END OF LAB4A_P1 – Part I *******

LAB 4A - PART II : Enhance the script

Test Requirements : Issue new ***separate*** checkpoints for creating a ***New Reservation***

1. Save *LAB4A_P1* script as **LAB4A_P2**
2. Create the following checkpoints while creating a ***new*** order :

   a) **Update, Delete, Insert** buttons are visible & disabled (*CheckPoint – Start*)
   b) Verify the Order was *created* : Look for message: **"Insert Done…"**
   c) Verify the Order was *deleted* : Look for message: **"Delete Done…"**

3. Modify the properties of an existing Checkpoint

Steps: (Part II)

1. Invoke the AUT & Login (on your own) → Verify you are in the main window ***Flight Reservation***

2. Save **C:\YOURNAME\QTP10\LVL1\LAB4A_P1 as LAB4A_P2** (if it is not already opened)

3. Highlight & Expand **Flight Reservation** → Highlight : √ ***Checkpoint "Flight Reservation"***

4. Start Recording (**F3**) → Swap to AUT → **File** → **New Order**

5. **Insert** (menu bar) → **Checkpoint** → **Standard Checkpoint** → cursor changes to a hand
6. Click on the **Update Order** button → Verify **WinButton : Update Order** → Click **OK**
7. Make the following changes → Click **OK**

   a) Uncheck all the properties
   b) Check **enabled** → Under **Configure Value** : Click **Constant** → Select **False**
   c) Check **text** → Under **Configure Value** : Click **Constant** → verify **&Update Order**

8. **Insert** (menu bar) → **Checkpoint** → **Standard Checkpoint** → cursor changes to a hand
9. Click on the **Delete Order** button → Verify **WinButton : Delete Order** → Click **OK**
10. Make the following changes → Click **OK**

    a) Uncheck all the properties
    b) Check **enabled** → Under **Configure Value** : Click **Constant** → Select **False**
    c) Check **text** → Under **Configure Value** : Click **Constant** → Verify **&Delete Order**

11. **Insert** (menu bar) → **Checkpoint** → **Standard Checkpoint** → cursor changes to a hand
12. Click on the **Insert Order** button → Verify **WinButton : Insert Order** → Click **OK**
13. Make the following changes → Click **OK**

    a) Uncheck all the properties
    b) Check **enabled** : Under **Configure Value** : Click **Constant** → Select **False**
    c) Check **text** Under **Configure Value** : Click **Constant** → Verify **&Insert Order**

14. Enter all the required *data* to **add** a reservation (on your own)

15. Click **Insert Order** button

16. **Insert** (menu bar) → **Checkpoint** → **Standard Checkpoint** → cursor changes to a hand

17. Click on the **Insert Done…** message → Verify **ActiveX:Threed Panel Control** → Click **OK**

18. Verify **ActiveX:Threed Panel Control** is displayed → Click **OK**

19. Make the following changes → Click **OK**

    a) Uncheck all the properties
    b) Check **enabled** : Under **Configure Value** : Click **Constant** → Select **True**
    c) Check **Caption** Under **Configure Value** : Click **Constant** → Verify **Insert Done...**

20. Click **Delete Order** → Click **Yes** (to confirm the **deletion** of the *order*)

21. **Insert** (menu bar) → **Checkpoint** → **Standard Checkpoint** → cursor changes to a hand

22. Click on the **Delete Done...** message

23. Verify **ActiveX:Threed Panel Control** → Click **OK**

24. Make the following changes → Click **OK**

    a) Uncheck all the properties
    b) Check **enabled** : Under **Configure Value** : Click **Constant** → Select **True**
    c) Check **Caption** Under **Configure Value** : Click **Constant** → Verify **Delete Done...**

25. Stop Recording (**F4**) → Review the script → Verify all your actions have been recorded

Item	Operation	Value	Documentation
Action1			
SystemUtil	Run	"C:\Program Files\HP\QuickTest ...	Open the "C:\Program File
Login	Check	CheckPoint("Login")	Check whether the "Login"
Flight Reservation	Check	CheckPoint("Flight Reservation")	Check whether the "Flight
Menu	Select	"File;New Order"	Select item "File;New Orde
Update Order	Check	CheckPoint("Update Order")	Check whether the "Upda
Delete Order	Check	CheckPoint("Delete Order")	Check whether the "Delete
Insert Order	Check	CheckPoint("Insert Order")	Check whether the "Insert
MaskEdBox	Type	"121010"	Type "121010" in the "Ma
Fly From:	Select	"Frankfurt"	Select the "Frankfurt" item
Fly To:	Select	"Los Angeles"	Select the "Los Angeles" i
FLIGHT	Click		Click the "FLIGHT" button
Flights Table			
Name:	Set	"Bobby Stewart"	Enter "Bobby Stewart" in t
Insert Order	Click		Click the "Insert Order" bu
Threed Panel Control	Check	CheckPoint("Threed Panel Control")	Check whether the "Three
Delete Order	Click		Click the "Delete Order" bu
Flight Reservations			
Threed Panel Control	Check	CheckPoint("Threed Panel Control...	Check whether the "Three
Menu	Select	"File;Exit"	Select item "File;Exit" from

26. Save the Script (**LAB4A_P2**)

27. Close the AUT

28. Playback & verify the **Test Results**

29. Close the **Test Results** window when finished

30. Highlight **CheckPoint ("Cancel")** and click until the Blue Check mark appears

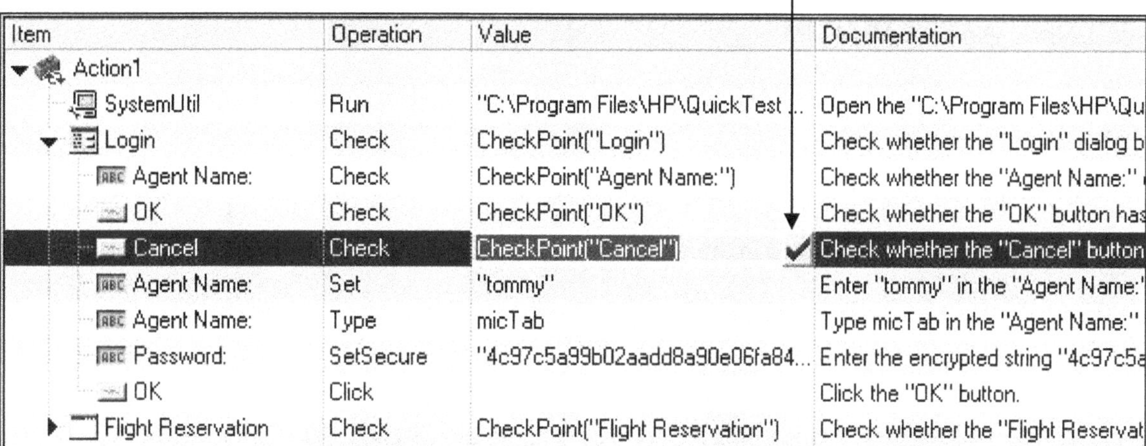

Item	Operation	Value	Documentation
Action1			
SystemUtil	Run	"C:\Program Files\HP\QuickTest ...	Open the "C:\Program Files\HP\Qu
Login	Check	CheckPoint("Login")	Check whether the "Login" dialog b
Agent Name:	Check	CheckPoint("Agent Name:")	Check whether the "Agent Name:"
OK	Check	CheckPoint("OK")	Check whether the "OK" button has
Cancel	Check	CheckPoint("Cancel")	Check whether the "Cancel" button
Agent Name:	Set	"tommy"	Enter "tommy" in the "Agent Name:"
Agent Name:	Type	micTab	Type micTab in the "Agent Name:"
Password:	SetSecure	"4c97c5a99b02aadd8a90e06fa84...	Enter the encrypted string "4c97c5a
OK	Click		Click the "OK" button.
Flight Reservation	Check	CheckPoint("Flight Reservation")	Check whether the "Flight Reservat

31. Scroll & Choose **text** property → verify the text contains **Cancel** → Click **OK**

32. Highlight **CheckPoint ("Flight Reservation")** and click until the Blue Check mark appears

33. Scroll & Choose **text** property → verify the text contains **Flight Reservation** → Click **OK**

34. Save the Script (**LAB4A_P2**) → Playback & Verify the **Test Results**

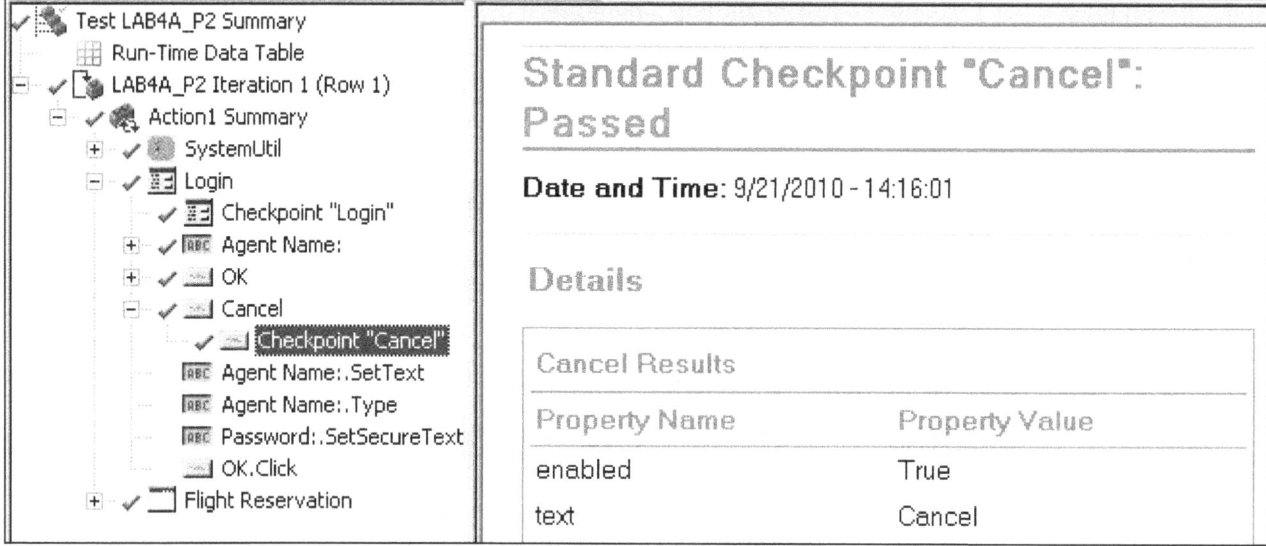

35. Close the **Test Results** window when finished

******* END OF LAB4A_P2 *******

LAB 4A - PART III : Further Enhancing the script

Test Requirements : Create a ***Bitmap*** Checkpoint

1. Save LAB4A_P2 as **LAB4A_P3**
2. Verify the picture (image) on the **ABOUT** window is the last **action** in the script (not the last line in script)
3. Where would you do that ?

   Answer: Review & Identify the ***line*** in the script before exiting the AUT

Steps: (Part III)

1. Invoke the AUT → Login (on your own)
2. Open **C:\YOURNAME\QTP10\LVL1\LAB4A_P2** → Save as **LAB4A_P3**
3. Highlight & Expand **Flight Reservation**
4. Highlight the **last** *checkpoint* in the script (the last action before *exiting* the AUT) :

   Threed Panel Control ***Check "Threed Panel Control "***

5. Start Recording (Press **F3**) → Swap to AUT → **Help** (menu bar) → **About ...**
6. **Insert** (menu bar) → **Checkpoint** → **Bitmap Checkpoint ...** → cursor changes to a hand
7. Point & Click on the airplane (image)
8. The **Bitmap Checkpoint Properties** window opens

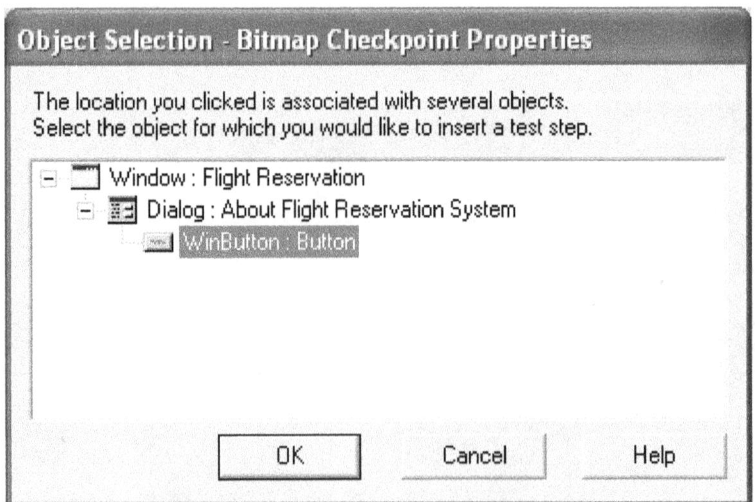

   Observe : The Hierarchy and the *object* (picture) is recognized as a **button**

9. Click **OK**

10. The **Bitmap Checkpoint Properties** window *opens*

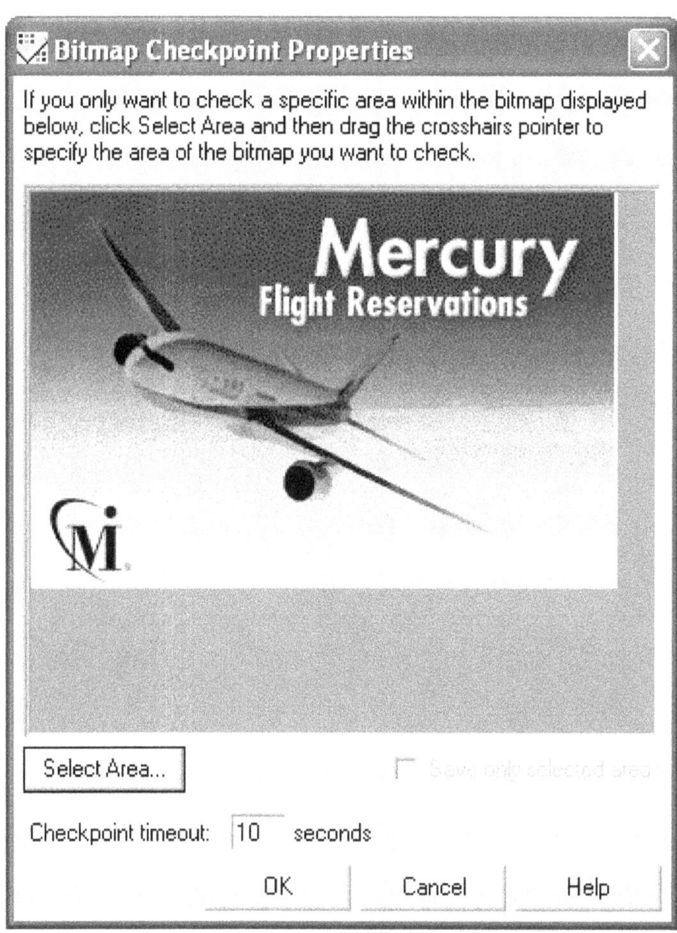

11. Click **OK** (Close the ***Bitmap Checkpoint Properties*** window)

12. Swap to AUT → Click **OK** (Close the ***About*** window)

13. Stop the recording & *Review* the script

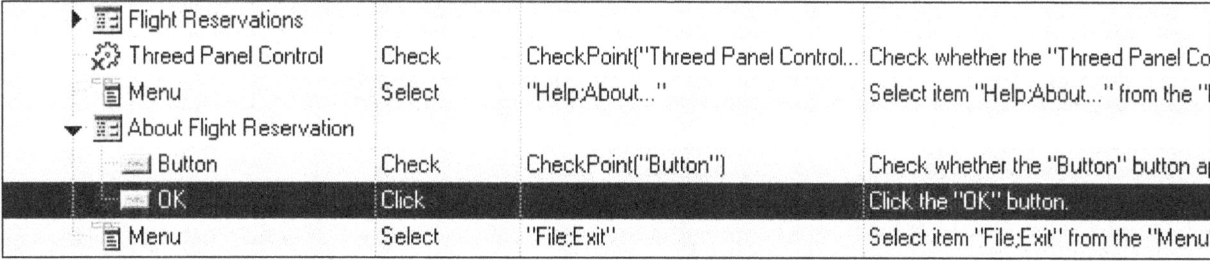

Flight Reservations			
Threed Panel Control	Check	CheckPoint("Threed Panel Control...	Check whether the "Threed Panel Co
Menu	Select	"Help;About..."	Select item "Help;About..." from the "
About Flight Reservation			
Button	Check	CheckPoint("Button")	Check whether the "Button" button a
OK	Click		Click the "OK" button.
Menu	Select	"File;Exit"	Select item "File;Exit" from the "Menu

14. Save the script **LAB4A_P3**

15. Close the AUT

16. Playback and Review the *Test Results* → Close the **Test Results** window when finished

***** **END OF LAB4A_P3** *****

LAB 4B : Text CheckPoints - (Part I, II)

Overall Process:

- Invoke Flight4A → Log-In → Open an *Order* → Verify *values* on the screen → Exit the AUT
- Create a ***new*** test script → Playback & Verify the results
- Add a new CheckPoint via the Active Screen (Flight Number)

PART I : Test Requirements : Create ***text*** checkpoint for an **existing** *order*

1. Open *existing* **Order # : 2**

2. Verify the following : (if different write the current values) : → Close the AUT

   a) Date of Flight : **mm/dd/yyyy** ______________________________

   b) Fly From: **Los Angeles** ______________________________

   c) Fly To: **San Francisco** ______________________________

   d) Name: **Fred Smith** ______________________________

Steps: (Part I)

1. **File** → **New** → **Test**
2. **Automation** (menu bar) → **Record F3**
3. Check **Record and run test on any open Windows-based application** → Click **OK**
4. **Start → Programs → QuickTest Professional → Sample Applications → Flight**
5. Login with ***yourname*** and use the password ***mercury*** → Click **OK →** Wait for the **Flight Reservation**
6. **File** → **Open Order** → Select **Order No.** check box → Type **2** Click **OK**
7. **Insert** (menu bar) → **Checkpoint** → **Standard Checkpoint** → Point to **Date of Flight**
8. Verify **ActiveX:MaskEdBox →** Click **OK** → Uncheck all properties
9. Check the following : ***cliptext***, ***default text*** , ***formatted text*** → Click **OK**
10. **Insert** (menu bar) → **Checkpoint** → **Standard Checkpoint** → Point to **Fly From :**
11. Verify **WinComboBox:Fly From →** Click **OK** → Uncheck all properties
12. Check the following : ***selection*** → Click **OK**
13. **Insert** (menu bar) → **Checkpoint** → **Standard Checkpoint** → Point to **Fly To :**
14. Verify **WinComboBox:Fly To →** Click **OK**
15. Check **Selection** that **San Francisco** is displayed → Click **OK**

16. **Insert** (menu bar) → **Checkpoint** → **Standard Checkpoint** → Cursor changes to a hand
17. Point & Click inside ***Name*** input field
18. Verify **WinEdit: Name** is diaplayed → Click **OK** → UnCheck all properties
19. Check that **Fred Smith** is displayed → Click **OK**
20. Swap to AUT → **File** → **Exit** (Close the AUT)
21. Stop the Recording (Press **F4**)
22. Save the script **C:\YOURNAME\QTP10\LVL1\LAB4B_P1**

Item	Operation	Value	Documentation
Action1			
SystemUtil	Run	"C:\Program Files\HP\QuickTest ...	Open the "C:\Program Files\HP\Quick
Login			
Flight Reservation			
Menu	Select	"File;Open Order..."	Select item "File;Open Order..." from th
Open Order			
MaskEdBox	Check	CheckPoint("MaskEdBox")	Check whether the "MaskEdBox" Acti
Fly From:	Check	CheckPoint("Fly From:")	Check whether the "Fly From:" list has
Fly To:	Check	CheckPoint("Fly To:")	Check whether the "Fly To:" list has th
Name:	Check	CheckPoint("Name:")	Check whether the "Name:" edit box h
Menu	Select	"File;Exit"	Select item "File;Exit" from the "Menu"

23. Run the *test* → It should pass ! → Close the ***Test Results*** window
24. Highlight **Name: Check CheckPoint ("Name:")** → In the Active Screen – Look for ***Flight No:*** **2824**

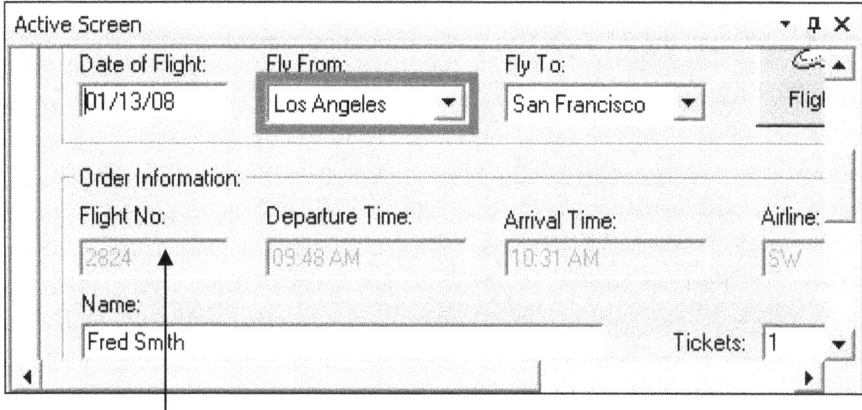

25. Move mouse over *Flight No:* **2824** → Right-Click → **Insert Standard Checkpoint …**
26. Verify **WinEdit: Flight No:** → Click **OK**
27. UnCheck all Properties → Check **Text** → Check **After current step** → Click **OK**
28. Save **LAB4B_P1** → Run the *test* → It should pass ! → Close the ***Test Results*** window

***** **END OF LAB4B_P1** *****

LAB 4B : CheckPoints - Part II (Lab will fail !)

Test Requirements : Modify the ***Date Of Flight*** data

a) Perform a partial recording in the middle of the script
b) After a user logins – verify the bitmap or *picture* exists in the **Flight Reservation** window
c) Modifying an existing *checkpoints* definition to check if it is *enabled*

Steps:

1. Save ***LAB4B_P1*** as **LAB4B_P2**

2. Start the AUT → Login → Wait for the Main Window (**Flight Reservation** window) to open [On your own]

3. In QTP → ***View*** (menu bar) → **Collapse All** → Click **Keyword View**

4. In **Keyword View** - Expand **Action1** → Highlight **Flight Reservation**

Item	Operation	Value	Documentation
Action1			
SystemUtil	Run	"C:\Program Files\HP\QuickTest ...	Open the "C:\Program Files\HP\QuickTest
Login			
Flight Reservation			

5. Start Recording (or Press **F3**)

6. **Insert** (menu bar) → **Checkpoint** → **Bitmap Checkpoint**

7. Point & Click on the picture (located on *right* side of the AUT's main window)

8. Verify **Static:static** → Click **OK** → A snapshot of the picture displays → Click **OK** (to accept the picture)

9. Stop the recording (or Press **F4**) → Close the AUT → The script should be similar to :

Item	Operation	Value	Documentation
Action1			
SystemUtil	Run	"C:\Program Files\HP\QuickTest ...	Open the "C:\Program Files\HP\Qui
Login			
Flight Reservation			
Menu	Select	"File;Open Order..."	Select item "File;Open Order..." from
Static	Check	CheckPoint("Static")	Check whether the "Static" text labe
Open Order			
MaskEdBox	Check	CheckPoint("MaskEdBox")	Check whether the "MaskEdBox" Ac
Fly From:	Check	CheckPoint("Fly From:")	Check whether the "Fly From:" list ha
Fly To:	Check	CheckPoint("Fly To:")	Check whether the "Fly To:" list has
Name:	Check	CheckPoint("Name:")	Check whether the "Name:" edit box
Flight No:	Check	CheckPoint("Flight No:")	Check whether the "Flight No:" edit b
Menu	Select	"File;Exit"	Select item "File;Exit" from the "Menu

a) Where is the first Checkpoint issued (taken) ? At Which *step* ?

b) Is the Checkpoint placed correctly in the *sequence* of actions ?

10. In the ***Keyword View*** mode → Expand All (if it is not already expanded)

11. Rearrange the **Static Check CheckPoint ("Static')** to before the ***Menu Select File;Open Order***

12. Highlight **Checkpoint ("MaskEdBox")** in the **Value** column (Date of Flight)

13. Click **Checkpoint ("MaskEdBox")** [in the **Value** column]

14. Click the **Blue** check mark → **Checkpoint properties ...**

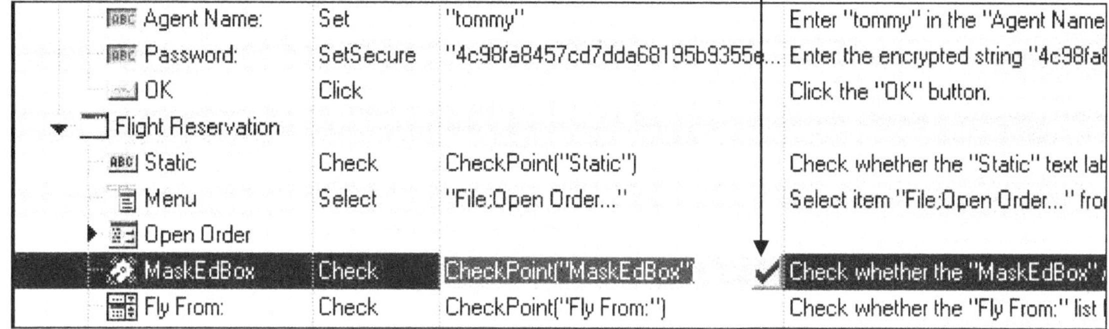

Agent Name:	Set	"tommy"	Enter "tommy" in the "Agent Name
Password:	SetSecure	"4c98fa8457cd7dda68195b9355e...	Enter the encrypted string "4c98fa
OK	Click		Click the "OK" button.
Flight Reservation			
Static	Check	CheckPoint("Static")	Check whether the "Static" text lab
Menu	Select	"File;Open Order..."	Select item "File;Open Order..." fro
Open Order			
MaskEdBox	Check	CheckPoint("MaskEdBox")	Check whether the "MaskEdBox"
Fly From:	Check	CheckPoint("Fly From:")	Check whether the "Fly From:" list

15. Uncheck ***cliptext*** and ***defaulttext*** → Check **enabled**

16. Under **Configure Value** → in **Constant** change to **False** → Click **OK** → Save **LAB4B_P2**

17. Run the *test* → It should Fail ! → Review the Results → Expand All

18. Highlight **CheckPoint MaskEdBox** → Why did it fail ! → Close the *Test Results* window

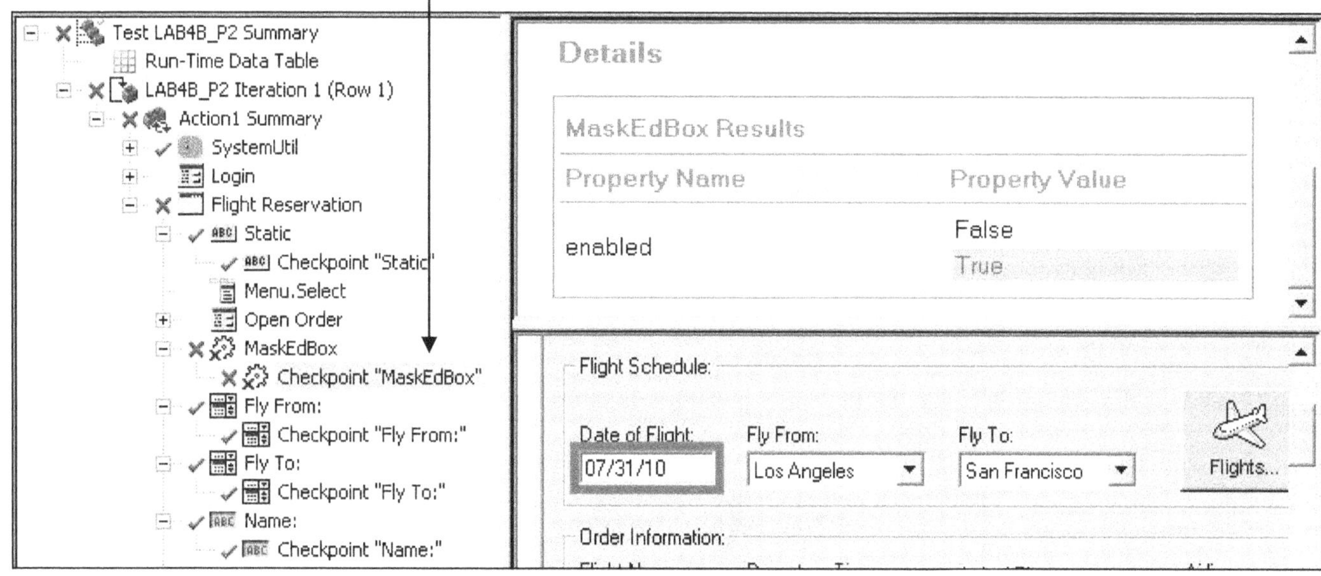

19. Highlight **MaskEdBox Check CheckPoint("MaskEdBox")**

20. Click **Checkpoint ("MaskEdBox")** [in the **Value** column] → Click the **Blue** check mark

21. Make the following changes and Click **OK**

- Uncheck all properties , Check : **formattedtext** , Check : **enabled : True**

22. Save **LAB4B_P2** → Run the *test* → It should Pass ! → Review & Close the ***Test Results*** window

Modify the Bitmap CheckPoint

23. Highlight **Static Check Checkpoint ("Static")**

24. Click **Blue** Check Mark → **Checkpoint properties …**

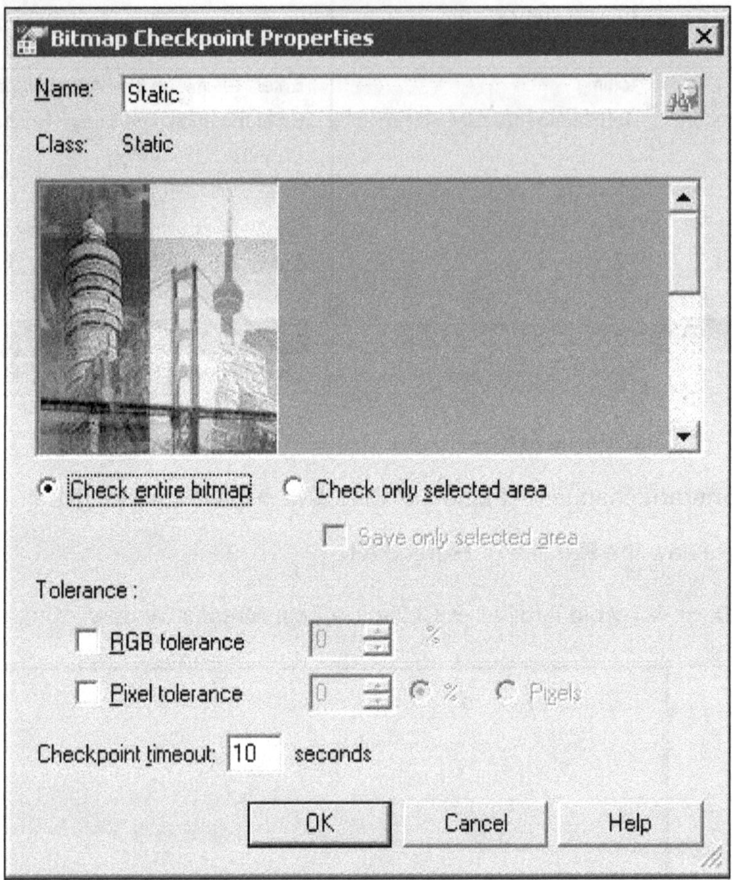

25. Click **Check only selected area …** button

26. Move *mouse* over the picture → cursor changes to plus sign (**+**)

27. Drag & draw an *area* inside the picture you want to capture

28. Select **Save only selected area** check box → Click **OK**

29. Change the checkpoint timeout to **5** seconds → Click **OK**

30. Highlight **Static Check CheckPoint("Static")** (in the **Item** column)

31. Right Click → **Checkpoint properties …** → Verify that a ***portion*** of the map is captured → Click **OK**

32. Save the LAB4B_P2 script → Playback & Verify the Results → Close the **Test Results** window

***** **END OF LAB 4B_P2** *****

LAB 4C : CheckPoints & Properties (Optional)

Requirements:

- Create a new script and save it as LAB4C
- Copy & replace the **objectrepository.bdb** from LAB4A into **LAB4C\Action1** sub-folder
- Perform the following :

1. Login

   a) Check for the following and issue a checkpoint:

   - Title Bar text **Login**
   - Agent Name & Password is ***enabled*** & text are *empty*
   - Bitmap picture of the airplane

2. Main Window (**Flight Reservation**)

   a) Class : **First , Business, Economy** : focused=false, enabled=false, checked=off
   b) Message area (bottom left) enabled=true , focused=false, text= ""
   c) Order No (bottom right) enabled=true , focused=false, text= ""
   d) The Flight Button is ***disabled***
   e) After entering: ***Date of Flight, Fly From, Fly To*** the Flight button is: enabled
   f) Issue a bitmap checkpoint of the *plane* only
   g) Insert an order and take a checkpoint to verify the order was created (*on your own*)

3. Updating

   a) Use the order that was added earlier in Step 2
   b) The **Insert Order** should be disabled when the order selected is populated (issue a checkpoint)
   c) Change any of the fields on the screen and Update the order

4. Faxing

a) Check for the following and issue a checkpoint:

- The Fax Number is editable (enabled)
- The Name, Order #, Flight and Date are visible , but not editable

b) Click the Preview & check & issue a checkpoint for the following:

- The **Send** is *enabled* and in *focus*
- The **Cancel** is *enabled*

5. Delete

   a) Delete the order created earlier
   b) Issue a checkpoint confirming the delete
   c) The **Insert Order** Button should be disabled
   d) The **Flight** button should be disabled

***** END OF LAB 4C *****

OBJECT REPOSITORY MAINTENANCE

1. The Object Repository must contain ***all*** the objects required for your *test*

2. Test Cases may not reference all the objects in the AUT

3. During recording, only the objects that you come in contact with are stored in the *Object Repository*

4. At Run-Time, the AUT may change the *appearance* & *properties* of an object

5. During the Software Development, the developers will be creating many builds

6. Each **build** *may* contain objects that have been modified :

   a) Additional properties
   b) Removed properties
   c) Change the value of an existing property
   d) Renamed the *property* name
   e) New Objects
   f) Renamed Objects
   g) Replace with a different object which provide the same function

7. During Playback (*Object Not Found* error) :

   a) QTP may not be able to identify or *locate* the object
   b) User intervention required and the test will be stopped at that *step*
   c) The tester must *evaluate* and *analyze* why the test failed

   Question : How can you *add* or *update* objects during the execution of your script ?

UPDATE RUN mode

1. During execution , QTP will flash the *Update Run* status and it will attempt to **update** the following :

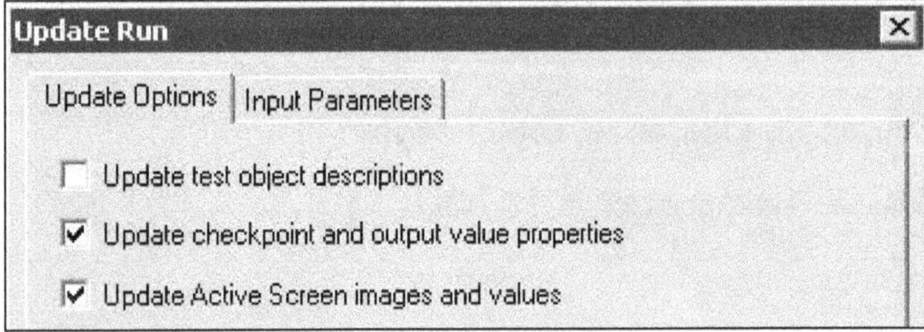

   a) **test object** descriptions
   b) Active Screen images & values
   c) Expected ***CheckPoint*** Values

2. Run only one (1) iteration of the test and one (1) iteration of each action

3. The new values will be saved & stored for future test *runs* in the Local Repository

4. The local repository has a higher priority than a Shared Repository (. ***TSR***)

5. It does NOT update any of the **parameterized** values (DataTable & Environment)

REASONS WHY A TEST FAILS ?

1. Object does NOT exist in the AUT

   a) The Mandatory Properties in the Repository do NOT match the AUT

2. The **parent** object has changed

   a) The Object's Hierarchy defined in the Repository do NOT match the AUT

3. The Object Description Property Values have changed

   a) The Object's Description property values has changed from the original values in the Repository

4. The Object does NOT exist in the Repository

   a) The Object is missing from either the Shared or Local repository

5. The Object Description Set (one or more properties) needs to be changed

   a) The current definition (to uniquely describe the object) is no longer unique

MAINTENANCE RUN WIZARD (MRW)

1. Useful when QTP is having problems or difficulties identifying an *Object* referenced in the script
2. Typically in subsequent Builds , the AUT may have changed
3. The Maintenance Run Wizard will attempt to help you re-identify & add missing object(s) to the Repository :

Problem (Error) Encountered	**Run Wizard Solution :**
1. Object Not Found in the AUT	Allowing you to Point and Click on the object in the AUT, and recommend one of many options
2. Object is missing from one of the Associative Repositories	Add the missing object to the Repository
3. Object exists in the AUT, but can only be *identified* thru ***Smart Identification***	Assist you to modify the description of the Object without requiring ***Smart Identification***

Note: *Smart Identification* is a feature allowing QTP to identify an object in the AUT by referencing & using Additional properties to uniquely locate the object. However, it takes more time to execute.

4. Comments can be included as part of the Test Step
5. It is valid while executing *tests* in the Normal mode (Not in the ***Fast*** mode)
6. A Summary Screen provides a count of objects added or modified
7. Requires a User Interface (will NOT work for web services or Non-Visuals Interface)
8. Alternate method is to use **Update from Application** (from the Object Repository)

LAB 4D : Update from Application

Requirements:

- Scenario : Run AUT → Login → Issue Several Checkpoints → Exit AUT
- Backup the repository (you will be making changes)
- Update the repository with new *property* values
- Update the repository with new *test object* description (Part 2)

Steps (Part 1)

1. **File → New → Test ...** (create a *new* script)
2. **Automation** (menu bar) → **Record and Run Settings ...**
3. Check **Record and run test on any open Window-based application** radio button → Click **OK**
4. **Automation** (menu bar) → **Record... F3** (***Record*** message is flashing in red)
5. **Start → Programs → QuickTest Professional → Sample Applications → Flight**
6. **Insert** (menu bar) → **Checkpoint → Standard Checkpoint** → Point to **Login** title bar
7. Verify **Dialog : Login →** Click **OK** → Uncheck all properties → Check **text** → Click **OK**
8. Enter your *user-id* & *password* → Click **OK**
9. **Insert** (menu bar) → **Checkpoint → Standard Checkpoint** → Point to **Flight Reservation** title bar
10. Verify **Window : Flight Reservation →** Click **OK** → Uncheck all properties → Check **text** → Click **OK**
11. **Insert** (menu bar) → **Checkpoint → Standard Checkpoint** → Point to **Fly From** como-box
12. Verify **Window : Flight Reservation →** Click **OK** → Uncheck all properties → Check **text** → Click **OK**
13. **Insert** (menu bar) → **Checkpoint → Standard Checkpoint** → Point to **Fly From** como-box
14. Verify **WinEdit : Name →** Click **OK** → Uncheck all properties → Check **text** → Click **OK**
15. **File → Exit** (Close the AUT)
16. Press **Stop (F4)** (Stop Recording)
17. Save as **LAB4D** → Playback → It Should Pass !
18. Review & Close the *Test Results* window

BACKUP THE REPOSITORY

19. **Resources → Object Repository**
20. **File → Export Local Objects...**
21. **C:\YOURNAME\QTP10\LVL1\REPO\LAB4D_ORIG.TSR** → Click **Save**
22. Close the ***Object Repository*** window
23. Verify the file exists (Windows Explorer) : **C:\YOURNAME\QTP10\LVL1\REPO\LAB4D_ORIG.TSR**

24. **Resources** → **Object Repository** → Expand **Checkpoint and Output Objects** (bottom left)

25. Highlight **Login** → In the Right Pane , select **enabled** → Change to **False**

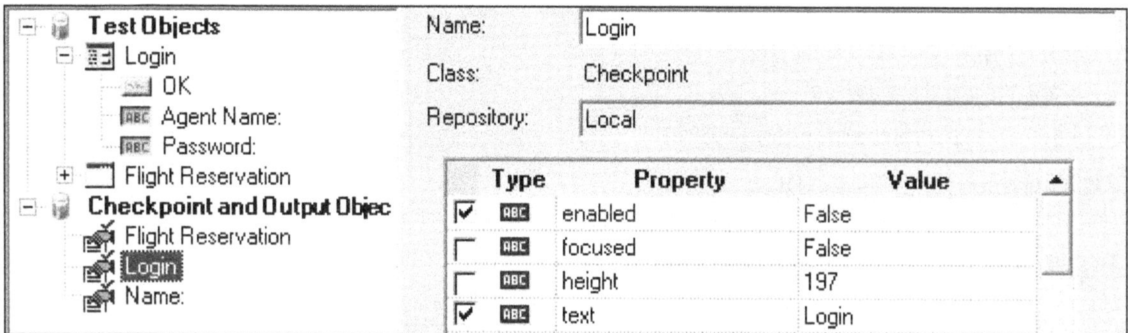

26. Highlight **Flight Reservation** → Check **text**

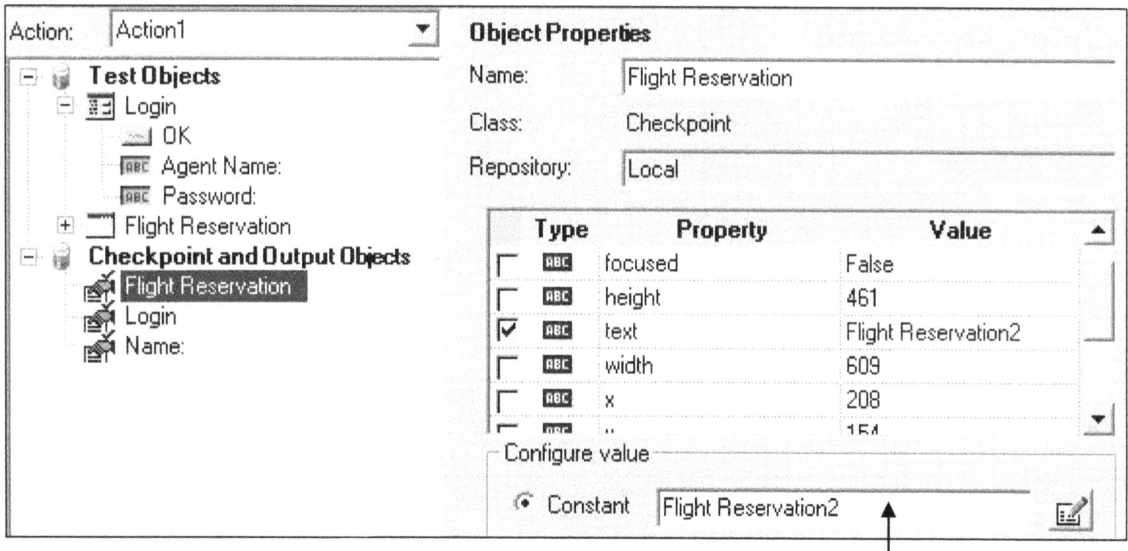

27. In the **Configure value** → Change to **Flight Reservation2**

28. **File** → **Close** (Close the *Object Repository* window)

29. Save **LAB4D**

30. Playback → It Should Fail on both CheckPoints !

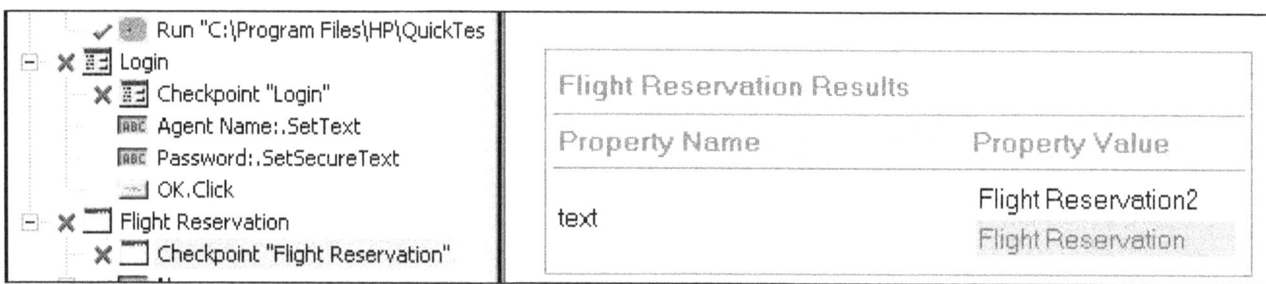

31. Close the **Test Results** window

32. **Automation** (menu bar) → **Update Run Mode** → Click **OK**

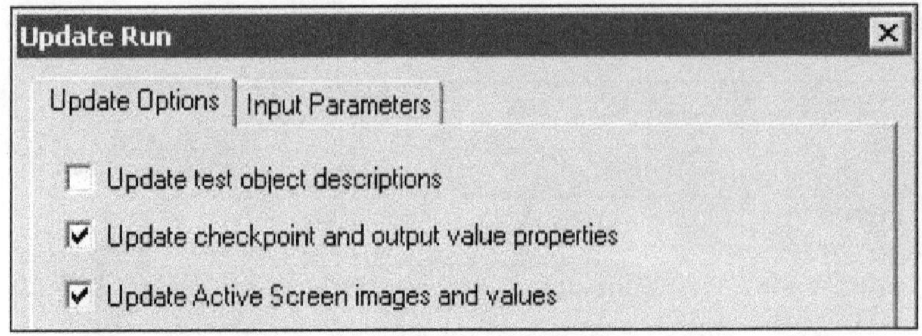

33. The Test should *Pass* ! → Review the **Test Results** window

34. Expand & Highlight **Checkpoint :Login**"

Test LAB4D Summary
Run-Time Data Table
LAB4D Iteration 1 (Row 1)
Action1 Summary
SystemUtil
Run "C:\Program Files
Login
Checkpoint "Login"
Agent Name:.SetText
Password:.SetSecure
OK.Click
Flight Reservation
Checkpoint "Flight Res

Date and Time: 9/27/2010 - 17:09:53

Details

Login Results

Property Name	Property Value	Updated Value
enabled	False	True
text	Login	<Unchanged>

35. Expand & Highlight **Checkpoint :Flight Reservation**"

36. Close the ***Test Results*** window

*** **END OF LAB4D – Part 1** ***

STEPS (Part 2) : Assume the Object Description has changed !

1. **Resources → Object Repository** → Highlight **Agent Name:** (Top Left)
2. Highlight **attached text Agent Name**: property → Click Red (**X**) → Remove the selected property
3. Click the green plus (+) sign, scroll & highlight property **x** (coordinate) → Click **OK**
4. Click the green plus (+) sign, scroll & highlight property **y** (coordinate) → Click **OK**
5. Assign the following values : x = 20 and y = 59

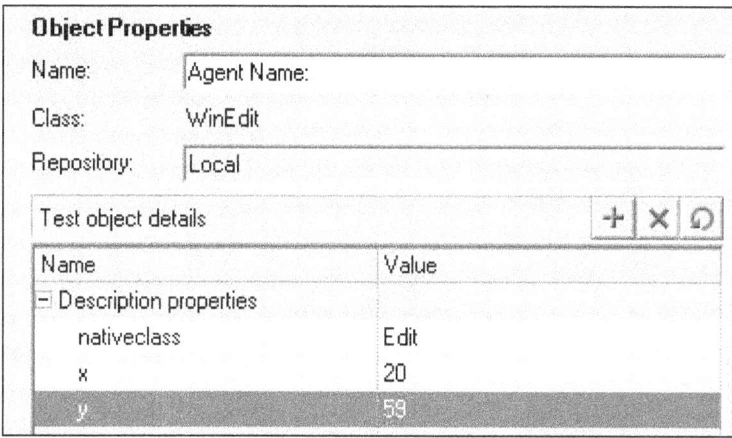

6. Expand **Flight Reservation** → Highlight object **Name:** (Passenger name)
7. Highlight **attached text Name**: property → Click Red (**X**) → Remove the selected property
8. Click the green plus (+) sign, scroll & highlight property **height** (coordinate) → Click **OK**
9. Click the green plus (+) sign, scroll & highlight property **width** (coordinate) → Click **OK**
10. Assign the following values : height = 20 and width = 270

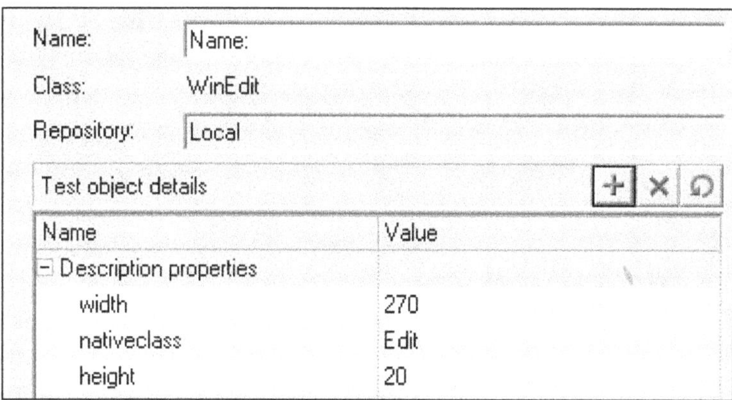

11. Close the ***Object Repository*** window
12. Save **LAB4D**

13. Highlight **Password: SetSecure** → Review the User-Id used (tommy)

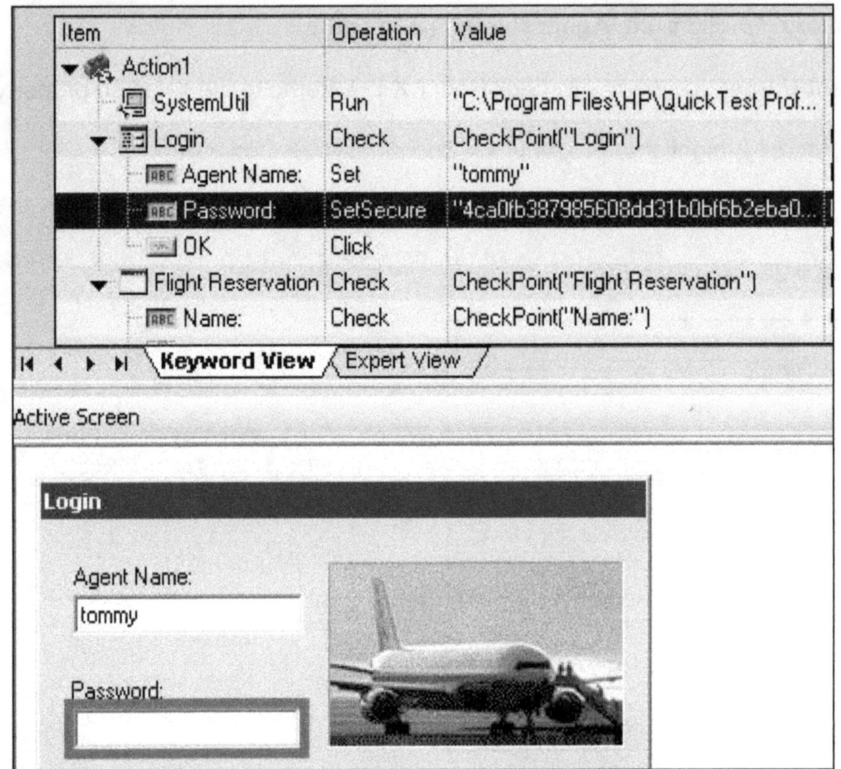

Item	Operation	Value
Action1		
SystemUtil	Run	"C:\Program Files\HP\QuickTest Prof...
Login	Check	CheckPoint("Login")
Agent Name:	Set	"tommy"
Password:	SetSecure	"4ca0fb387985608dd31b0bf6b2eba0...
OK	Click	
Flight Reservation	Check	CheckPoint("Flight Reservation")
Name:	Check	CheckPoint("Name:")

14. Change your username (Do NOT use the same User-ID you specified in the original recording)

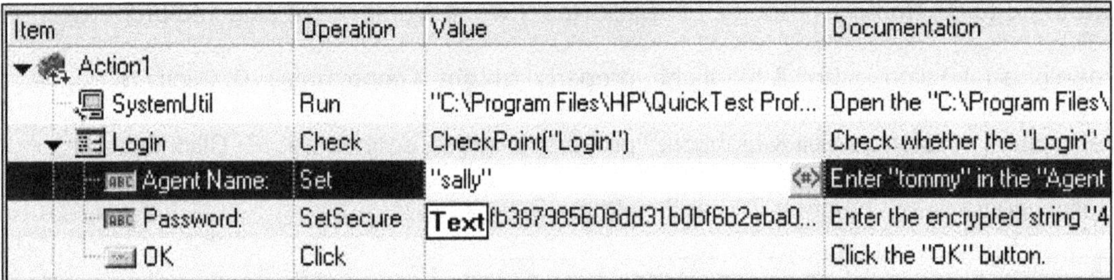

Item	Operation	Value	Documentation
Action1			
SystemUtil	Run	"C:\Program Files\HP\QuickTest Prof...	Open the "C:\Program Files\
Login	Check	CheckPoint("Login")	Check whether the "Login" c
Agent Name:	Set	"sally"	Enter "tommy" in the "Agent
Password:	SetSecure	Text fb387985608dd31b0bf6b2eba0...	Enter the encrypted string "4
OK	Click		Click the "OK" button.

Note: The user-id is different (**sally**)

15. Save **LAB4D**

16. **Automation** (menu bar) → **Update Run Mode** → Click **OK**

17. Choose all the *Checkboxes* → Click **OK**

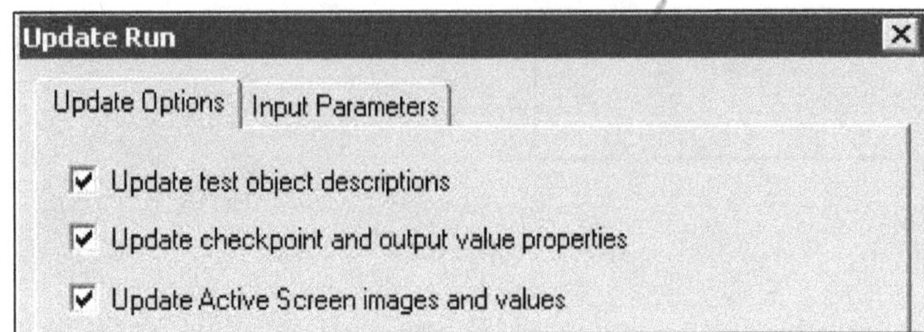

18. Review the **Test Results** window → Highlight **Name:**

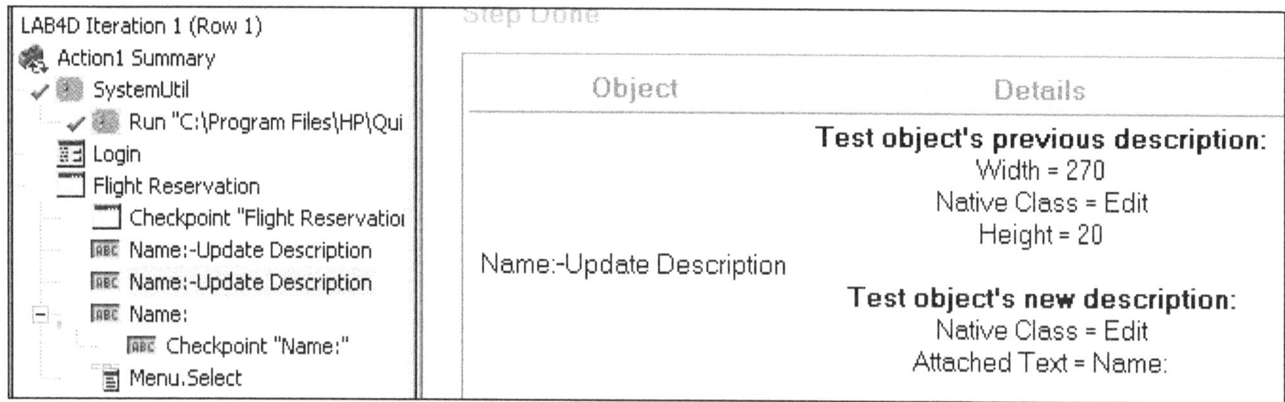

19. In the left pane, Highlight **Name:-Update description**

20. Right-Click → **Jump to Step in HP Quicktest** → It will bring you to the referenced position in the QTP script

    a) You can review the object that has changed !

21. Swap to & Close the *Test Results* window

22. Highlight **Password: SetSecure** → Review the User-Id used (sally)

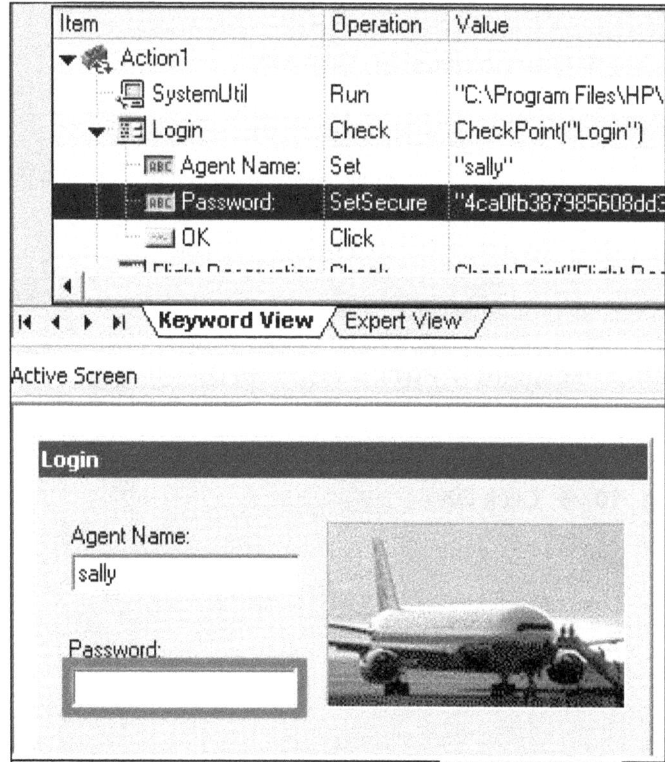

Note: The Active Screen has been updated with the new values

*** **END OF LAB 4D – Part 2** ***

LAB 4E : Maintenance Run Wizard (M R W)

Requirements:

- Scenario : Run AUT → Login → Open an Order →Issue several Checkpoints → Exit AUT
- Backup the repository
- Delete objects from the repository & Change property values
- Add objects with the Maintenance Run Wizard

Steps (Part 1)

1. **File → New → Test ...** (create a *new* script)
2. **Automation** (menu bar) → **Record and Run Settings ...**
3. Check **Record and run test on any open Window-based application** radio button → Click **OK**
4. **Automation** (menu bar) → **Record... F3** (***Record*** message is flashing in red)
5. Run Flight4A
6. **Insert** (menu bar) → **Checkpoint** → **Bitmap Checkpoint** → Click inside the **Airplane** Picture
7. Verify **Static : Static** → Click **OK** → Select **Check entire bitmap** → Click **OK**
8. Login with your User-Id & Password → Click **OK**
9. **File** → **Open Order** → Select **Order No** check-box → Type **2** → Click **OK**
10. **Insert** (menu bar) → **Checkpoint** → **Standard Checkpoint** → Point to **Name** Edit field
11. Verify **WinEdit : Name:** → Click **OK** → Uncheck all properties → Check **text** → Click **OK**
12. **Insert** (menu bar) → **Checkpoint** → **Standard Checkpoint** → Point to **Fly From** combo-box
13. Verify **WinCombo : FlyFrom** → Click **OK**
14. Uncheck all properties → Check **Items Count 10** → Click **OK**
15. **File** → **Exit** (Close AUT)
16. **Stop the Recording**
17. Save **LAB4E**
18. Playback → It Should Pass !
19. Review & Close the **Test Results** window
20. **Resources** → **Object Repository** → **File** → **Export Local Objects...**

21. **C:\YOURNAME\QTP10\LVL1\REPO\LAB4E_ORIG.TSR** → Click **Save**

22. Close the ***Object Repository*** window

23. Verify the file **C:\YOURNAME\QTP10\LVL1\REPO\LAB4E_ORIG.TSR** exists

Item	Operation	Value	Documentation
Action1			
SystemUtil	Run	"C:\Program Files\HP\QuickTest Prof...	Open the "C:\Program Files\HP\(
Login			
Static	Check	CheckPoint("Static")	Check whether the "Static" text la
Agent Name:	Set	"tommy"	Enter "tommy" in the "Agent Nam
Agent Name:	Type	micTab	Type micTab in the "Agent Name:
Password:	SetSecure	"4ca158221c92ffa1dccd50c490b75d...	Enter the encrypted string "4ca15
OK	Click		Click the "OK" button.
Flight Reservation			
Menu	Select	"File;Open Order..."	Select item "File;Open Order..." fr
Open Order			
Flight Reservation	Activate		Make the "Flight Reservation" wir
Name:	Check	CheckPoint("Name:")	Check whether the "Name:" edit b
Flight Reservation	Activate		Make the "Flight Reservation" wir
Fly From:	Check	CheckPoint("Fly From:")	Check whether the "Fly From:" list
Menu	Select	"File;Exit"	Select item "File;Exit" from the "M

24. **Resources** → **Object Repository** → In **Test Objects**

25. Highlight **Static** → Right-Click → **Delete** → Click **Yes**

26. Highlight **Name:** → Right-Click → **Delete** → Click **Yes**

27. In *CheckPoint and Output Objects* : Highlight **Fly From:** → Change *items count 10* to **items count 12**

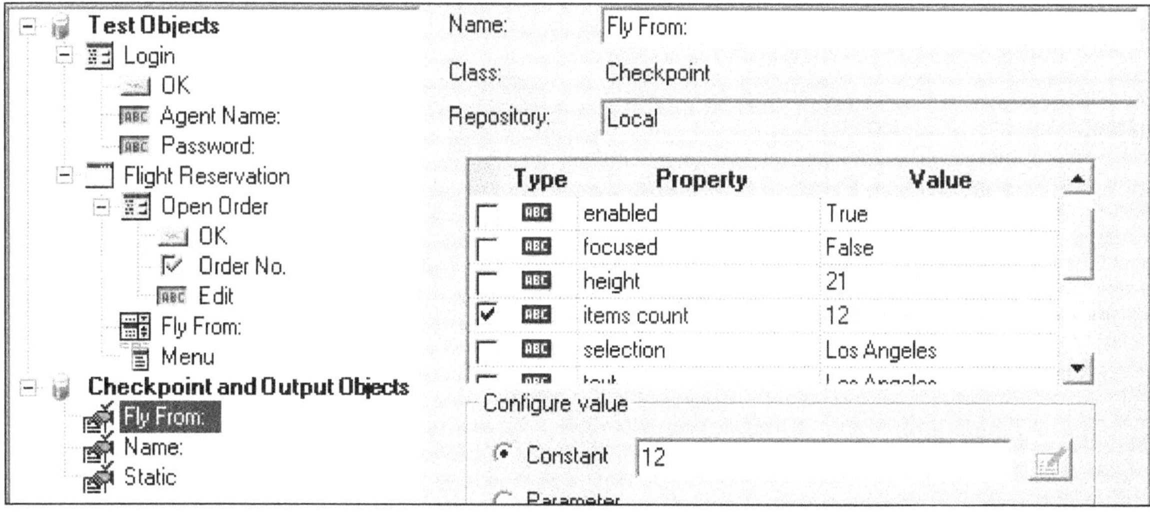

28. Close the ***Object Repository***

29. Save **LAB4E** → Playback → It should Fail with :

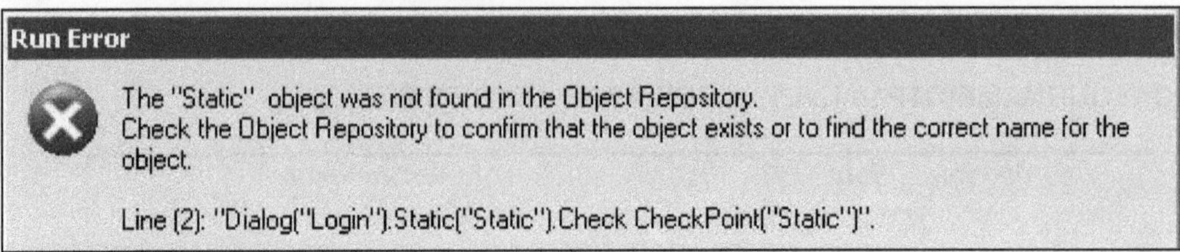

Run Error
The "Static" object was not found in the Object Repository.
Check the Object Repository to confirm that the object exists or to find the correct name for the object.
Line (2): "Dialog("Login").Static("Static").Check CheckPoint("Static")".

30. Click **Skip** → It Should Fail with :

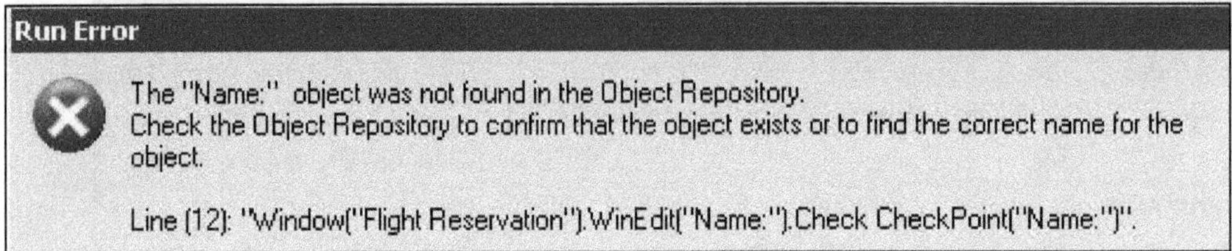

Run Error
The "Name:" object was not found in the Object Repository.
Check the Object Repository to confirm that the object exists or to find the correct name for the object.
Line (12): "Window("Flight Reservation").WinEdit("Name:").Check CheckPoint("Name:")".

31. Click **Skip** → The Test will continue with errors and Review the ***Test Results*** window

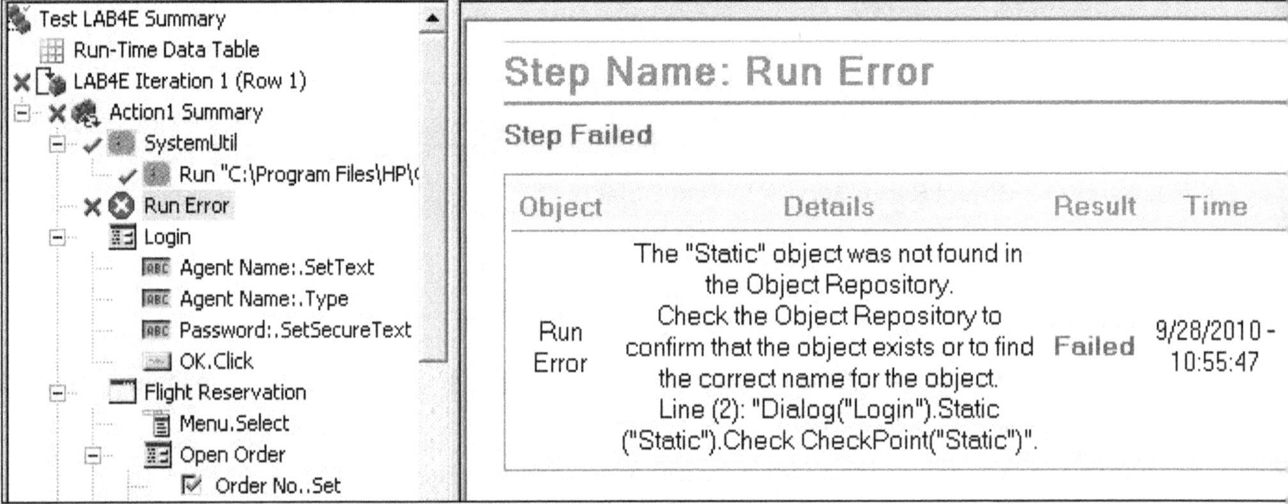

32. Highlight the first **Run Error** → Evaluate the error message

33. Highlight the second **Run Error** → Evaluate the error message

- The Object is NOT in the repository

34. Highlight the **Checkpoint "Fly From"** → Evaluate the error message

    a) The **items count** property value is different

35. Close the **Test Results** window

*** **END OF LAB4E – Part 1** ***

STEPS (Part 2) : Maintenance Run Wizard [MRW]

1. **Automation** (menu bar) → **Maintenance Run Mode ...** → The MRW will display :

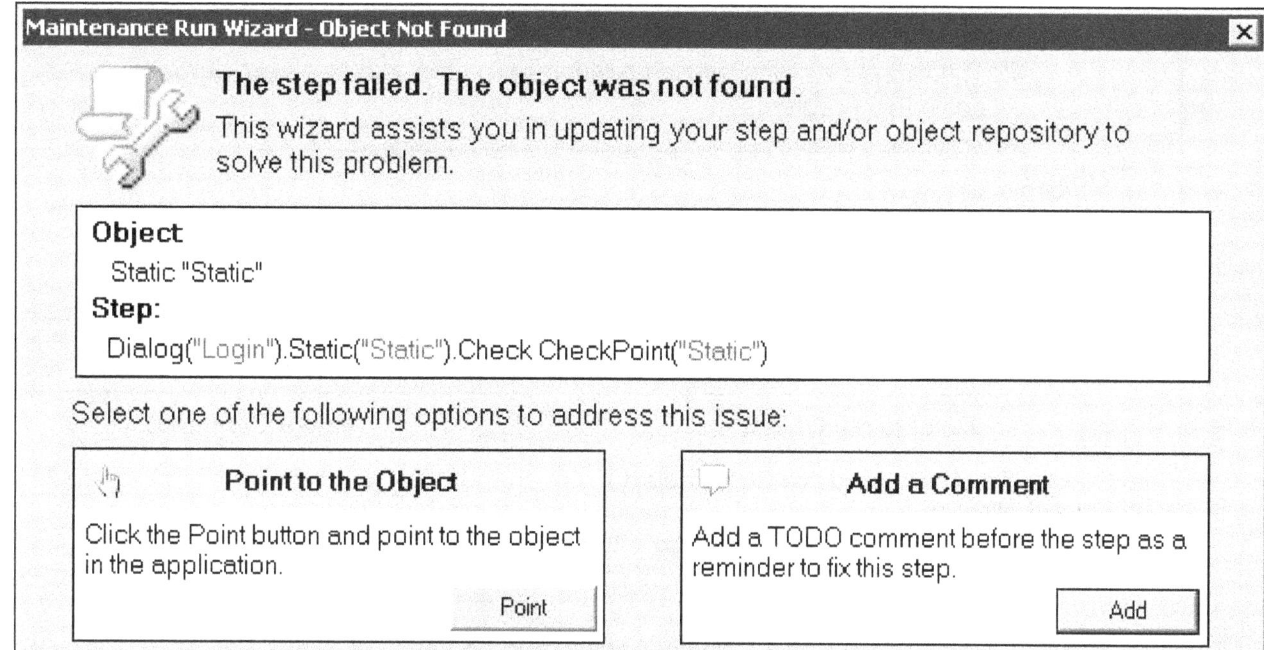

2. Click **Point** → Point and Click on the ***Airplane*** picture → Verify **Static : Static** → Click **OK**

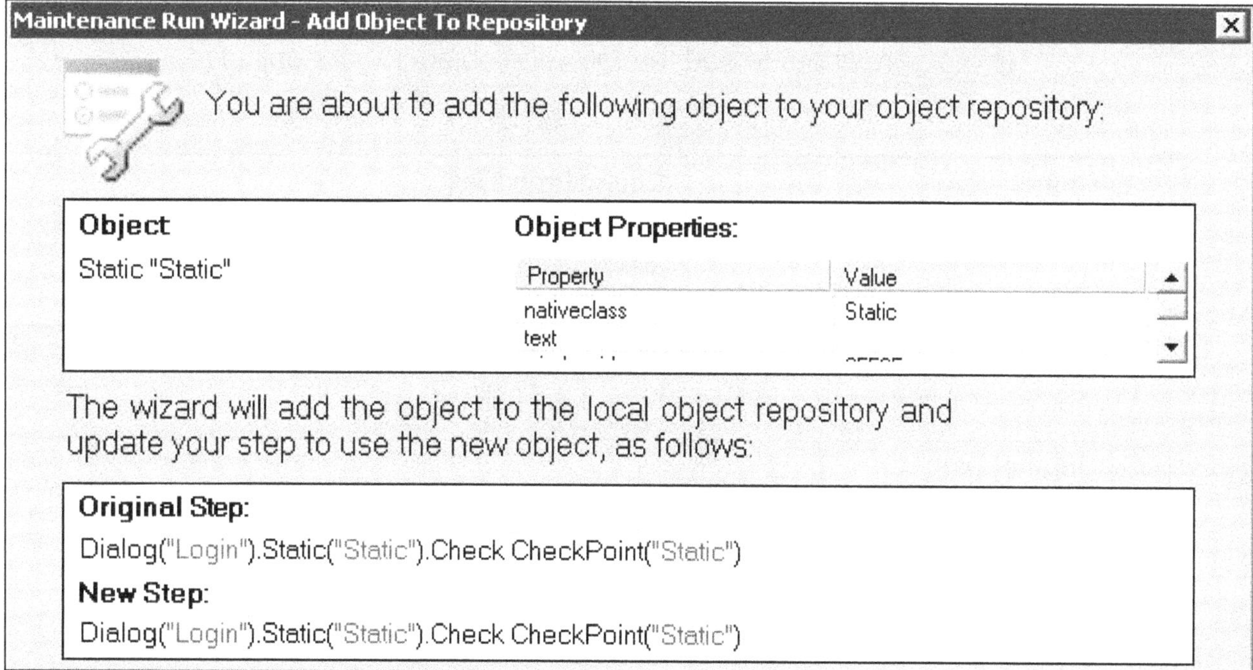

3. Review and Verify the recommended action :

   a) Add the missing object and keep the original step

4. Choose **Add the object and then update and rerun the step** → Click **OK**

5. MRW will recommend the following :

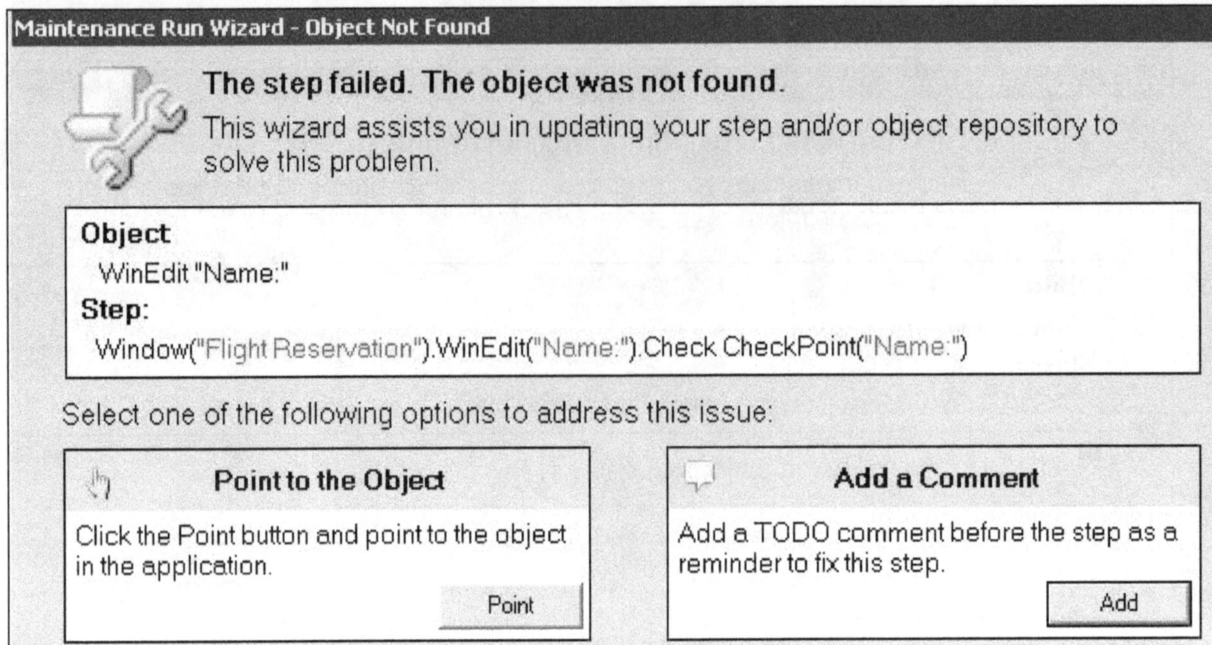
Maintenance Run Wizard - Object Not Found

The step failed. The object was not found.

This wizard assists you in updating your step and/or object repository to solve this problem.

Object
WinEdit "Name:"
Step:
Window("Flight Reservation").WinEdit("Name:").Check CheckPoint("Name:")

Select one of the following options to address this issue:

Point to the Object
Click the Point button and point to the object in the application.
Point

Add a Comment
Add a TODO comment before the step as a reminder to fix this step.
Add

6. Click **Point** → Point and Click on the ***Name:*** editable field → Verify **WinEdit : Name:** → Click **OK**

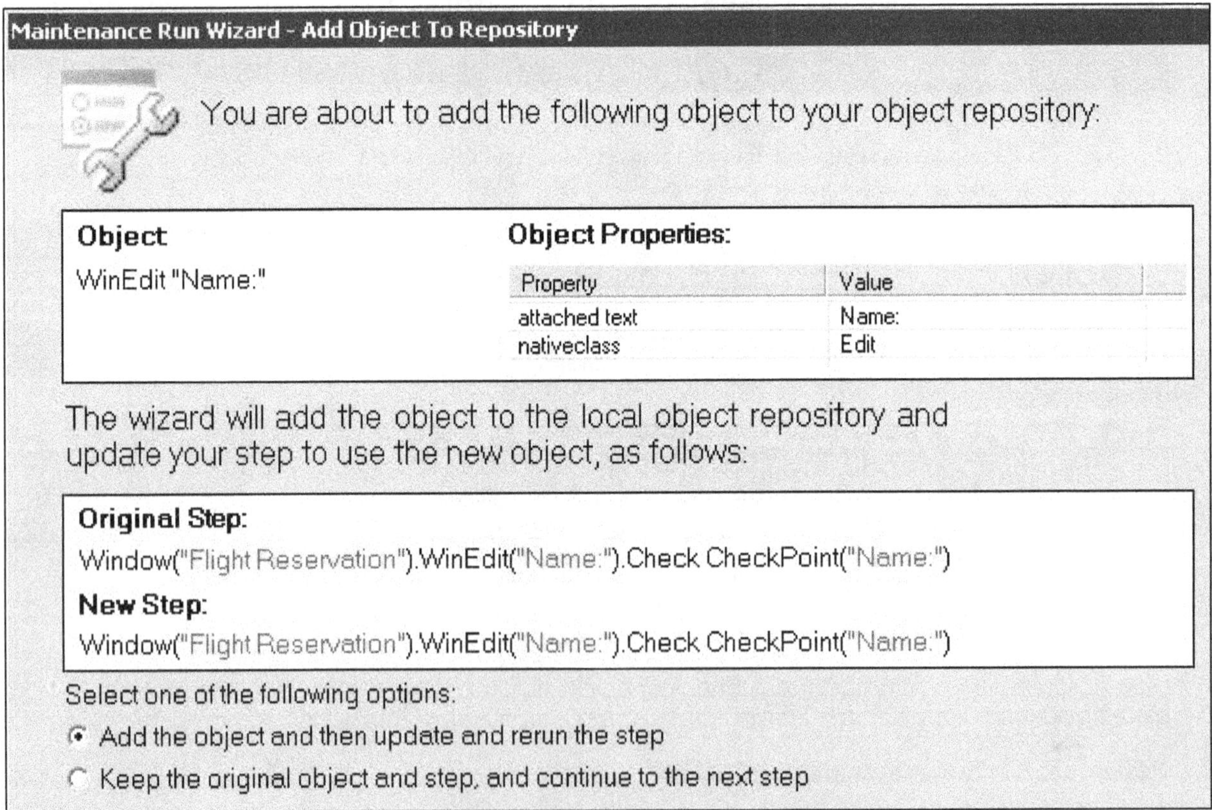
Maintenance Run Wizard - Add Object To Repository

You are about to add the following object to your object repository:

Object
WinEdit "Name:"

Object Properties:

Property	Value
attached text	Name:
nativeclass	Edit

The wizard will add the object to the local object repository and update your step to use the new object, as follows:

Original Step:
Window("Flight Reservation").WinEdit("Name:").Check CheckPoint("Name:")
New Step:
Window("Flight Reservation").WinEdit("Name:").Check CheckPoint("Name:")

Select one of the following options:
- (•) Add the object and then update and rerun the step
- ( ) Keep the original object and step, and continue to the next step

7. Choose **Add the object and then update and rerun the step** → Click **OK**

8. The Test will continue and MRW will generate a summary :

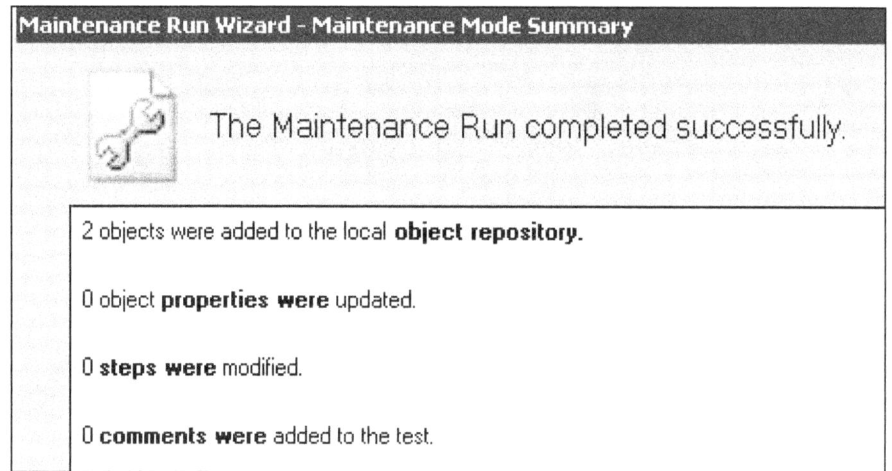

9. Click **Finish** → Review the **Test Results** window → The **FlyFrom** ***CheckPoint*** has failed :
   a) The **FlyFrom** object exists in the Repository
   b) MRW will refer to the current definition
10. Close the **Test Results** window
11. **Resources** → **Object Repository** → Observe the Objects have been added to the Repository
12. Expand **Flight Reservation** → Highlight **FlyFrom** → Right-Click → **Delete** → Click **Yes**
13. Close the **Object Repository** window
14. **Automation** (menu bar) → **Maintenance Run Mode ...** → The MRW will display :

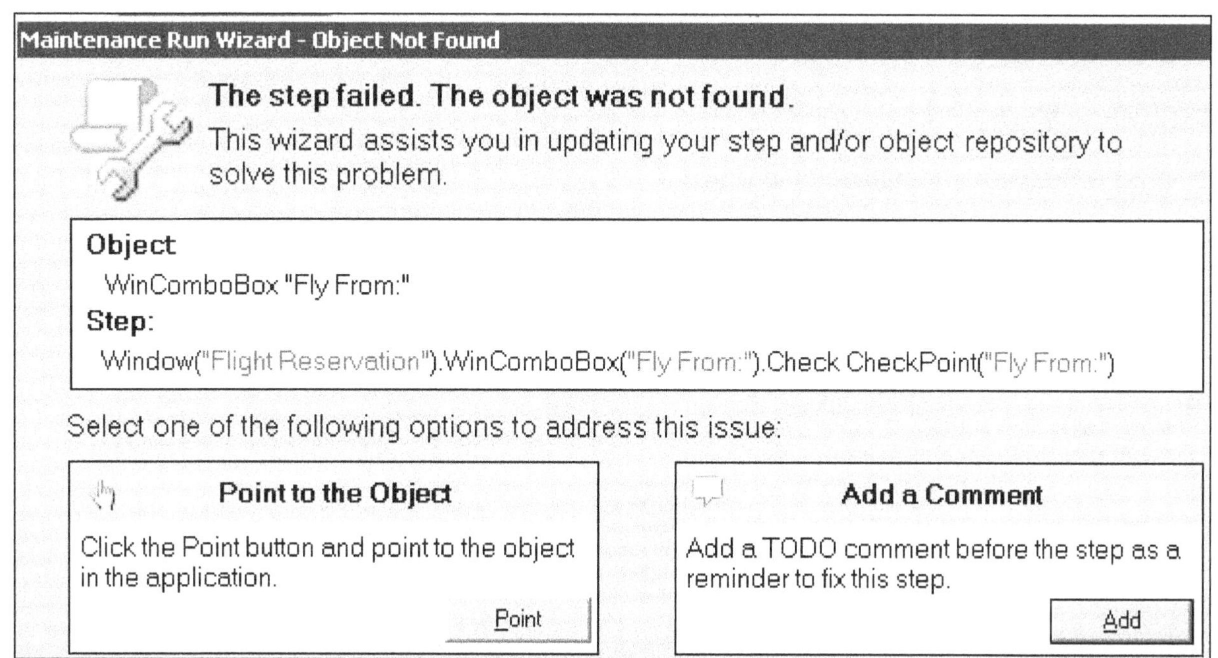

15. Click **Point** → Point and Click on the ***FlyFrom:*** Combo-Box

16. Verify **WinCombobox : Fly From:** → Click **OK** → MRW will display :

Maintenance Run Wizard - Add Object To Repository

You are about to add the following object to your object repository:

Object

WinComboBox "Fly From:"

Object Properties:

Property	Value
attached text	Fly From:
nativeclass	ComboBox

The wizard will add the object to the local object repository and update your step to use the new object, as follows:

Original Step:

Window("Flight Reservation").WinComboBox("Fly From:").Check CheckPoint("Fly From:")

New Step:

Window("Flight Reservation").WinComboBox("Fly From:").Check CheckPoint("Fly From:")

Select one of the following options:

- (•) Add the object and then update and rerun the step
- ( ) Keep the original object and step, and continue to the next step

17. Choose **Add the object and then update and rerun the step** → Click **OK**

18. Click **Finish**

19. Close & Review the **Test Results** window

20. Playback the Test → It Should Pass !

21. Close & Review the **Test Results** window

***** END OF LAB 4E *****

LAB 4F : Using the Shared & Local Repositories

Requirements:

- Scenario : Run AUT → Login → Open an Order →Issue several Checkpoints → Exit AUT
- Create a shared repository
- Delete objects from the local repository
- Use the *Maintenance Run Wizard* to add comments
- Copy objects to the Local thru the Shared Repository

Steps (Part 1)

1. **File → New → Test ...** (create a *new* script)
2. **Automation** (menu bar) → **Record and Run Settings ...**
3. Check **Record and run test on any open Window-based application** radio button → Click **OK**
4. **Automation** (menu bar) → **Record... F3** (***Record*** message is flashing in red)
5. Run Flight4A
6. **Insert** (menu bar) → **Checkpoint** → **Bitmap Checkpoint** → Click inside the **Airplane** Picture
7. Verify **Static : Static →** Click **OK** → Select **Check entire bitmap** → Click **OK**
8. Login with your User-Id & Password → Click **OK**
9. **File → Open Order →** Select **Order No** check-box → Type **4** → Click **OK**
10. **Insert** (menu bar) → **Checkpoint** → **Standard Checkpoint** → Point to **Name** Edit field
11. Verify **WinEdit : Name: →** Click **OK** → Uncheck all properties → Check **text** → Click **OK**
12. **Insert** (menu bar) → **Checkpoint** → **Standard Checkpoint** → Point to **Fly To** combo-box
13. Verify **WinCombo : FlyTo →** Click **OK**
14. Uncheck all properties → Check **selection** → Click **OK**
15. **File → Exit** (Close AUT)
16. **Stop the Recording**
17. Save **LAB4F**
18. Playback → It Should Pass !
19. Review & Close the ***Test Results*** window
20. **Resources → Object Repository** → **File → Export Local Objects...**
21. Save **C:\YOURNAME\QTP10\LVL1\REPO\LAB4F.TSR**

22. Delete the following objects :

Window	Object
Login	Password
Open Order	Order No (Check Box)
Flight Reservation	Fly To (Combo-Box)
CheckPoint	Name :

23. Close the **Object Repository**

24. Save **LAB4F** → Playback → It should Fail with :

25. Click **Stop** → Review & Close the *Test Results* window

26. **Automation** (menu bar) → **Maintenance Run Mode ...** → The MRW will display :

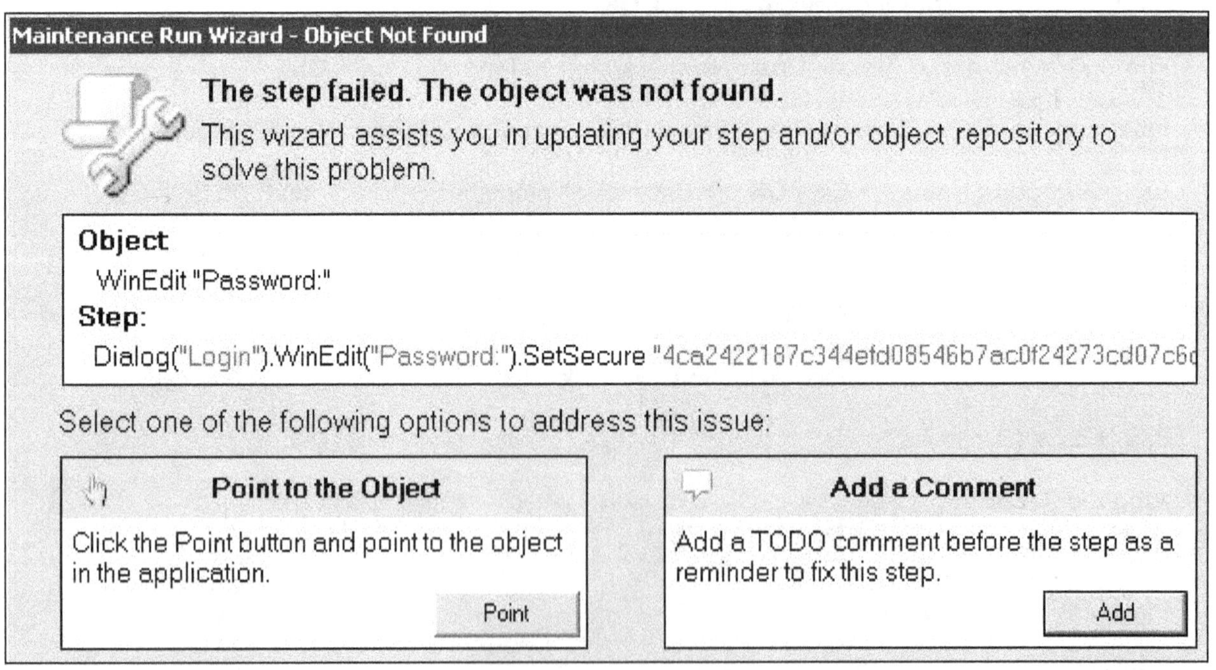

27. Click **Point** → Point and Click on the ***Password:*** Input field

28. Verify **WinEdit : Password** → Click **OK** → MRW will display :

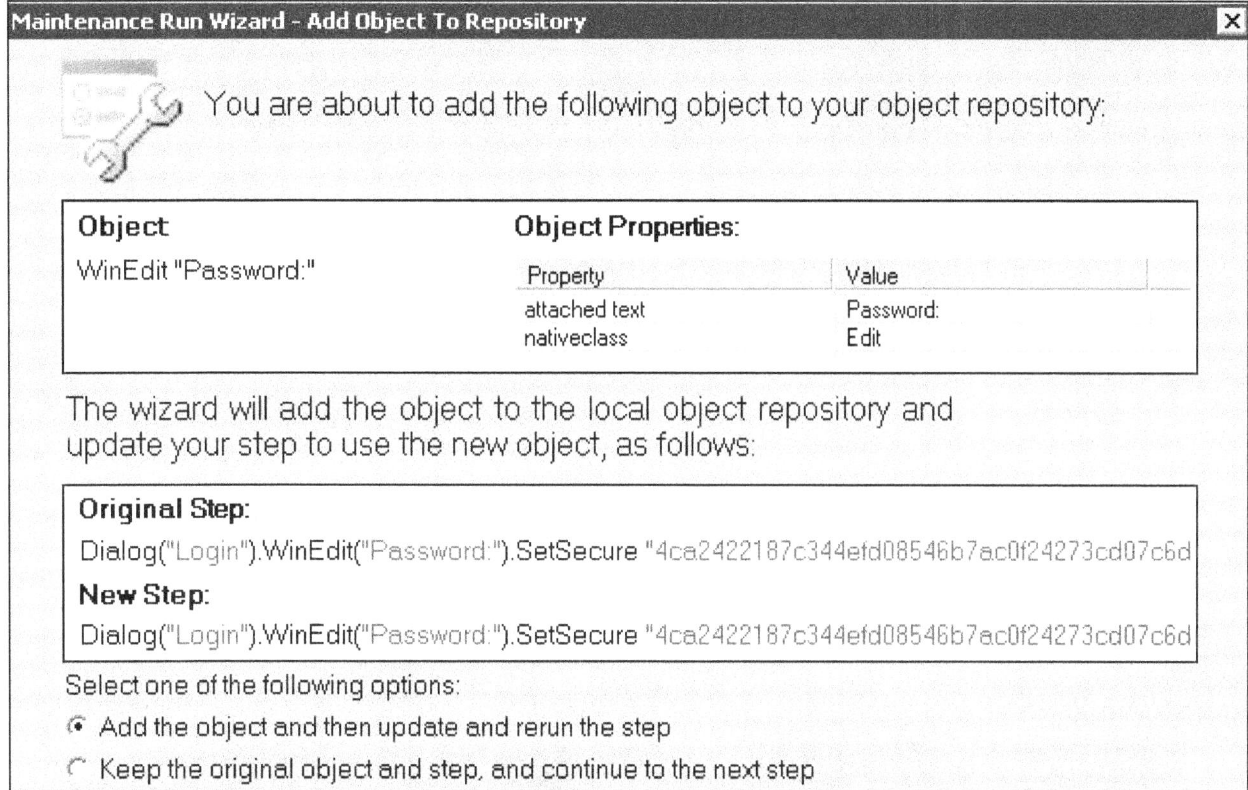

29. Click **OK** → MRW will display :

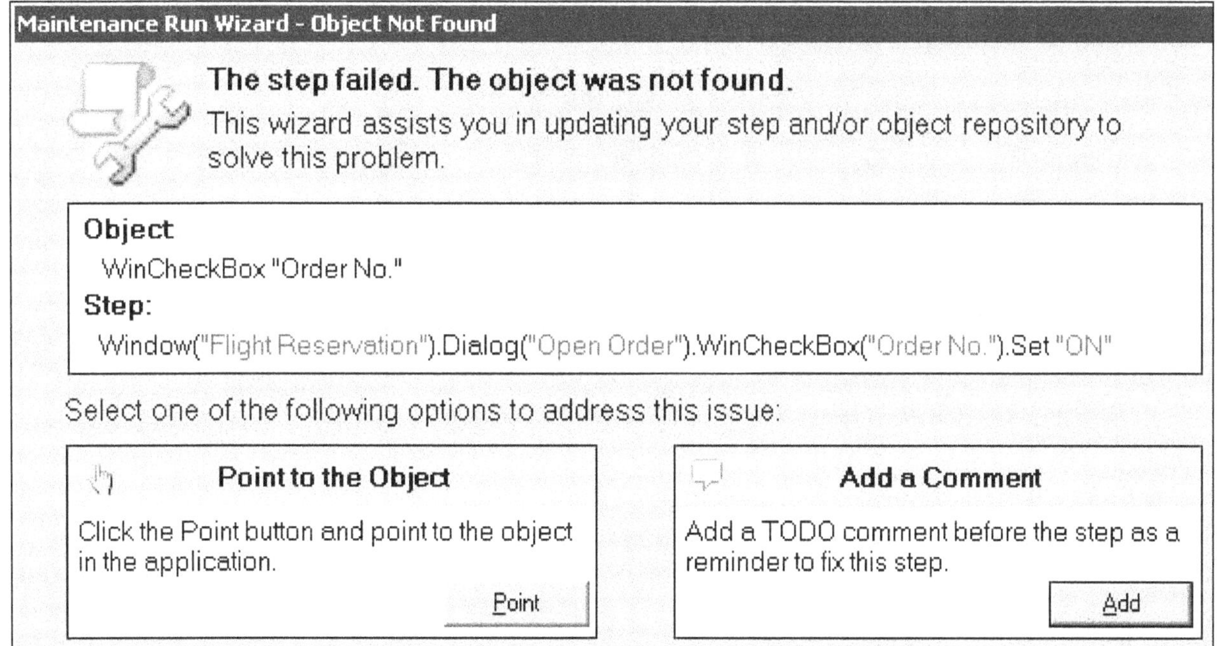

30. Click **Add** (You will ad a comment) as a reminder

31. Add the following comment :

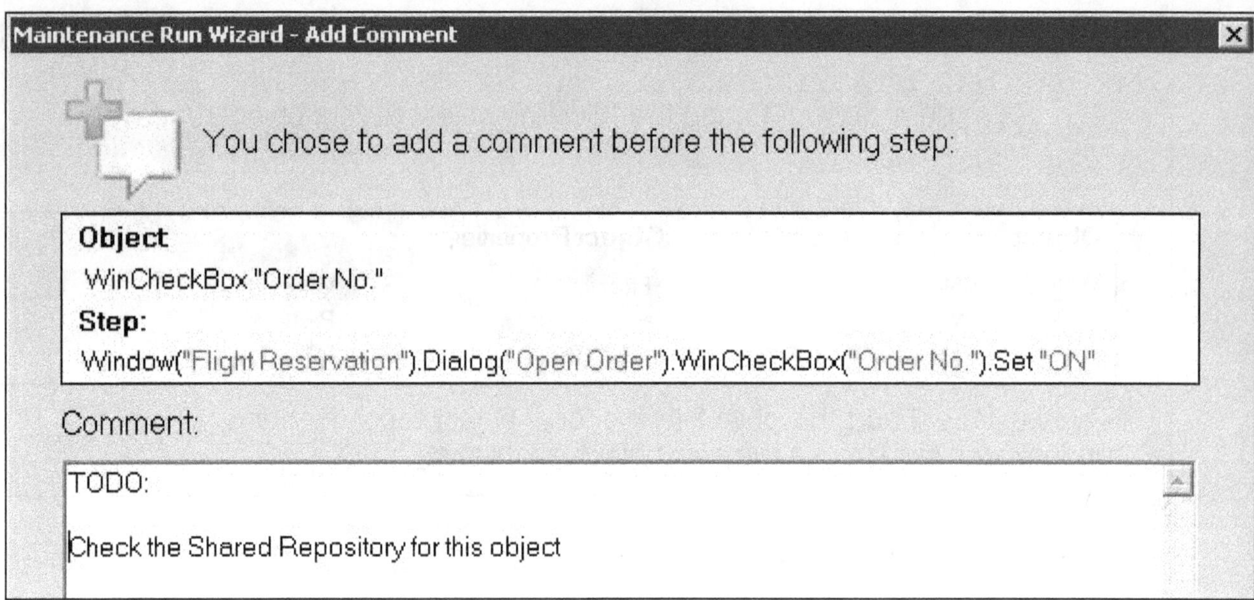

32. In the ***Comment*** type: **Check the Shared Repository for this object** → Click **OK**

33. Check the **Order No** check box → The AUT will generate the following error :

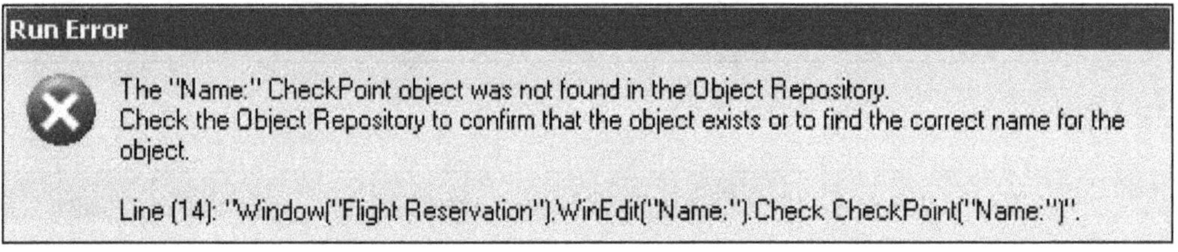

34. Click **Stop** → Click **Finish** → → Review & Close the *Test Results* window

35. Close the **AUT** → Click **Expert View** → Observe the Comments (Lines 8 thru 10)

```
1:  SystemUtil.Run "C:\Program Files\HP\QuickTest Professional\samples\flight\app\flight4a.exe","",""
2:  Dialog("Login").Static("Static").Check CheckPoint("Static")
3:  Dialog("Login").WinEdit("Agent Name:").Set "tommy"
4:  Dialog("Login").WinEdit("Agent Name:").Type micTab
5:  Dialog("Login").WinEdit("Password:").SetSecure "4ca2422187c344efd08546b7ac0f24273cd07c6d"
6:  Dialog("Login").WinButton("OK").Click
7:  Window("Flight Reservation").WinMenu("Menu").Select "File;Open Order..."
8:  'TODO:
9:  '
10: 'Check the Shared Repository for this object
11: Window("Flight Reservation").Dialog("Open Order").WinCheckBox("Order No.").Set "ON"
12: Window("Flight Reservation").Dialog("Open Order").WinEdit("Edit").Set "4"
13: Window("Flight Reservation").Dialog("Open Order").WinButton("OK").Click
14: Window("Flight Reservation").WinEdit("Name:").Check CheckPoint("Name:")
15: Window("Flight Reservation").WinComboBox("Fly To:").Check CheckPoint("Fly To:")
16: Window("Flight Reservation").WinMenu("Menu").Select "File;Exit"
17:
```

36. Click **Keyword View**

37. **Resources → Associate Repositories…** → Click on the Green Plus Sign **(+)**

38. Point to **C:\YOURNAME\QTP10\LVL1\LAB4F.TSR** → Click **Open**

39. Highlight **Action1** → Click the **>** button → Click **OK**

40. **Resources → Object Repository → Expand Flight Reservation** → It should similar to :

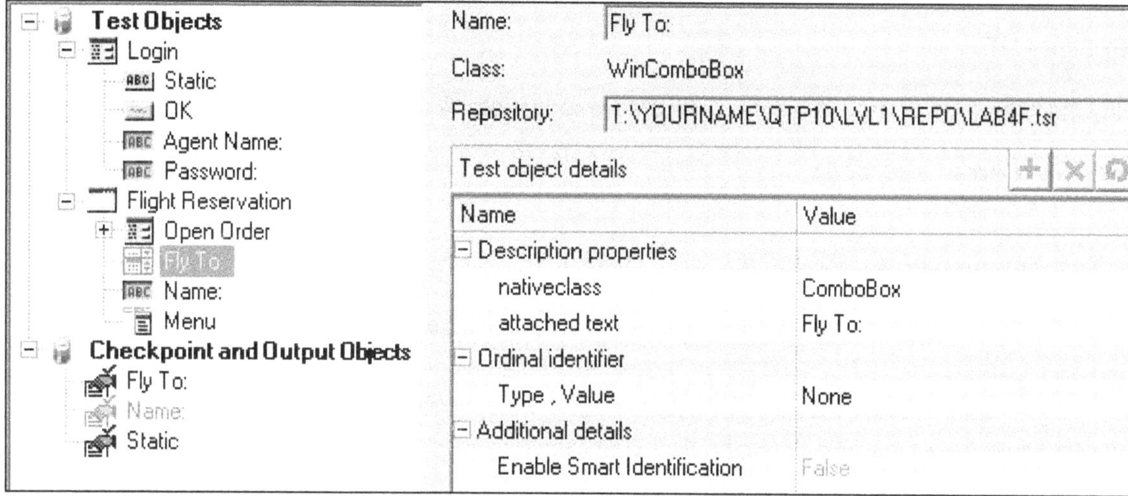

a) The **Fly To:** is greyed and is stored in the Shared Repository on the **T** Drive

b) The **Name:** Checkpoint is greyed and is stored in the Shared Repository on the **T** Drive

41. Highlight **Fly To:** → Right-Click → **Copy to Local**

42. Expand ***Open Order*** → Highlight **Order No:** → Right-Click → **Copy to Local**

43. Highlight **Name:** CheckPoint → Right-Click → **Copy to Local**

44. Highlight **Fly To:** → Observe the object is now in the Local repository

45. Expand ***Open Order*** → Highlight **Order No:** → Observe the object is now in the Local repository

46. Highlight **Name:** CheckPoint → Observe the object is now in the Local repository

47. Close the **Object Repository** window

48. **Resources → Associate Repositories…**

49. Highlight **C:\YOURNAME\QTP10\LVL1\LAB4F.TSR** → Click Red (**X**)

50. Click **Yes** (to confirm the delete) → Click **OK** (Close the **Associate Repositories** window)

51. Save **LAB4F** → Playback → It Should Pass → Review and Close the ***Test Results*** window

***** END OF LAB 4F *****

LAB 4G : TEXT AREA CHECKPOINTS & RECORD ADDITIONAL STEPS

PART 1 : Test Requirements : Create ***text area*** checkpoint (via recording)

1) Scenario:

   a) Invoke Flight4A → Log-In → Add a new *Order* → Verify *text values* on the screen

   b) Change the number of *tickets* → Update the Reservation → Verify *text values* on the screen

   c) Exit the AUT

2) Identify the ***message*** or *text* to verify : **Flight Reservation** & **Insert Done...**

3) Create a ***Text Area*** Checkpoint

PART 2 : Test Requirements : Create ***text area*** checkpoint (via Active Screen)

1) Add New Steps to : *Change the number of tickets → Update the Reservation*

2) Verify *text values* on the screen **Update Done...**

3) Add a new Text Area CheckPoint via the Active Screen

Steps: (Part I)

1. **File → New → Test**
2. **Automation** (menu bar) → **Record F3**
3. Check **Record and run test on any open Windows-based application** → Click **OK**
4. **Start → Programs → QuickTest Professional → Sample Applications → Flight**
5. Login with ***yourname*** and use the password ***mercury*** → Click **OK**
6. Wait for the **Flight Reservation** (main menu)
7. Enter *positive data* in all the fields to ***Add*** a new order
8. Click **Insert Order** button → Wait for the Confirmation message
9. Write down the Order Number: _____________ (You will delete this *Order* later !)
10. **Insert → Checkpoint → Text Area Checkpoint...** → cursor changes to a *cross-hair*
11. Click & Hold the Left button of the mouse (Do NOT let go) *and*
12. Drag & Draw the rectangle area around the message : **Insert Done...** and *then* release the left ***Button***
13. Verify **ActiveX: Threed Panel Control** → Click **OK**

14. Verify that the *text* **Insert Done...** is in Red ! → Click **OK**

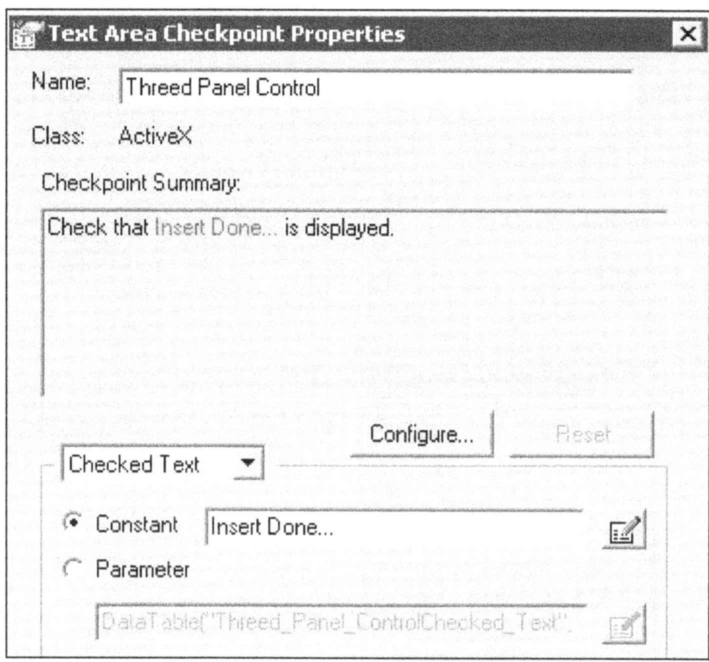

15. **File → Exit** (Close the AUT)

16. Press **F4** (Stop the Recording) and Review the Script :

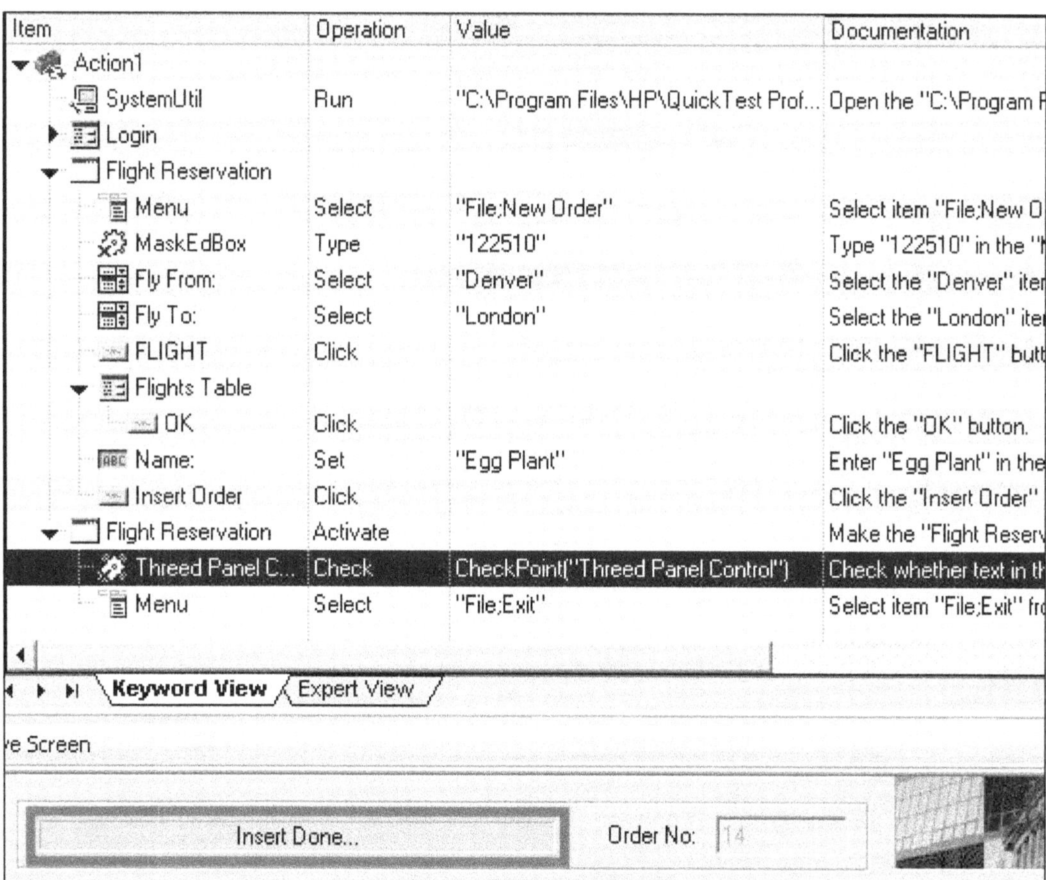

17. Save Script as **LAB4G**

18. Manually Start the AUT → Login → Open Order Number : _________ (Created from the earlier *step*)

19. Click **Delete Order** buttton → Click **Yes** (to *confirm* the delete) → **File** → **Exit** (Close the AUT)

20. Playback **LAB4G** → Watch & Write down the Order Number : _______________

21. Review the ***Text Area*** checkpoint in the *Test Results* window

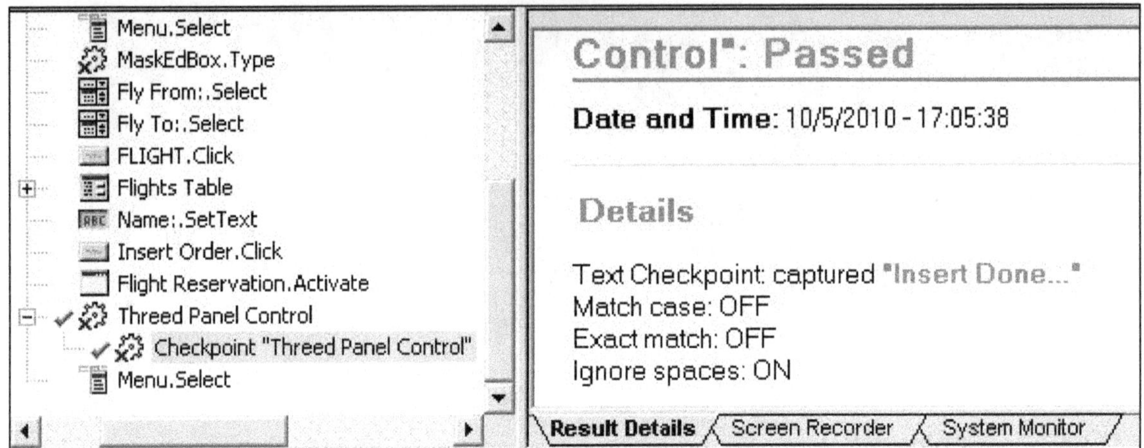

Add a *Text Area Checkpoint* to the LOGIN window (via the Active Screen)

22. Highlight **Login** (in the Keyword view)

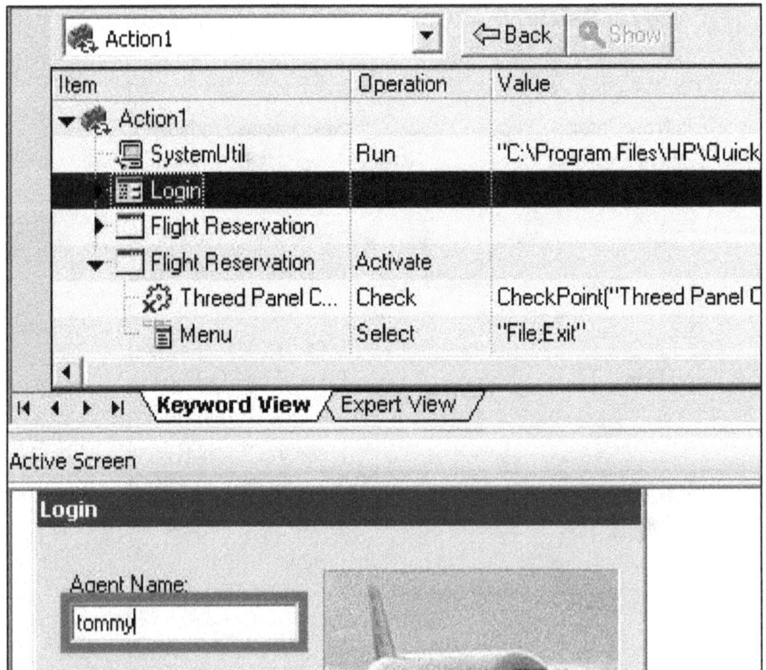

23. Move the mouse over the **Login** title bar until the Logical name ***Login*** is shown

24. Right-Click → **Insert Text Checkpoint...**

25. Verify the Text is correct and **Login** is displayed in Red !

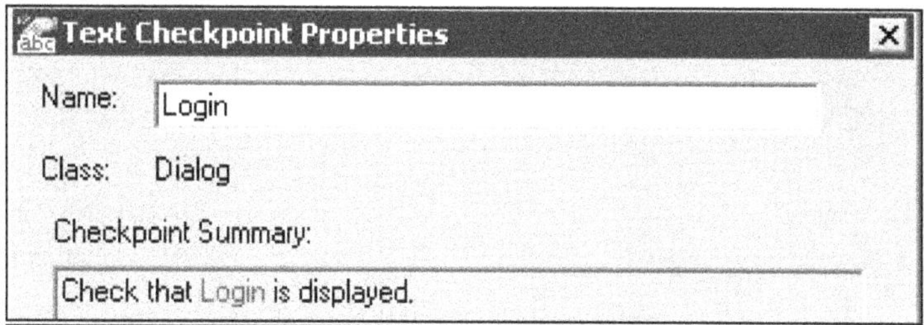

26. Check **After the current step** → Click **OK**

27. Sample Script :

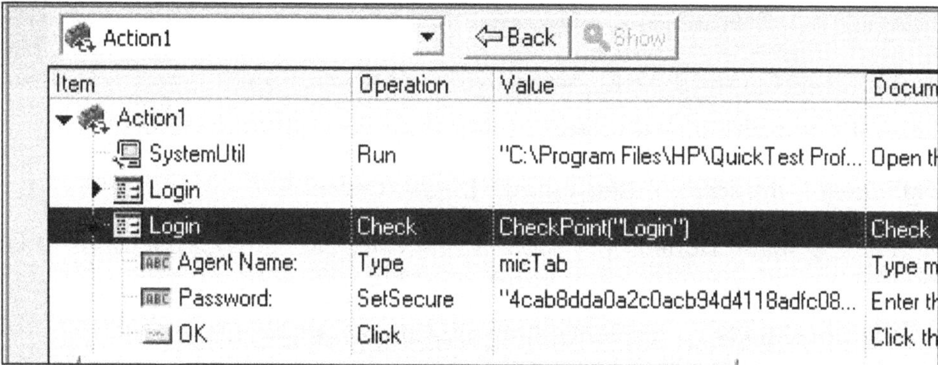

Item	Operation	Value	Docum
Action1			
SystemUtil	Run	"C:\Program Files\HP\QuickTest Prof...	Open tl
Login			
Login	Check	CheckPoint("Login")	Check
Agent Name:	Type	micTab	Type m
Password:	SetSecure	"4cab8dda0a2c0acb94d4118adfc08...	Enter tl
OK	Click		Click th

28. Save **LAB4G** → Playback → Review & Close the *Test Results* window

STEPS: (Part 2) : Add Steps to Delete the Order

1. Manually Start the *AUT* → *Login* → *Open Order Number* : _________ (Created from the earlier *step*)
2. In Keyword View, highlight the ***Last Checkpoint*** in the script (Step before **Menu Select File;Exit**)

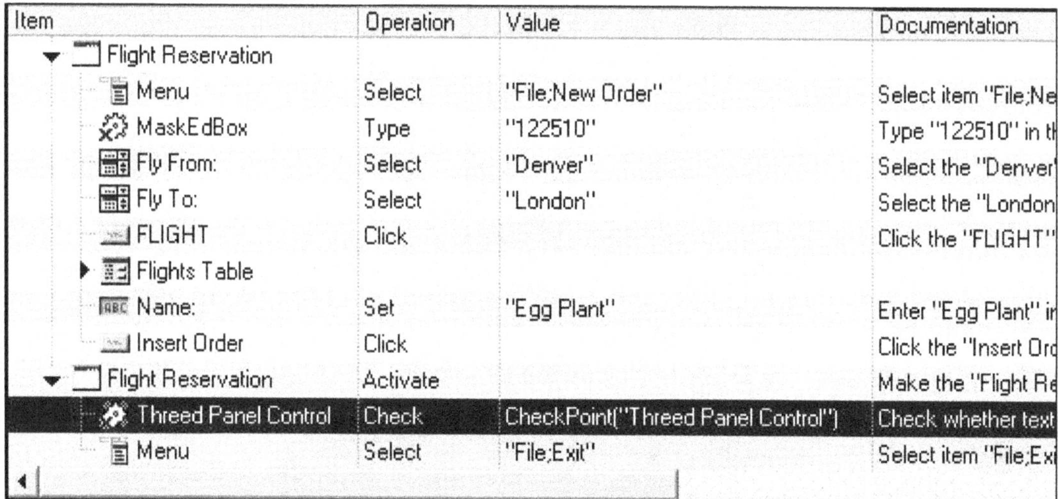

Item	Operation	Value	Documentation
Flight Reservation			
Menu	Select	"File;New Order"	Select item "File;Ne
MaskEdBox	Type	"122510"	Type "122510" in tl
Fly From:	Select	"Denver"	Select the "Denver'
Fly To:	Select	"London"	Select the "London
FLIGHT	Click		Click the "FLIGHT"
Flights Table			
Name:	Set	"Egg Plant"	Enter "Egg Plant" i
Insert Order	Click		Click the "Insert Ord
Flight Reservation	Activate		Make the "Flight Re
Threed Panel Control	Check	CheckPoint("Threed Panel Control")	Check whether text
Menu	Select	"File;Exit"	Select item "File;Exi

3. Press **F3** (Start Recording)
4. Click **Delete Order** button → Click **Yes** (to *confirm* the delete) → *Stop* the Recording

5. Close the ***AUT*** → Review the Script

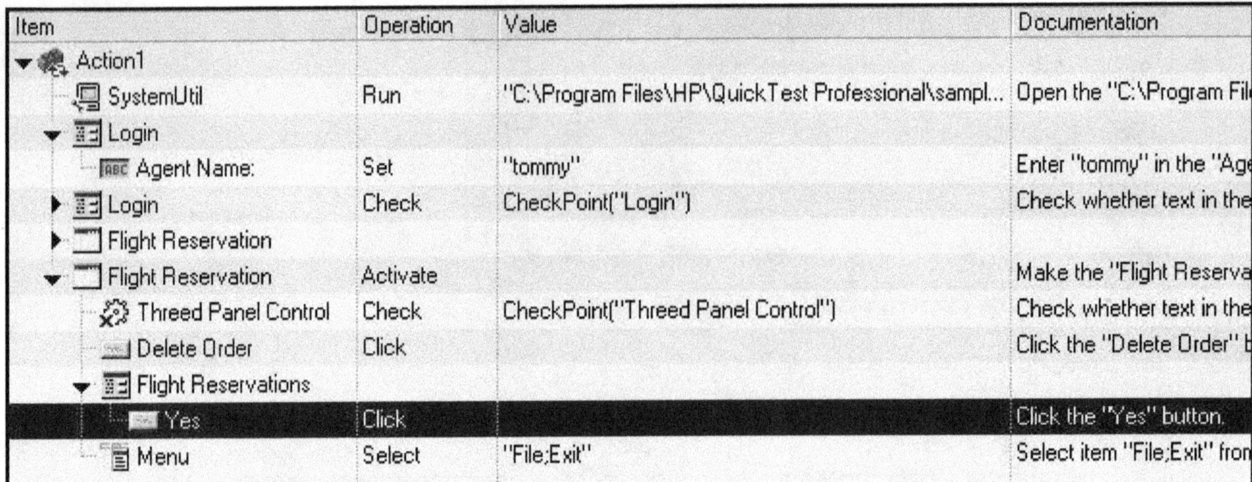

Item	Operation	Value	Documentation
Action1			
SystemUtil	Run	"C:\Program Files\HP\QuickTest Professional\sampl...	Open the "C:\Program Fil
Login			
Agent Name:	Set	"tommy"	Enter "tommy" in the "Ag
Login	Check	CheckPoint("Login")	Check whether text in the
Flight Reservation			
Flight Reservation	Activate		Make the "Flight Reserva
Threed Panel Control	Check	CheckPoint("Threed Panel Control")	Check whether text in the
Delete Order	Click		Click the "Delete Order" b
Flight Reservations			
Yes	Click		Click the "Yes" button.
Menu	Select	"File;Exit"	Select item "File;Exit" fron

6. Save **LAB4G** → Ensure the *Active Screen* is visible

7. Highlight the ***Last Checkpoint*** in the script (Step before **Delete Order**)

8. Move mouse over the *Message* **Insert Done…** (in Active Screen) until the ***ThreedPanelControl*** is visible

9. Right-Click → **Insert Text Checkpoint…** → Verify **ActiveX:ActiveX** → Click **OK**

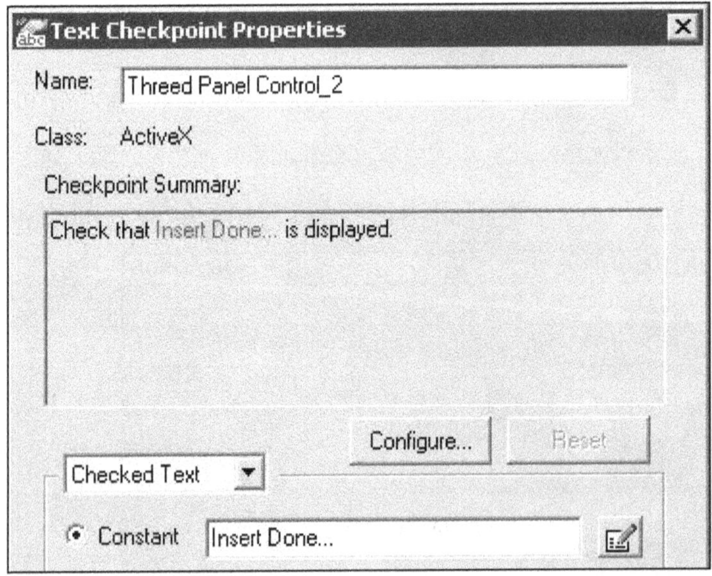

10. Highlight only the **Insert Done** → Do NOT include the three dots

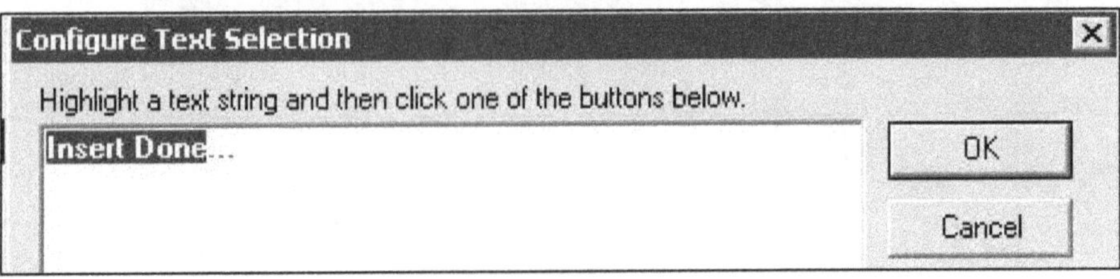

11. Click **Checked Text** → Observe the three dots are not included → Click **OK**

12. Check the following : ***After current step*** → Click **OK**

13. Drag & Drop the Second Checkpoint before **Menu Select File;Exit**

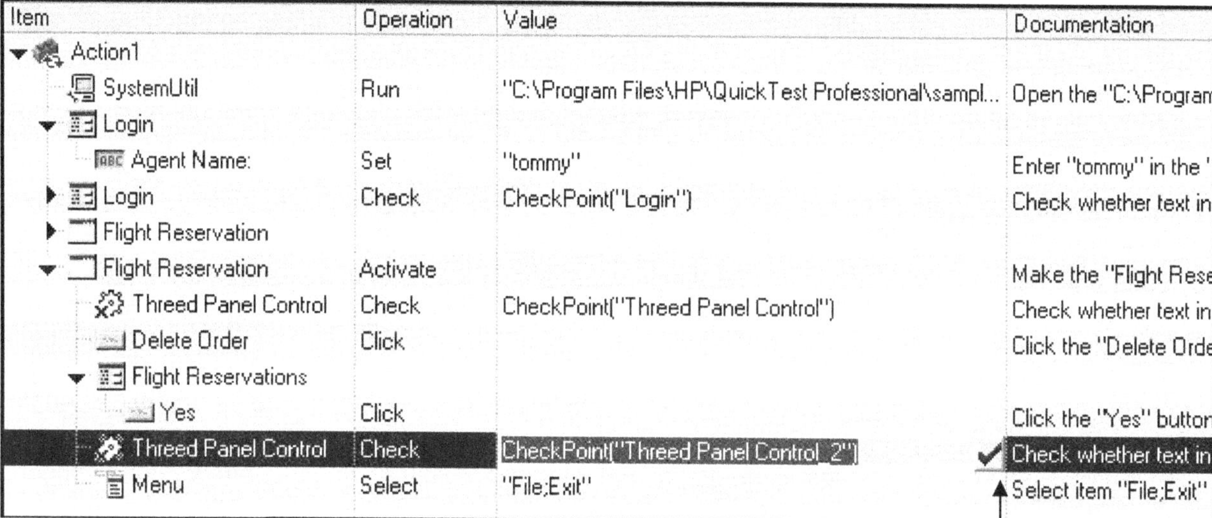

Item	Operation	Value	Documentation
Action1			
SystemUtil	Run	"C:\Program Files\HP\QuickTest Professional\sampl...	Open the "C:\Program
Login			
Agent Name:	Set	"tommy"	Enter "tommy" in the "
Login	Check	CheckPoint("Login")	Check whether text in
Flight Reservation			
Flight Reservation	Activate		Make the "Flight Rese
Threed Panel Control	Check	CheckPoint("Threed Panel Control")	Check whether text in
Delete Order	Click		Click the "Delete Orde
Flight Reservations			
Yes	Click		Click the "Yes" button
Threed Panel Control	Check	CheckPoint("Threed Panel Control_2")	Check whether text in
Menu	Select	"File;Exit"	Select item "File;Exit"

14. In the **Threed Panel Control_2** *value* column , Click **Configure the value icon**

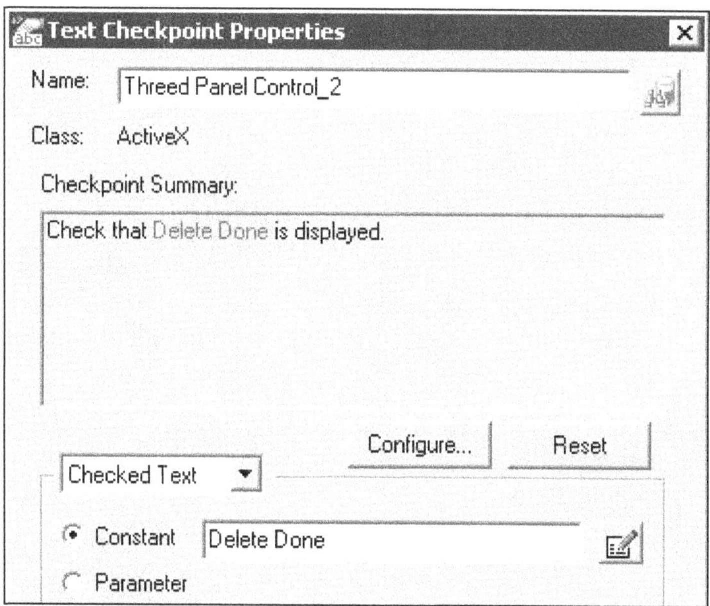

15. Check the Constant radio button → Change **Delete Done** (from Insert Done) → Click **OK**

16. Save **LAB4G**

17. Playback → Review the **Test Results** window

18. Verify the second Checkpoint contains **Delete Done** (with the three dots)

19. Close the ***Test Results*** window

STEPS : (Part 3)

20. Highlight **Check Checkpoint ("Threed Panel Control_2")** → Observe the Active Screen :

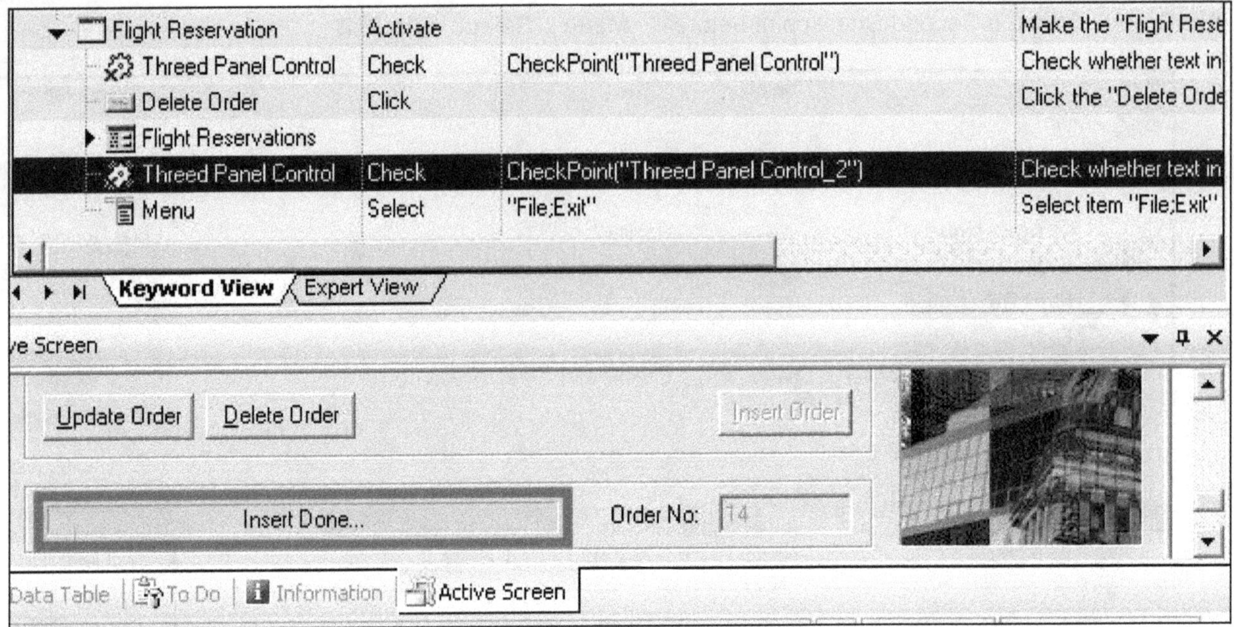

Question : Why does the ***Active Screen*** NOT contain the correct message : **Delete Done...** ?

21. **Automation** (menu bar) → **Update Run Mode**
22. Check all the options → Click **OK**
23. Review & Close the ***Test Results*** window → Review the Script :

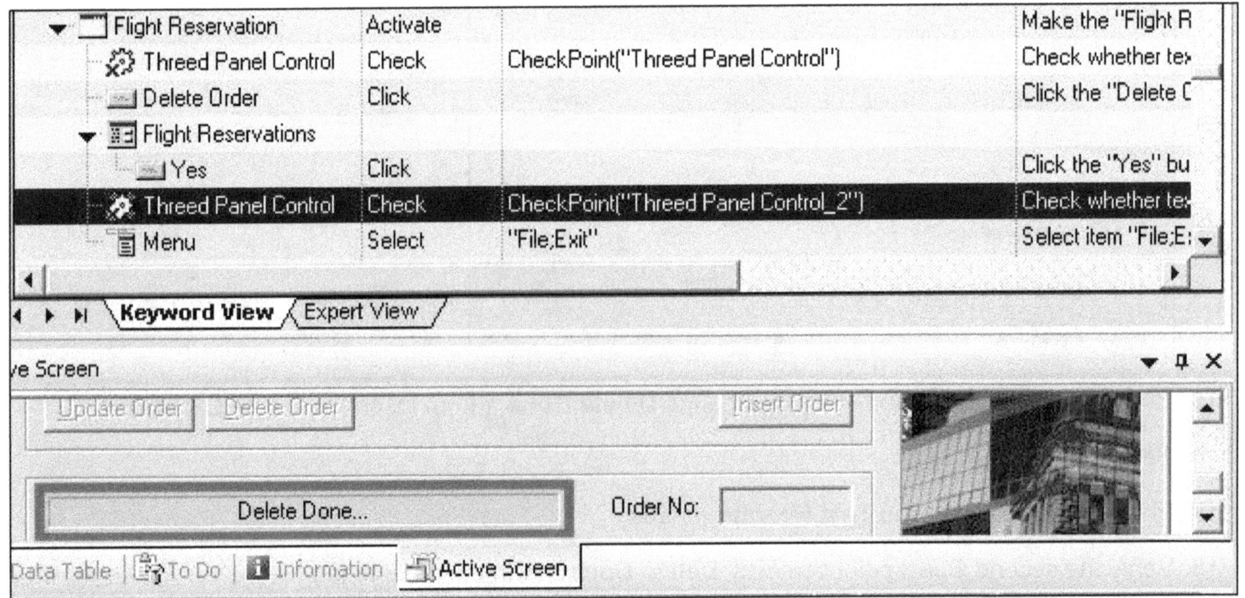

***** END OF LAB 4G *****

LAB 4H : Object Repository (Optional)

Requirements:

- Test 1 : Add a New reservation
- Test 2 : Update a existing Reservation
- Use the Object Spy to learn the properties & values
- Delete Objects
- Use the *Maintenance Run Wizard* and/or *Update from Application*

1. Export all objects from **LAB1D** and save as **C:\YOURNAME\QTP10\LVL1\REPO\LAB4G**

2. Test 1 : Script **LAB4H_Test1**

   a) Add a New order

   b) Create Checkpoints : *Login, Flight Reservation* window and a the Confirmation Message

      - **OK** button, **Flights Table** button , Confirmation
      - Include properties : X, Y, Height, Width, etc ...
      - Change the property values on several objects

   c) Delete Objects :

      - **OK** button, **Flights Table** button , Confirmation Message

   d) The Local repository should be missing the above objects

   e) Playback the test , Fix & correct for any missing objects

3. Test 2 : Script **LAB4G_Test2**

   a) Update an existing order (refer to the order added in Test 1)

   b) Create Checkpoints : *Login, Flight Reservation* window and a the Confirmation Message

      - **OK** button, **Flights Table** button , Confirmation

   c) Delete Objects :

      - **Tickets** (Number of), **Economy** button , **Date of Flight**

   d) The Local repository should be missing the above objects

   e) Playback the test , Fix & correct for any missing objects

   d) Change any of the fields on the screen and Update the order

***** **END OF LAB 4H** *****

LAB 4I : Zip File

Requirements:

- Use ***LAB4A_P1***
- *Zip* a QTP script file : ***Compress*** your HP QuickTest files
- Export & Import your tests

Steps

1. Create the following folder : **C:\YOURNAME\QTP10\LVL1\BKUP** (on your own)
2. Open **C:\YOURNAME\QTP10\LVL1\LAB4A_P1**

Export the file

3. **File** (menu bar) → **Export Test to Zip File …** → the **Export to Zip File** window opens
4. Change the destination ***Zip file:*** to **C:\YOURNAME\QTP10\LVL1\BKUP\LAB4A_P1** → Click **OK**
5. Verify in Windows Explorer the LAB4A_P1.ZIP exists

Import the file : The target folder must not exist !

6. **File** (menu bar) → **Import Test from Zip File …** → Make the following changes → Click **OK**

Zip file: **C:\yourname\QTP10\LVL1\bkup\lab4a_p1.zip**
Extract to: **C:\yourname\QTP10\LVL1\bkup**

7. After the **Import** – QTP opens the imported file (you are positioned at the target folder)
8. In the **Windows Explorer** verify that **C:\yourname\QTP10\LVL1\bkup\lab4a.zip** file still exists

Delete the main test *LAB4A* :

9. Delete the following folder & sub-folders : **C:\YOURNAME\QTP10\LVL1\LAB4A_P1**
10. **File** (menu bar) → **Import Test from Zip File …** → Make the following changes → Click **OK**

Zip file: **C:\yourname\QTP10\LVL1\bkup\lab4a_p1.zip**
Extract to: **C:\yourname\QTP10\LVL1**

11. You have the Back Up file restored to the original source folder

***** **END OF LAB4I** *****

NOTES:

CHAPTER 3

Objectives of this section

- **Creating Keyword View script (without recording)**
- **LAB 5A - Export Object Repository & Create a Keyword Script**
- **LAB 5A - Optional**
- **The SYSTEMUTIL & REPORTER Object**
- **VBScript (Introduction to Programming)**
- **Data Types**
- **Debugging**
- **LAB 5B – Debugging**
- **LAB 5C – Block AUT Input**
- **Expert View (Programming)**
- **Retrieving Properties**
- **LAB 5D , LAB 5E, LAB 5F (Optional)**
- **DataTables & Data Driven Tests**
- **LAB 6A : Login – DDT (Data Driven Test)**
- **LAB 6B : Exporting Data from RDBMS (MS SQL Server)**
- **Loop Statements (FOR... NEXT , DO WHILE ...)**
- **DataTable Functions**
- **LAB 6C : Importing Data from MS Excel**
- **LAB 6D : Reverse Transaction Processing (Optional)**
- **LAB 6E : Verifying Existing Orders (DDT)**

Object Not Found ?

Describe the reasons as to why the message **Object Not Found** is displayed ?

- The word ***Object*** refers to the actual Object (WinEdit, WinComboBox, WinList, Push Button, etc …)

1. The AUT has changed

   a) The physical characteristics is different than the previous build (*version*)

   b) The number of properties have been increased

   c) Some properties *removed* or *modified*

   d) The *name* of Object was changed

2. Synchronization

   a) QTP is running faster than the AUT

   b) The AUT is running slow

3. Not the *Current* or *Active* window

   a) QTP is attempting to execute an *Action* in a specific window (or *page*)

   b) The AUT is another window (or *page*)

4. Wrong or InCorrect Global Repository

   d) The **Test Shared Repository (TSR)** does not contain the objects required for the current test

   e) The current test is not linked to the correct TSR file

5. The *logical* name of the Object is *Misspelled* or *Incorrect* in :

   a) Script

   b) Object Repository

6. Wrong AUT is running

Keyword View & Object Repository

1. What is a Test Case ? ____________________

2. List, describe or *name* several characteristics of a **test case** :

3. What does the **Keyword View** display ?

   a) Actions required to perform/conduct/execute a test

   b) The actions are logically organized in a sequence

   c) The **entire** series of steps *required* to complete / finish a single **Business Transaction**

   d) Data Validation Level (Basic / Minimum):

      - Basic : (Match the data type : *integer, decimal, string, date, datetime, etc ...*)
      - Does the field format (on the screen match with the actual input data)

   e) Business Rule Validation (BRV) :

      - Required Fields (Minimum) to complete the entire **Business Transaction** (Process)

      - The verifications can be specific to a single *or* multiple fields

      - Dependent / Linked fields *determine* the policy or rules of the company

      - BRV's occur thru out the entire business process (steps)

   f) After the last data input – a **user** clicks on a button to indicate this is end of the business process

   g) The AUT/SUT should **confirm** the success or failure of the business process

   h) Typically a final **checkpoint** is performed after the user has completed the business process

   i) A secondary **checkpoint** (validation) to query the database may be required

4. Customer / End User / Departments

   a) Discuss / review with the customer the Functional Requirements
   b) Involve them from the beginning thru the end
   c) Interview the users :

      - What documents do they have before they use the AUT
      - Have them explain or walk you thru the entire process (if possible)
      - Review the AUT/SUT – observe & document problems in MS Word
      - The outcome of their demonstration (success or failure) encountered will be your test cases

5. Review Your Work (Make friends !!) :

   - First check your work, next have your colleague review your work (teamwork)
   - Always be prepared with a **Pen** & **Paper** before going to any meetings

Exporting Local Objects

1. Creating a **Keyword View** without recording <u>*requires*</u> the Object Repository

2. How would you *create* objects in the **Object Repository** ?

   a) Recording
   b) Learn the Objects in a *window / web page*
   c) Active Screen
   d) Update **Object Repository** with the *new / modified* objects

3. QTP allows you to **export** objects from a <u>repository</u> to a file from an <u>existing</u> script

4. It will create a file with the extension of : **.TSR**

5. A QTP repository can be defined to either be **shared** or **local** or ***both***

6. Associative Repositories

   a) Shared repositories are stored as **.TSR** extension
   b) Referenced by other QTP Tests
   c) Not Permanent to an **Action** file
   d) If objects are missing in one Test – you can associate it with other shared repositories

7. Sample File Structure :

C:\FRS					**Description**
	Build22				
	Build23				
		FRS_REPO			
			FRS1.TSR		*Shared Repository*
			FRS2.TSR		*Shared Repository*
		Fax_Order			*QTP Test*
			Action 0		
				Snapshot	
				Resource.mtr Script.mts *ObjectRepository.bdb*	*Local Repository*
			Action 1		
				Snapshot	
				Resource.mtr Script.mts *ObjectRepository.bdb*	*Local Repository*
			Default.xls		*DataTable*
			Default.*	.cfg, .usp	*Internal files*
			Parameters.mtr		*Parameter file*
		Delete Order			*QTP Test*

8. Associative Repository : (**Object Repository** : Tools → Associative Repositories…)

Fax_Order	**Repository**	QTP Test
	ObjectRepository.bdb	Local
	C:\FRS\FRS_REPO\FRS1.TSR	Shared (Associative)
	C:\FRS\FRS_REPO\FRS2.TSR	Shared (Associative)

LAB 5A : Export Object Repository to create a Shared Repository

PART I : Test Requirements :

a) Export Objects from **LAB4B_P2** to a file **LAB5A.TSR**
b) Create two new scripts using the new file (repository)
c) Perform tests: **Login** without recording
d) Review the Script & Steps → Execute the test
e) Review the Test Results

STEPS: (Part I) : Perform the Login verification

1. Create **C:\YOURNAME\QTP10\LVL1\REPO** folder (if not already created)
2. **File** (menu bar) → **Open** → **C:\YOURNAME\QTP10\LVL1\LAB4B_P2** → Click **Open**
3. **Resources** (menu bar) → **Object Repository...** → Review the Objects in the repository
4. **File** → **Export Local Objects ...** → Point to **C:\YOURNAME\QTP10\LVL1\REPO**
5. Type **LAB5A.TSR** → Click **Save**
6. **File → Close** (Close the Object Repository)
7. **File → New** → **Test** (Create a New Script)
8. **Resources** (menu bar) → **Object Repository...** → There are **NO** Objects in the repository
9. **File → Close** (Close the Object Repository)
10. **File → Save** → **C:\YOURNAME\QTP10\LVL1\LAB5A** → Click **Save**
11. **Resources** (menu bar) → **Object Repository...**
12. **Tools** (menu bar) → **Associate Repositories...** → Click **Green Plus Sign (+)**

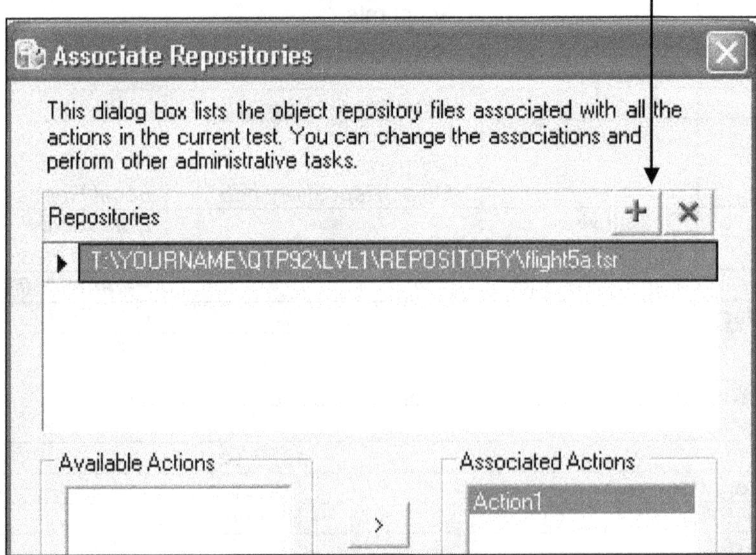

13. Point to **C:\YOURNAME\QTP10\LVL1\REPOSITORY\LAB5A.TSR** → Click **Open**
14. Highlight **Action1** → Click Right Arrow (**>**) → Click **OK** → Observe the Repository is in **Read-Only** mode

15. **File → Close** (Close the Object Repository) → **File → Save** (Save LAB5A)

16. Click on the empty line below **Action1** → Wait for the *Combo/List* box to appear

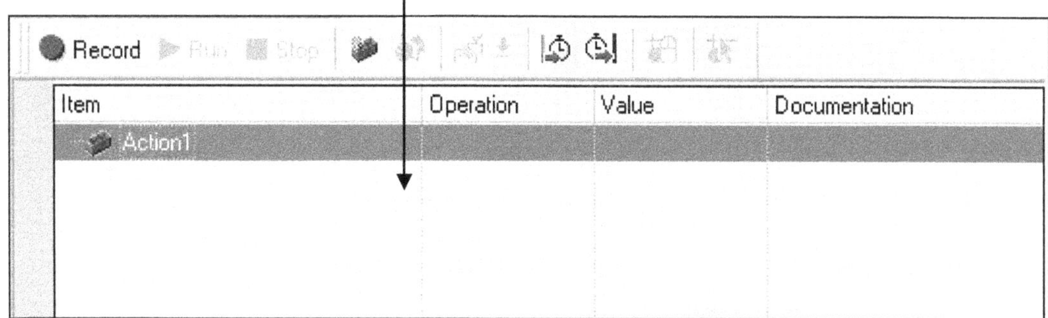

17. Select **Step Generator…** from the **< Select an item>** combo box

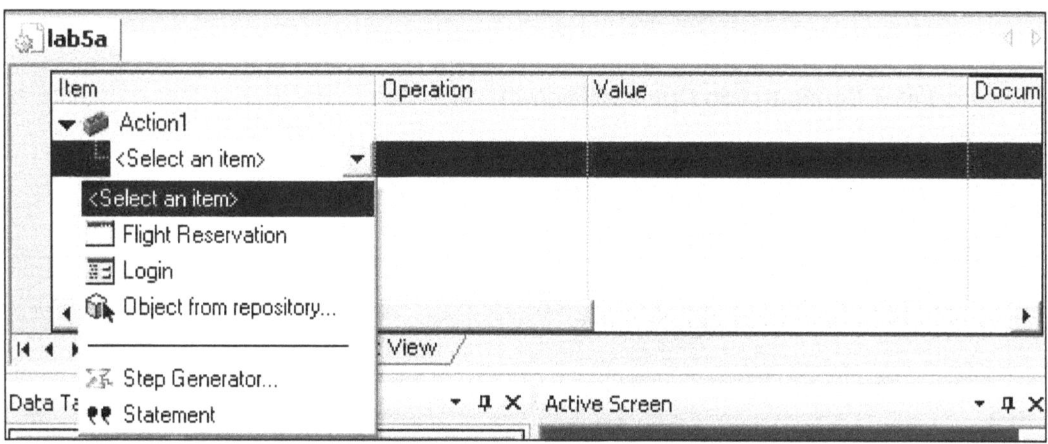

18. Select the following : **Utility Objects → SystemUtil → Run** → Move cursor line **file*** in the ***Value*** field

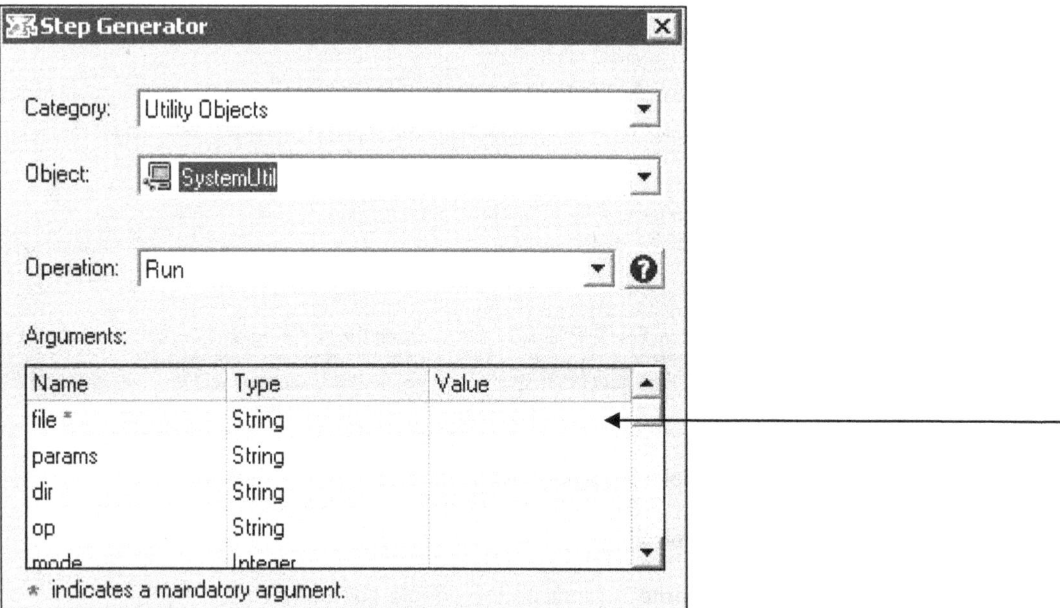

19. Click the Hash symbol **< # >**

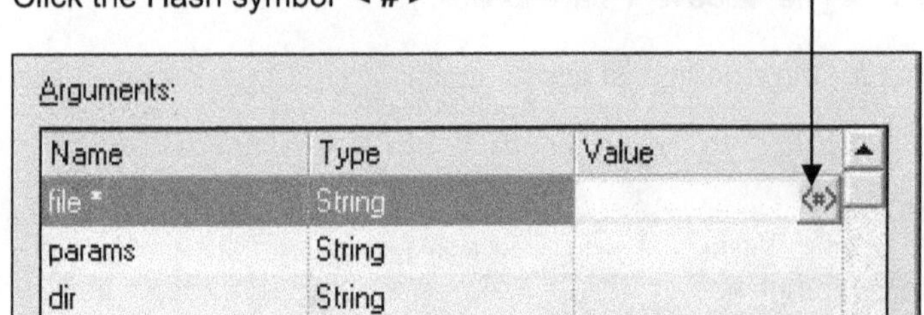

20. Choose **Constant** and type the following : (or the pathname on your machine)

    C:\Program Files\HP\QuickTest Professional\samples\flight\app\flight4a.exe

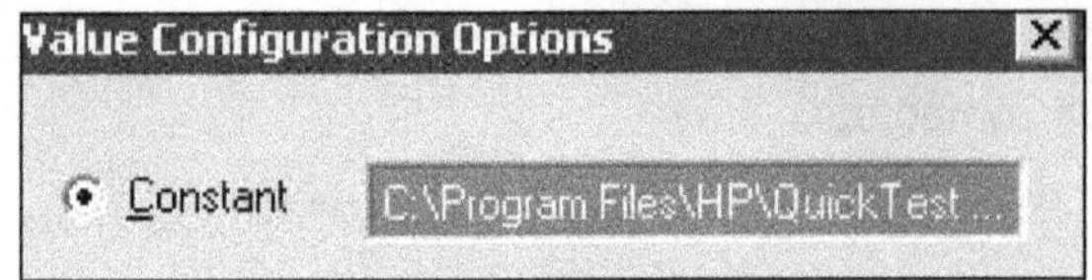

21. Click **OK** (Close the **Value Configuration Options** window)

22. Click **OK** (Close the **Step Generator** window)

23. The first step is created :

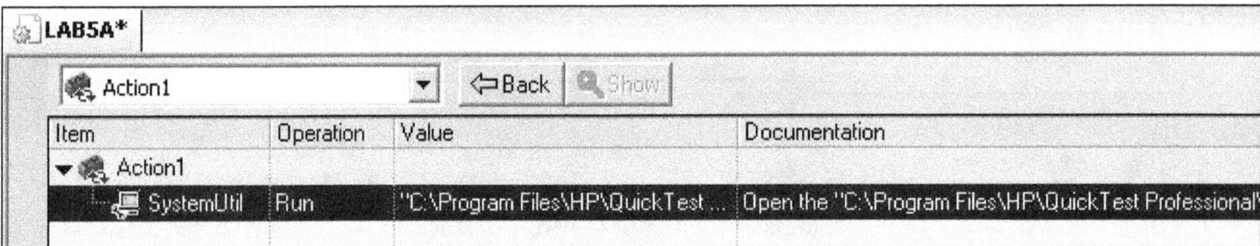

24. Move cursor inside **Item** column and click below **SystemUtil** until < **Select an Item** > appears

25. Click on arrow → Select **Login** (It will display **Activate** in the Operation column)

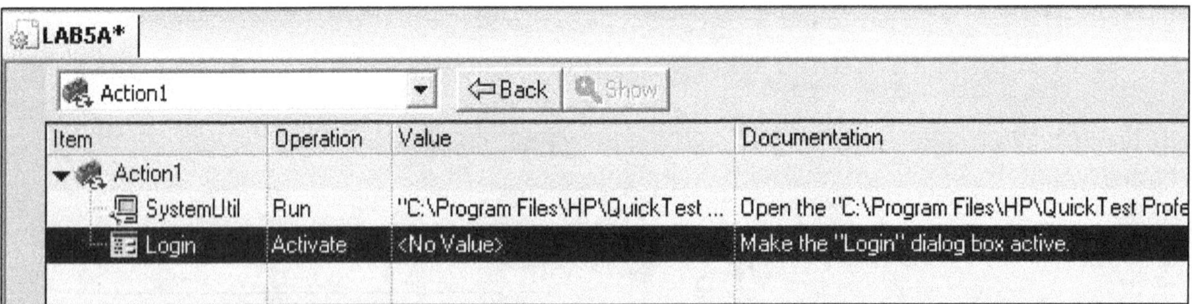

26. Move cursor inside **Item** column and click below **Login** until < **Select an Item** > appears

27. Click on arrow → Select **Agent Name:** (It will display : **Set** in the Operation column)

28. Move cursor to **Value** field and type **"yourname"** (Include the double quotes")

29. Move cursor to another field (the **< Tab >** might NOT work all the time !)

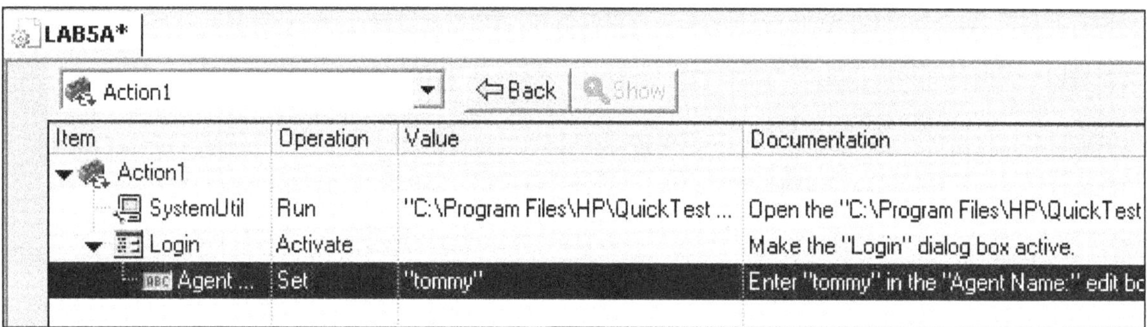

30. Move cursor inside **Item** column and click below **Agent Name:** until < **Select an Item** > appears

31. Click on arrow → Select **Password:** (It will display : **Set** in the Operation column)

32. Move cursor to **Value** field and type **"mercury"** (Include the double quotes")

33. Move cursor to another field (the **< Tab >** might NOT work all the time !)

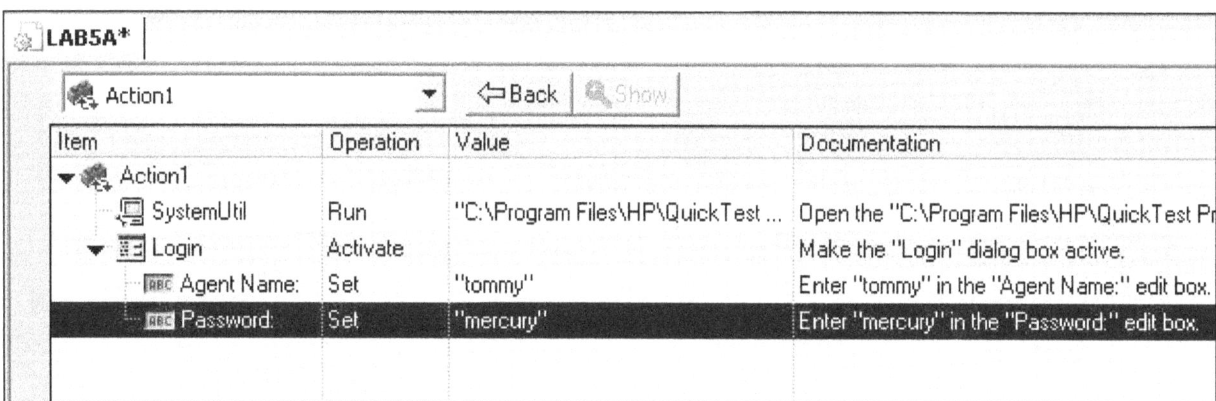

34. Move cursor inside **Item** column and click below **Password:** until < **Select an Item** > appears

35. Click on arrow → Select **OK** (It will display : **Click** in the Operation column)

36. Verify the **Keyword View** appears as follows:

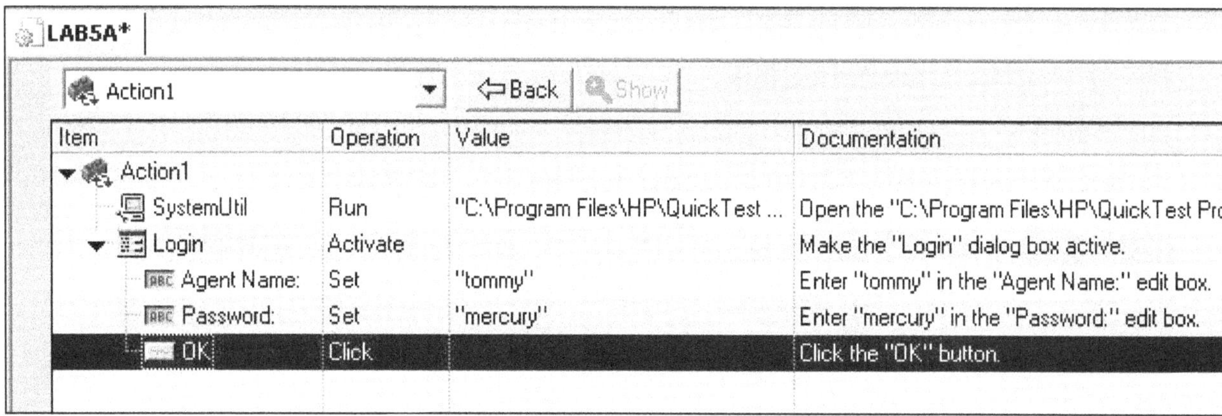

37. Move cursor inside **Item** column and click below **OK** until < **Select an Item** > appears

38. Click on arrow → Select **Object from repository…**

39. Highlight **Flight Reservation** window (in Select ***Object for Step*** window) → Click **OK**

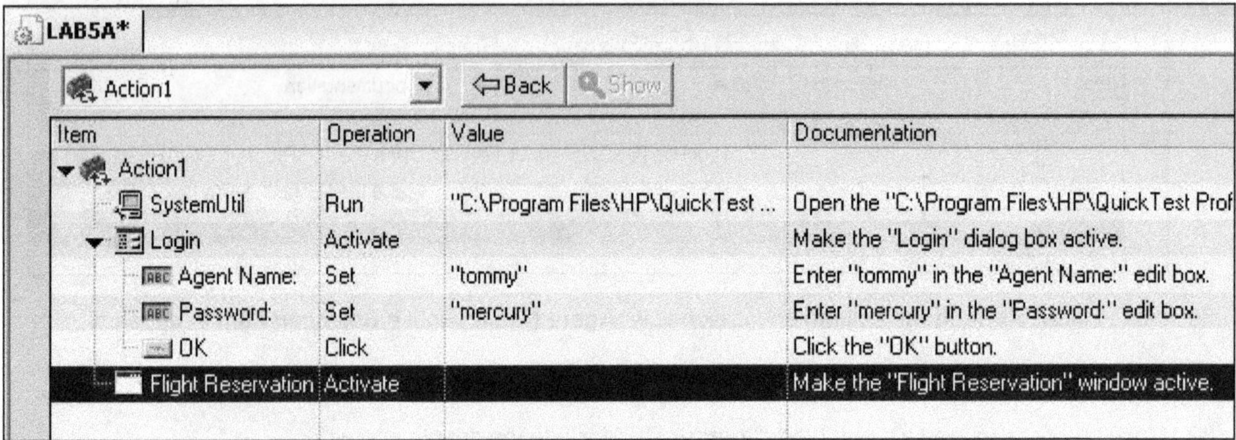

40. Move cursor inside **Item** column and click below **Flight Reservation:** until < **Select an Item** > appears

41. Click on arrow → Select **Menu** (It will display : **Select** in the Operation column)

42. Move cursor to **Value** field and type **"File;Exit"** (Include the double quotes)

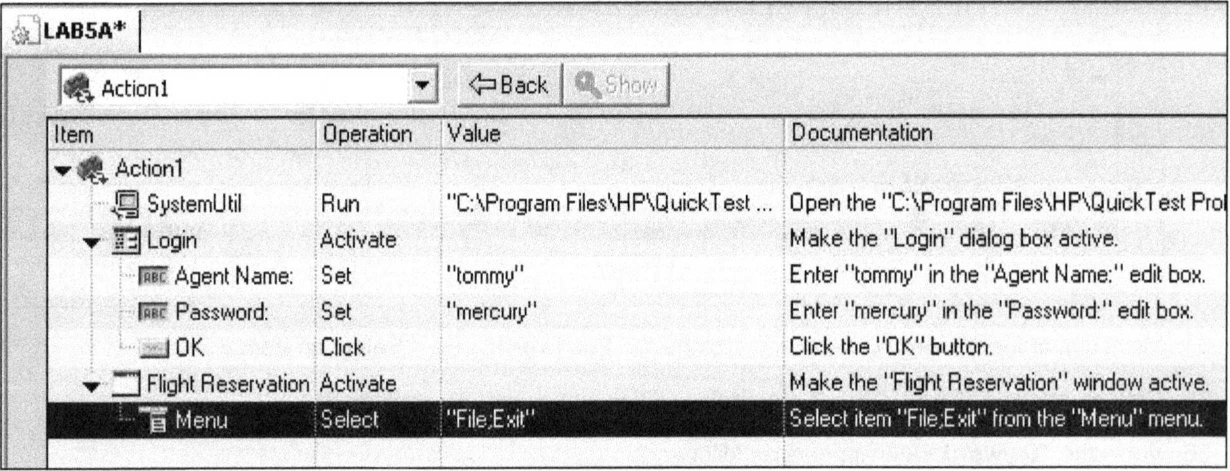

43. **File** → **Settings** → Click **Run** tab

44. Change *Object synchronization* to : **5** seconds → Click **OK**

45. **File** → **Save** → **C:\YOURNAME\QTP10\LVL1\LAB5A**

46. Playback → It should fail (with the following error message) : !

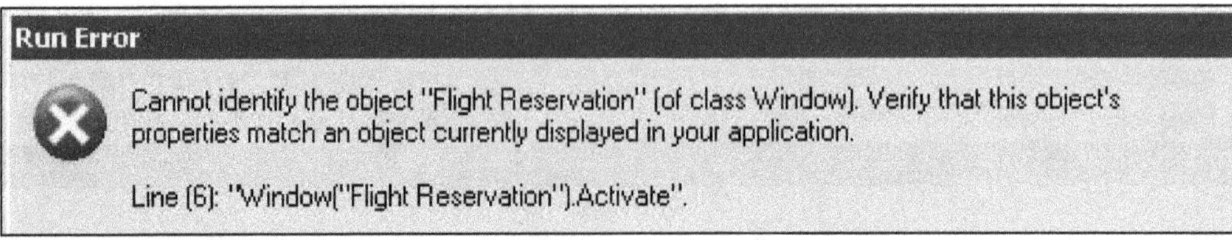

47. Click **Stop** (Stop the Playback)

48. Review the **Test Results** window → *Close* the **Test Results** window (when finished)

49. Close the AUT → How would you correct the problem ?

    Answer : Insert a **Synchronization** statement

50. Highlight **Flight Reservation** → Move cursor to field **Activate** → Click on the arrow

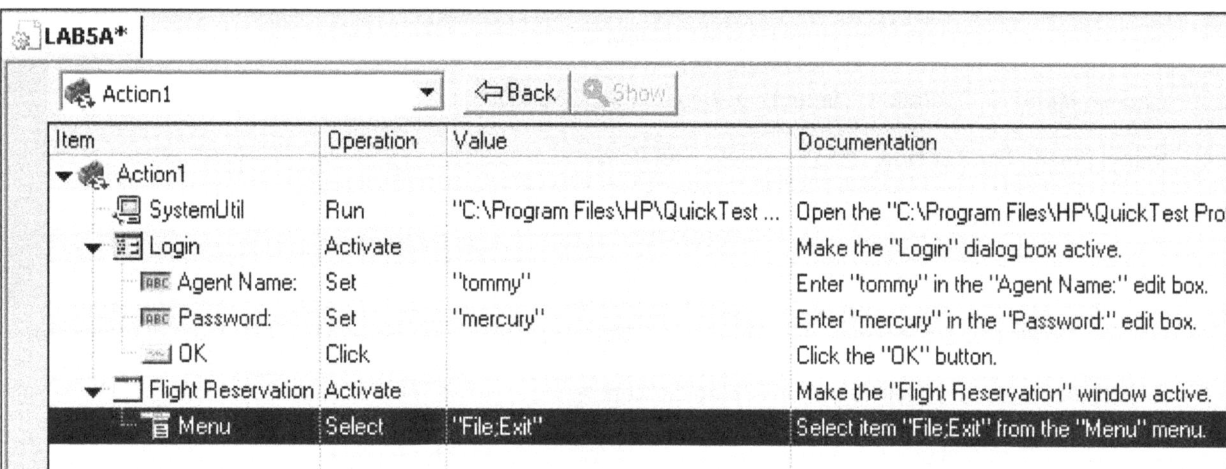

51. Scroll & select ***Activate*** with **WaitProperty** in ***Operation*** column

52. Move cursor to next empty field in **Value** column

53. Click on the left-most **Configure the Value < # >** icon (within the Value column)

54. In <u>Constant</u> type the following (including the double quotes) : **"enabled",1,12** → Click **OK**

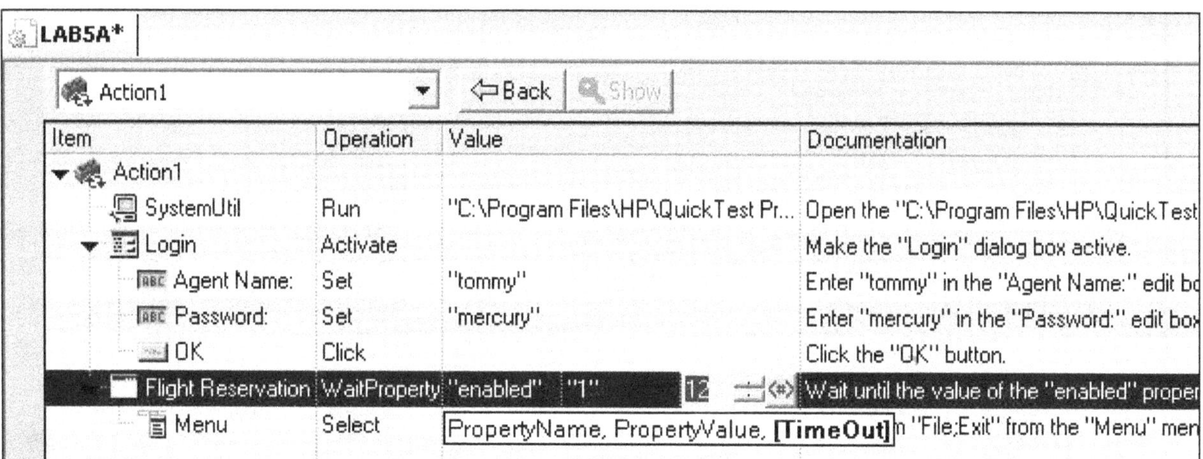

55. **File → Save → C:\YOURNAME\QTP10\LVL1\LAB5A**

56. Playback → It should Pass (with a warning message !) :

57. Playback → Correct & Fix any Synchronization issues (Increase the number of seconds !)

58. Verify the results → It should pass (you may get warnings – its OK !)

****** END OF LAB5A - Part 1 ******

LAB 5A : Part 2 – Complete the remaining Test Case — OPTIONAL

Requirements : Open an existing order

Steps:

1. Highlight **Flight Reservation WaitProperty** **Make the "Flight Reservation" window**
2. **Insert** (menu bar) → **New Step** → < Select an Item >
3. Choose **Menu** → Operation : **Select** → Value **"File;Open Order…"** → Move cursor back to **Menu**
4. **Insert** (menu bar) → **New Step** → < Select an Item >
5. Choose **Open Order** → Operation : **Activate**
6. **Insert** (menu bar) → **New Step** → < Select an Item >
7. Choose **Order No.** (checkmark) → Operation : **Set** → Value **"On"** → Move cursor back to **Order No.**
8. **Insert** (menu bar) → **New Step** → < Select an Item >
9. Choose **Edit** → Operation : **Set** → Value **"3"** → Move cursor back to **Edit**
10. Press **"F8" →** < Select an Item >
11. Choose **OK** → Operation : **Click** → Value **""** → Move cursor back to **OK**
12. Press **"F8" → Object from Repository** → Highlight **Flight Reservation** → Click **OK**

Item	Operation	Value	Documentation
Action1			
SystemUtil	Run	"C:\Program Files\HP\QuickTest Pr...	Open the "C:\Program Files\HP\QuickTest Pr
Login	Activate		Make the "Login" dialog box active.
Agent Name:	Set	"tommy"	Enter "tommy" in the "Agent Name:" edit box.
Password:	Set	"mercury"	Enter "mercury" in the "Password:" edit box.
OK	Click		Click the "OK" button.
Flight Reservation	WaitProperty	"enabled","1",12	Wait until the value of the "enabled" property
Menu	Select	"File;Open Order..."	Select item "File;Open Order..." from the "Men
Open Order	Activate		Make the "Open Order" dialog box active.
Order No.	Set	"On"	Set the state of the "Order No." check box to
Edit	Set	"3"	Enter "3" in the "Edit" edit box.
OK	Click		Click the "OK" button.
Flight Reservation	Activate		Make the "Flight Reservation" window active.
Menu	Select	"File;Exit"	Select item "File;Exit" from the "Menu" menu.

13. Save **LAB5A**
14. Playback and Verify the test → Correct & Fix any timing (synchronization) issues

***** END OF LAB 5A - Part 2 ******

LAB 5A : Create a QTP Script without recording (OPTIONAL)

REQUIREMENTS :

Part 1 : Create the TSR

1. Create a new folder **C:\YOURNAME\QTP10\REPO_SHARED**
2. Start the AUT
3. Create a single **Shared Repository** file containing the following windows via the AUT :
   a) LOGIN
   b) FLIGHT RESERVATION
   c) OPEN ORDER
   d) FAX ORDER NUMBER
4. Save the **Shared Repository C:\YOURNAME\QTP10\REPO_SHARED\FLIGHT.TSR**

   Note : Refer to LAB 2B - for help

Part 2 : Create the following two (2)Tests

1. Each Test Case is Separate script :
2. Each Script should reference the Shared Repository
3. Test Case (Test Scripts) :
   a) Create a New Reservation : **LAB5A_OPT1**
      - Enter Mandatory data for the required fields
      - Follow the Business Rules (A Valid Flight Date)

   b) Fax an order : **LAB5A_OPT2**
      - Pre-Conditions : The Order Must Exist

****** END OF LAB 5A – OPTIONAL ******

EXPERT VIEW

HP Quicktest has _two_ formats for viewing your test :

1. **Keyword View** : the object hierarchy in an ***icon***-based tree
2. **ExpertView** : the object hierarchy as a ***VBScript*** (a subset of the VB Language)
3. You may need to expand the window to see the ***VBScript*** statement (in the *left* pane)
4. Each line of VBScript represents a *step* in the test
5. HP Quicktest *records* the operations your *perform* on your application
6. An object's description is displayed parentheses following the object type
7. The script contains the objects ***logical name*** and is ***not*** case sensitive
8. The Object Repository contains both the **logical** and **physical description**

Example 1: Sample Script from the **Notepad** demonstration

```
01. SystemUtil.Run "notepad","","C:\Documents and Settings\ITM",""
02. Window("Notepad").WinEditor("Edit").Type "Line 1"
03. Window("Notepad").WinEditor("Edit").Type  micReturn
04. Window("Notepad").WinEditor("Edit").Type "Line 2"
05. Window("Notepad").WinEditor("Edit").Type  micReturn
06. Window("Notepad").WinEditor("Edit").Type "Line 3"
07. Window("Notepad").WinEditor("Edit").Type  micReturn
08. Window("Notepad").WinMenu("Menu").Select "File;Exit"
09. Window("Notepad").Dialog("Notepad").WinButton("Yes").Click
10. Window("Notepad").Dialog("Save As").WinEdit("File name:").Set  "C:\qa\test1"
11. Window("Notepad").Dialog("Save As").WinButton("Save").Click
```

9. The ***objects*** in the hierarchy are separated by a dot
10. The ***method*** performed on the object is always displayed at the end of the line
11. Follows the **OMD** model : **Parent.Object .Method.Data**

ENHANCING TESTS VIA PROGRAMMING

1. The easiest way to create a script (*test*) is to ***record*** a business process

2. After recording – a ***Script*** and ***Object Repository*** are created

3. To increase your test's power and flexibility you add programming statements to the *recorded* script

4. Programming statements can contain:

   a. **Recordable** test object methods :

      - Operations that a user can perform on AUT/SUT

   b. **Non-Recordable** test object methods :

      - Operations that a user **cannot** perform on AUT/SUT
      - To ***Retrieve*** or ***Set*** information to or from the AUT/SUT
      - Perform ***operations*** triggered by an *event*

   c. Modify or *manipulate* data from within the script (calculations, extracting text, etc …)

   d. ***Test Object*** methods of the object being tested (at *design* time)

   e. ***Run-time*** methods of the object being tested (while the test is *running*)

   f. Various VBScript programming statements & functions that *affect* the way a test executes

      - Variables & Constants
      - Arrays
      - IF Conditions
      - Loops
      - Action files
      - etc …

   g. Supplemental statements: Comments, customized messages and using **With** statements

ADDING METHODS

1. Is a *programming* tool that helps you to quickly and easily add **recordable** & **non-recordable** methods

2. You can add a step that :

   a) waits for an object
   b) checks to see if an *object* exists
   c) verify or insert a checkpoint
   d) return the value of an objects property
   e) etc …

3. To add a method (from the KeyWord View)

   a. First select the *object* (point & click)
   b. Choose a *method* from a list
   c. Enter the appropriate values for that method

DIM statement

1. Declares *variables* and allocates storage space in memory

2. Allows you to **compute** and/or store *data* values and reference *variables* within the script

3. Default datatype is ***variant*** (accepts any *datatype* value: string, integer, decimal, date, time, etc ...)

Syntax: DIM *variable*

Example 2: Definition of variables as **variant** data type

```
DIM  vs_name              ' Declares  a single variable
DIM  vs_names ( 10 )      ' Declares  an array with ten elements
DIM  vs_state  ( )        ' Declares  a dynamic array
DIM  vs_first, vs_last    ' Declares  two variables
```

EXPLICIT Option

1. You must declare all variables before using the **DIM**, **PRIVATE**, **PUBLIC** or **REDIM** statement
2. Used to clarify the name of an existing *variable* or to avoid confusion in code where the scope of the variable is *not* clear

Syntax: Option Explicit

Example 3: Explicit Declaration

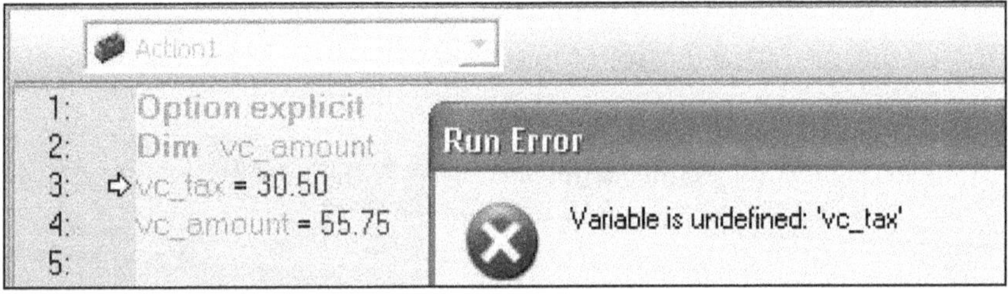

Note:

1. If you don't include an **Option Explicit** – then you can define any variables without errors
2. However it will not generate any errors if the variable name are misspelled
3. You still need to include a **DIM** statement

Quotes

1. Double Quotes to specify values in assignment

2. Single or *Double* Quote for comment

3. Be Consistent in the symbol used

CONDITIONAL STATEMENTS

1. You can control the flow of execution by using conditional statements

2. Conditional Statements used for decision-making (within the script)

   a) **IF** *condition* **Then** *Statements* [**ELSE** *elsestatements*] **END IF**

      - Used to evaluate whether a *condition* is **True** of **False**
      - Based on the results – specify one or more statements to *run*
      - You can **nest** as many *levels* as you need

   b) **IF** *condition* **Then**
      [*Statements*]
      [**ELSEIF** *condition-x*] **THEN**
      [Elseif*Statements*]
      ELSE
      [Else*Statements*]
      END IF

3. **Conditional** statements use operators

   a) = : Is equal to
   b) < : Less Than
   c) > : Greater than
   d) <= : Less than or equal to
   e) >= : Greater than or equal to
   f) < > : Not equal to

4. In the **Keyword View** mode – you can *insert* an **IF** statement :

   a) Determine the line in the script (the *step*) where you need to make a decision
   b) **Insert → Step → Conditional Statement → If … Then**
   c) Highlight the Newly created Step: **"" Statement True Check whether (True) is true. If so** *then*
   d) Click the **Expert View** → The *following* statement is displayed:

      If True Then
      End If

5. To complete the **THEN** statement you can :

   a) Copy an existing step and paste it in your **Then** statement *or*

   b) Enter the statement manually :

      Example 4:

      DIM vs_name

      If vs_name = "Fred Smith" Then
      action 1 (VB Statement)
      else
      action 2 (VB Statement)
      End If

SYSTEMUTIL Test Object

SystemUtil OBJECT

1. The **SystemUtil** object contains information about processes running in the system
2. Typically used to execute or ***run*** the Application Under Test (AUT) as the first line in the script
3. The **SystemUtil.Run** is automatically recorded / added to your test when you run from the ***Desktop***

Format : **Object.Run File, [Params], Directory, Operation, Mode**

Argument	Description
File	Required : Absolute / Relative PathName of the executable to run
Params	Optional : Parameters passed to the application
Dir	Optional : The default Directory / Folder of the application
Operation	Optional :
	Open (default) , Edit, Explore, Find, Print
Mode	How the Application is displayed when opened
	0 = Hide
	1 = Visible [Default]

Example 5

```
SystemUtil.Run "PAYROLL1.EXE" ," ","C:\YOURNAME\BUILD12", "Open"    or

SystemUtil.Run "C:\YOURNAME\PAYROLL1.EXE" ," ","", "Open"
```

SYSTEMUTIL. (Methods / Functions)

Methods	Description
BlockInput	Prevents keyboard and mouse input events from reaching applications
CloseDescendendantProcesses	Closes all processes opened by QTP
CloseProcessByHwnd	Closes a process that that is the owner of a window with the specified ID
CloseProcessById	Closes a process with the specified Process ID
CloseProcessByName	Closes a process according to a name
CloseProcessByWndTitle	Closes all processes that are owners of windows the specified Title
Run	Runs a file or application
UnBlockInput	Re-enables keyboard and mouse input events after a BlockInput

For detailed information: **QTP Help → Object Model Reference → Standard Windows → SystemUtil Object**

WRITING MESSAGES TO THE LOG

REPORTER OBJECT

1. The **LOG** is a permanent record that is viewable in the *Test Results* window

2. Sends a Customized message to the **TEST RESULTS** window (specific to the AUT or general messages)

Format : **Reporter.ReportEvent Eventstatus, Stepname, Message**

Argument	Description
Eventstatus	Status of the ReportStep
	0 or **micPass**
	1 or **micFail**
	2 or **micDone** : Does not affect the test (similar to **report_msg**)
	3 or **micWarning**
Stepname	Name of the intended *step*
Message	The error or message to be reported

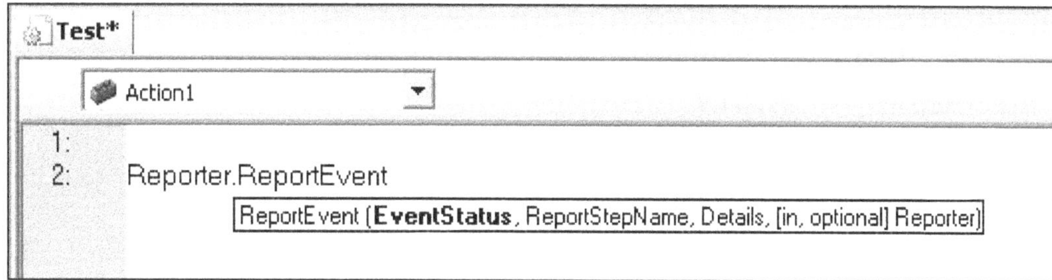

Example 6

```
Reporter.ReportEvent  1,  "Step 5",  "This step has failed"        or

Reporter.ReportEvent  micFail,  "Step 5",  "This step has failed"
```

Test Result :

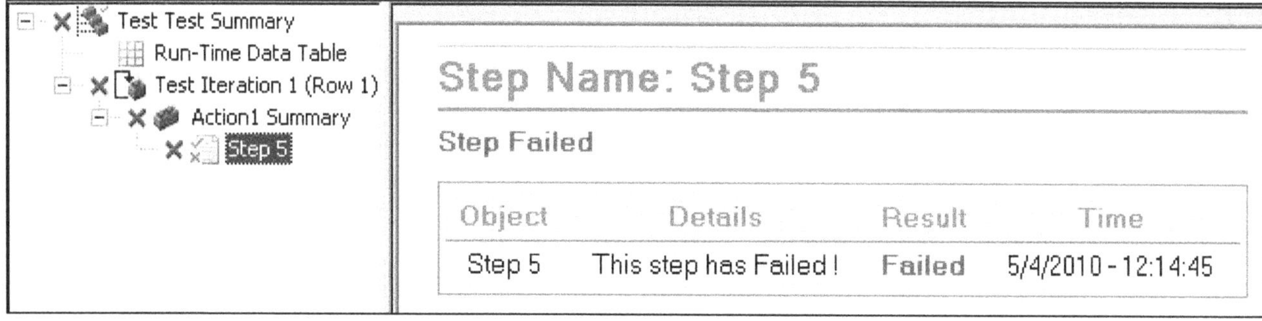

REPORTER FILTER

1. *To Retrieve* the current mode setting *currentmode* = **Reporter.Filter**
2. To *Set* the current mode for displaying events Reporter.Filter = *newmode*
3. You can *disable* or *enable* reporting of *steps* by setting a **Filter**
4. Any subsequent messages will be filtered based on the current settings specified (like a *FireWall*)

Mode	Description
0 or **rfEnableAll**	Default. All reported events *are* reported in the Test Results (TR)
1 or **rfEnableErrorsAndWarnings**	Only events with a ***warning*** or fail status are displayed in the TR
2 or **rfEnableErrorsOnly**	Only events with a **fail** status are displayed in the TR
3 or **rfDisableAll**	No events are displayed in the Test Results

Example 7

a) **Reporter.Filter = rfDisableAll**

b) **Reporter.Filter = rfEnableAll**

QTP Script

```
Dim vs_last, vs_first

Reporter.Filter = rfEnableErrorsAndWarnings
Reporter.ReportEvent  micDone, "Step 1",  "My own Message ... "
Reporter.ReportEvent  micWarning, "Step 2",  "A Warning Message ... "
Reporter.ReportEvent  micPass, "Step 3",  "Verification Passed ... "
Reporter.ReportEvent  micFail, "Step 4",  "Verification ! "
```

Test Results

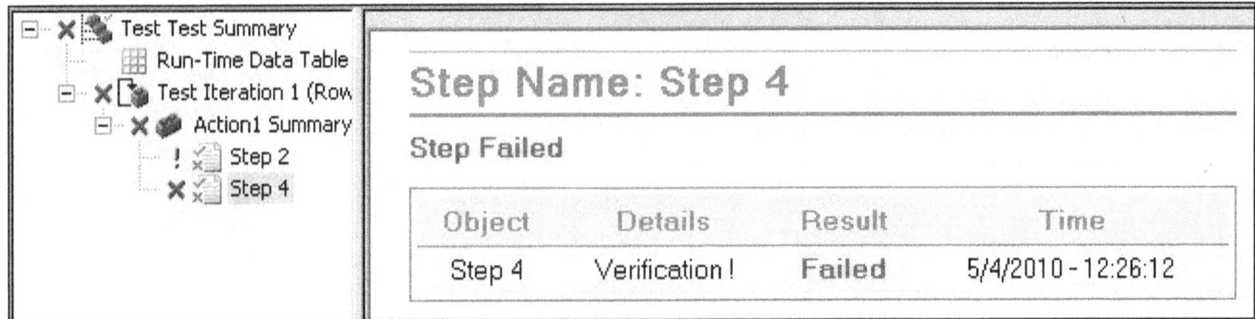

MSGBOX statement

1. Displays a message in a **dialog** box (response window)
2. Allows for user response (YES, NO, CANCEL , etc …)
3. You can use it to *display* information (variables) while debugging or for *decision-making* purposes

Format : **MsgBox (** ***message [buttons, title,])***

a) Message : Text string

b) buttons (Optional)

Constant	Value	Description
VbOKOnly	0	Display **OK** button only
VbOKCancel	1	Display **OK** and **CANCEL** buttons
VbAbortRetryIgnore	2	Display **ABORT, RETRY** and **IGNORE** buttons
VbYesNoCancel	3	Display **YES, NO** and **CANCEL** buttons
VbYesNo	4	Display **YES** and **NO** buttons

c) Return Values

Constant	Value	Description
VbOK	1	**OK**
VbCancel	2	**CANCEL**
VbAbort	3	Display **ABORT, RETRY** and **IGNORE** buttons
VbRetry	4	**RETRY**
VbIgnore	5	**IGNORE**
VbYes	6	**YES**
VbNo	7	**NO**

<u>Example</u> 8: Sample Script

```
DIM   vs_name , vs_rc                         ‘ Declares  two variables

Msgbox ( “Hello” )                            ‘ A dialog box to display any message within quotes

vs_name = “IBM”                               ‘ Assign a value to the variable

Msgbox ( “vs_name : “ & vs_name )             ‘ Display contents of the variable

Msgbox  “Name Incorrect ! Must be ITM “,vbOKOnly      ‘ Both statements are equivalent

vs_rc = MsgBox ( “Name Incorrect ! Do you want to continue “,vbYesNo )

IF   vs_rc =  7  THEN
     Reporter.ReportEvent  1,"Flight", "Could not open application"
     ExitTest
ELSE
     Reporter.ReportEvent  0,"Flight", "Opened App Successfully"
END IF

Reporter.ReportEvent  2,"Flight", "Last line in script"
```

EXIT statement

You can :

- *stop* or halt the execution of a script
- return a value to the parent script (or *action*)
- view the return value in the :

  a) Test Summary
  b) Iteration
  c) Action

STATEMENT	DESCRIPTION
ExitAction (*retvalue*)	Exits the **current** action, regardless of its iteration attributes
ExitActionIteration (*retvalue*)	Exits the current **iteration** of the action & returns to the top
	of the current iteration
ExitTestIteration (*retvalue*)	Exits the current global **iteration** of the action & begins the next
	row of the *Global* Data Table
ExitTest (*retvalue*)	Exits the script (*test*) regardless of its **iteration** attributes

Note:

a) retvalue : The return value which is a ***variant*** data type

LAB*XX* (QTP script)

Parent	Child	Iteration
Action 0		Global
	Action 1	Local Iteration

CHECKING THE SYNTAX

Before running (executing) the script – you can **check** the syntax to make sure it is correct

Tools → Check Syntax F7

- Missing : Quotes, Parentheses, Commas, etc …
- Invalid / Misspelled Method (function) *name*
- Actions and Associative Repositories are invalid or missing

LAB 5B : DEBUGGING

Test Requirements :

- Define breakpoints
- Single Step execution
- Examine the contents of *variable*
- Use *MsgBox* statement

Steps : (Part 1)

1. **File → New → Test** (Create a *new* script) → Click **ExpertView** tab

2. **View** (menu bar) → **DebugViewer** (Ensure the ***Debug Viewer*** window is displayed at bottom left)

3. Type the following Script for **Action1**

```
01.  DIM  vi_x, vi_y, vi_z
02.  DIM  vs_name, vs_value

03.  vi_x = 5
04.  vi_y = 14
05.  vs_name = "Abe Lincoln"
06.  vi_z = vi_x + vi_y

07.  MsgBox  "Name is:  "  &  vs_name
08.  Reporter.ReportEvent  micDone , " Name ",  vs_name
```

4. Move cursor to Line 3 → Press **F9** (a ***red*** stop sign displays)

5. Click **Run** → Click **OK** (to **Run** window)

6. Wait for the ***yellow*** arrow to display on top of the red hand

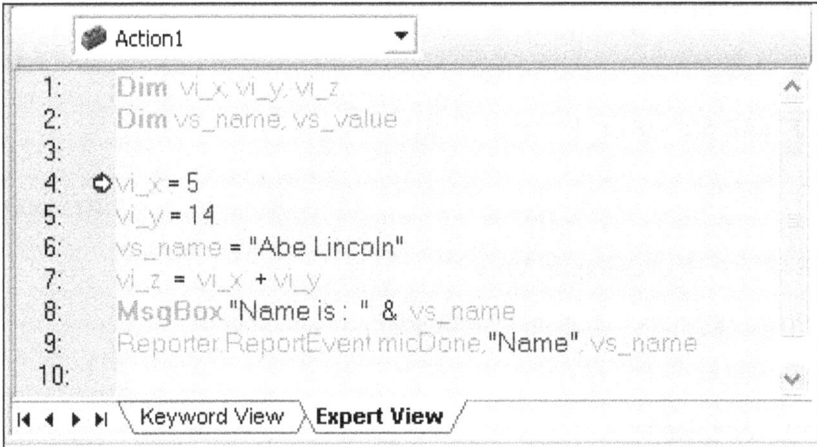

7. Click on the **DeBug Viewer** (located at the bottom Left)

8. Review the **DeBug Viewer** pane → Observe the three tabs (**Watch, Variables Command**)

9. Click **Variables** tab

10. Observe the **Variables** tab

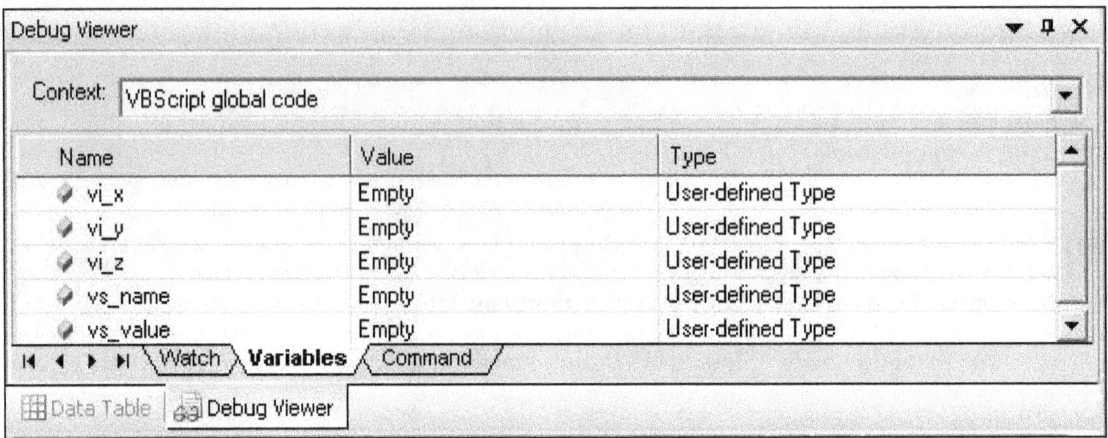

11. Press **F11** (Single Step) - Review the contents of **DeBug Viewer** vi_x = 5
12. Press **F11** (Single Step) - Review the contents of **DeBug Viewer** vi_y = 14
13. Press **F11** (Single Step) - Review the contents of **DeBug Viewer** vs_name = "Abe Lincoln"
14. Press **F11** (Single Step) - Review the contents of **DeBug Viewer** vi_z = 19
15. A Message Box appears → Click **OK**

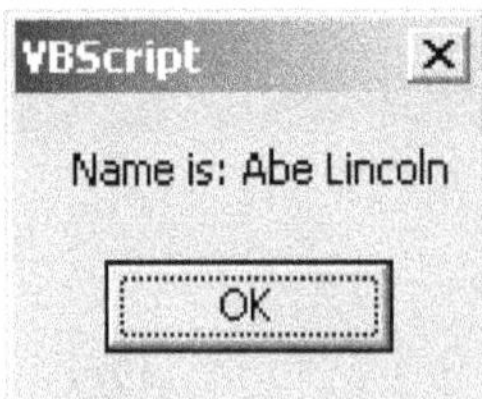

16. Save the script **C:\YOURNAME\QTP10\LVL1\LAB5B**
17. Close the **Test Results** window

***** **END OF LAB 5B - Part 1** *****

STEPS (Part 2) : **OPTIONAL**

1. Open either **LAB4B_P1** or **LAB4B_P2** → Click **Expert View**
2. Choose and mark breakpoints (Press **F9**) in the script after the **CheckPoint** statements
3. Run the Script (Playback) → it will pause
4. Click the **Debug Viewer** → Select **Variables** tab
5. Press **F11** to *single step* and observe the variables is memory being populated

***** **END OF LAB 5B - Part 2** *****

LAB 5C : Block AUT Input

Test Requirements :

- Use LAB1D (Open an Order & Fax)
- Use *MsgBox* statement to determine if the AUT should be blocked
- Use *SystemUtil* and *Reporter* object
- Choose a different *order number* & phone number than what was used in the original script
- Update the *Active Screen* & *Object Repository* via ***Update Run Mode***

STEPS : (Part 1)

1. **File → Open → Test** → Click **LAB1D**

2. **View** (menu bar) → **DebugViewer** (Ensure the ***Debug Viewer*** window is displayed at bottom left)

3. Playback and verify the Script runs with no errors ! If required, Re-Record in Analog mode

4. Click **Expert View** → Make the following changes (lines in **bold**) :

```
Dim   vi_answer
vi_answer = MsgBox  ("Do you want to Block AUT Input ?", vbYesNo)

' 6 = Yes   7 = No

If  vi_answer = 6  Then
     MsgBox  "AUT Input will be BLOCKED !", vbOKOnly
     SystemUtil.BlockInput
End If

SystemUtil.Run "C:\Program Files\HP\QuickTest Professional\samples\flight\app\flight4a.exe"
Dialog("Login").WinEdit("Agent Name:").Set "tommy"
Dialog("Login").WinEdit("Agent Name:").Type  micTab
Dialog("Login").WinEdit("Password:").SetSecure "4c925f67c92bb34105c734e6486354f1ce1b6887"
Dialog("Login").WinButton("OK").Click

.... Remain Code of the script
```

5. **Tools → Check Syntax** (Correct and Fix any Syntax errors)

6. Save **LAB1D** → Playback → The following message appears → Click **No**

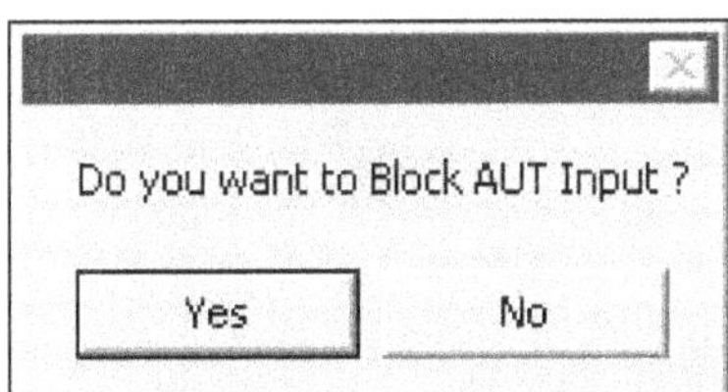

7. Wait for the ***Flight Reservation*** window to Open

8. Try and Click inside the AUT (Don't change the data)

9. Review and Close the ***Test Results*** window

10. Playback the Test again → The following message appears → Click **Yes**

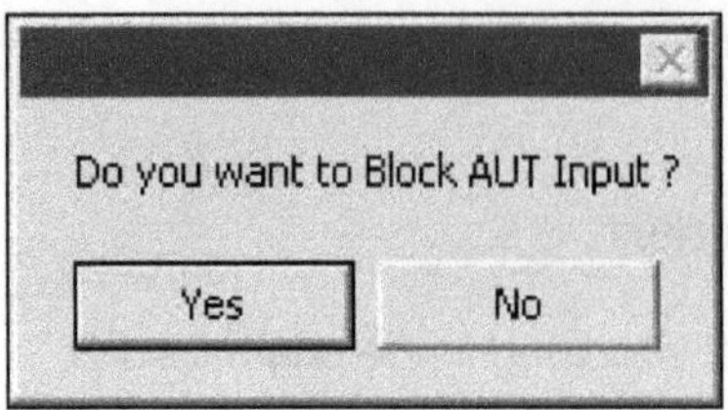

11. Wait for the ***Flight Reservation*** window to Open

12. Try and Click inside the AUT (The Keyboard input is blocked)

13. Review and Close the ***Test Results*** window

Steps (Part 2) : Use the Debugger

14. **View** → **Debug Viewer** (Ensure the Debug Viewer is visible)

15. Click **Expert View** → Specify a breakpoint in Double-Clicking **Line 9** until the ***red*** stop sign appears

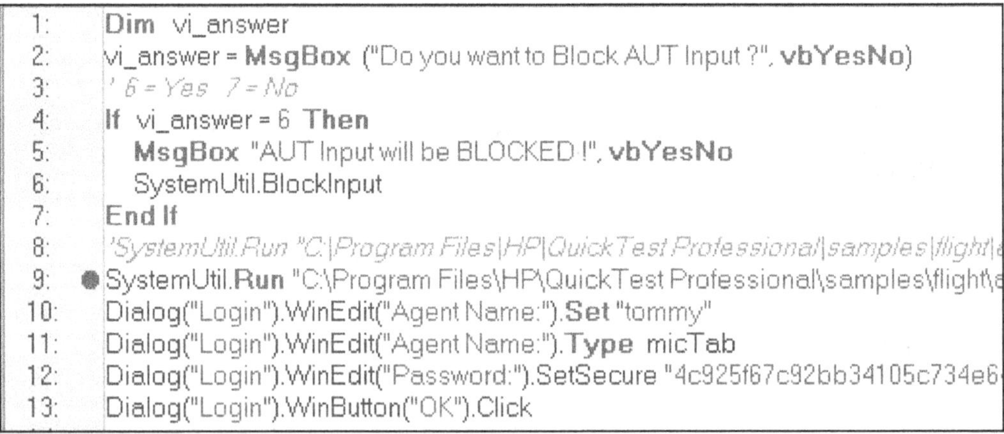

```
1:  Dim vi_answer
2:  vi_answer = MsgBox ("Do you want to Block AUT Input ?", vbYesNo)
3:  ' 6 = Yes  7 = No
4:  If vi_answer = 6 Then
5:     MsgBox "AUT Input will be BLOCKED !", vbYesNo
6:     SystemUtil.BlockInput
7:  End If
8:  'SystemUtil.Run "C:\Program Files\HP\QuickTest Professional\samples\flight\
9:  SystemUtil.Run "C:\Program Files\HP\QuickTest Professional\samples\flight\
10: Dialog("Login").WinEdit("Agent Name:").Set "tommy"
11: Dialog("Login").WinEdit("Agent Name:").Type micTab
12: Dialog("Login").WinEdit("Password:").SetSecure "4c925f67c92bb34105c734e6
13: Dialog("Login").WinButton("OK").Click
```

16. Playback → Click **Yes** (To Block Input) → Click **OK** (To verify)

17. QTP stops in *Line 9* → Click **Variables** tab Review the current value of the variable **vi_answer**

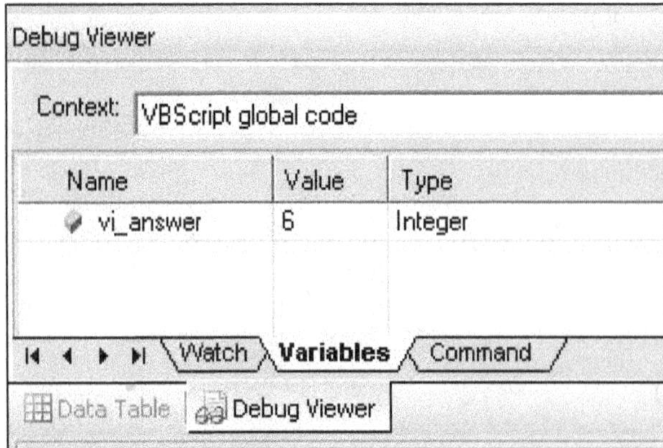

18. Press **F11** – Single Step → The AUT launches

19. Review and Close the ***Test Results*** window

20. Remove the breakpoint (Double-Click ***Line 6***) – **Stop Sign** is not visible

21. Make the following changes and use a different order number in *Line 18* only

```
01. Dim   vi_answer
02. vi_answer = MsgBox  ("Do you want to Block AUT Input ?", vbYesNo)
03. '  6 = Yes   7 = No
04. If  vi_answer = 6  Then
05.             Reporter.ReportEvent micDone,"AUT","Input will be be BLOCKED!"
06.     SystemUtil.BlockInput
07. Else
08.             Reporter.ReportEvent micDone,"AUT","Input NOT BLOCKED ..."
09.     SystemUtil.UnBlockInput
10. End If

11. SystemUtil.Run "C:\Program Files\HP\QuickTest Professional\samples\flight\app\flight4a.exe"
12. Dialog("Login").WinEdit("Agent Name:").Set "tommy"
13. Dialog("Login").WinEdit("Agent Name:").Type  micTab
14. Dialog("Login").WinEdit("Password:").SetSecure "4c925f67c92bb34105c734e6486354f1ce1b6887"
15. Dialog("Login").WinButton("OK").Click

16. Window("Flight Reservation").WinMenu("Menu").Select "File;Open Order..."
17. Window("Flight Reservation").Dialog("Open Order").WinCheckBox("Order No.").Set "ON"
18. Window("Flight Reservation").Dialog("Open Order").WinEdit("Edit").Set "2"
19. Window("Flight Reservation").Dialog("Open Order").WinButton("OK").Click
20. Window("Flight Reservation").WinMenu("Menu").Select "File;Fax Order..."
21. Window("Flight Reservation").Dialog("Fax Order No. 2").ActiveX("MaskEdBox").Type "7324445555"

22. Desktop.RunAnalog "Track1"

23. Window("Flight Reservation").Dialog("Fax Order No. 2").WinButton("Send").Click
24. Window("Flight Reservation").WinMenu("Menu").Select "File;Exit"
```

22. Click **Keyword View** → Correct & Fix any syntax errors

23. Save **LAB5C** → Expand **Fax Order No 2** → Highlight **MaskEdBox** → Review the phone number in script

24. **Automation** (menu bar) → **Update Run Mode** → Mark all the Check Boxes → Click **OK**

25. Click **Yes** (To Block Input) → Click **OK** (To verify) → Review the Test Results → Verify the LOG

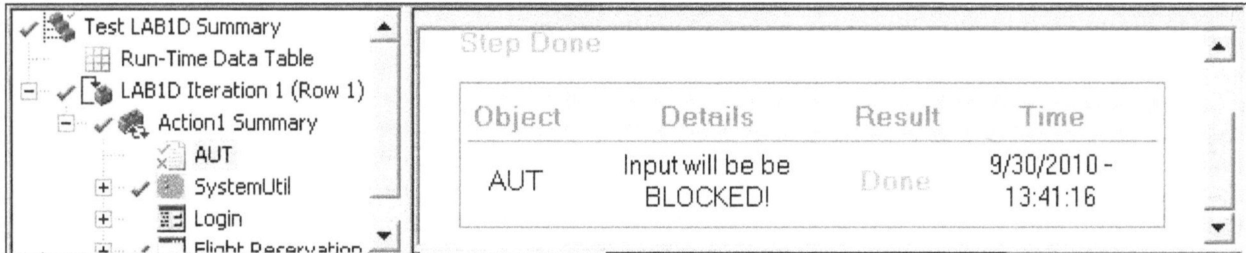

26. Close the **Test Results** window

27. Expand **Fax Order No 2** → Highlight **MaskEdBox** → Review the phone number in script has been updated

*** **END OF LAB 5C** ***

GetObject method

- Retrieves an existing object (in Memory)
- Allows you to *reference* an *Automated Application* inside the *script*
- Requires you to use the **set** statement
- The **set** statement allocates space in memory for an *object*
- The **dim** statement allocates space in memory to store *data* values

1. Format : GetObject (*pathname, class*)

   Pathname is *optional*
   Class is the Object's Class

2. Example : Reference *Automated Applications* in QTP

```
DIM  v_qtp , v_excel

set v_qtp   =  GetObject ( " ", "QuickTest.Application" )

set v_excel =  GetObject ( " ", "Excel.Application" )

v_qtp.WindowState = "Minimized"

set v_excel =  GetObject ( " ", "Excel.Application" )

v_excel.Save ("T:\YOURNAME\EXCEL\CUST.XLS " )

set v_qtp   =  Nothing

set v_excel =  Nothing
```

3. Each Automated Application has **Object Model**

   - Excel
   - Word
   - Outlook

4. QTP **Automation Object Model** helps you automate QuickTest operations :

   - Access Methods & Properties
   - Customize QTP to run a series of consecutive tests

LAB 5D : Minimize QTP Window

Test Requirements :

- Use LAB2D (Analog Recording)
- Use *GetObject* statement to access QTP
- Modify the script to Minimize the QTP window

Steps : (Part 1)

1. **File → Open → Test** → Click **LAB2D**

2. **File → Save As → LAB5D**

3. Click **ExpertView** → Make the following changes : (Lines ***1*** thru ***6***)

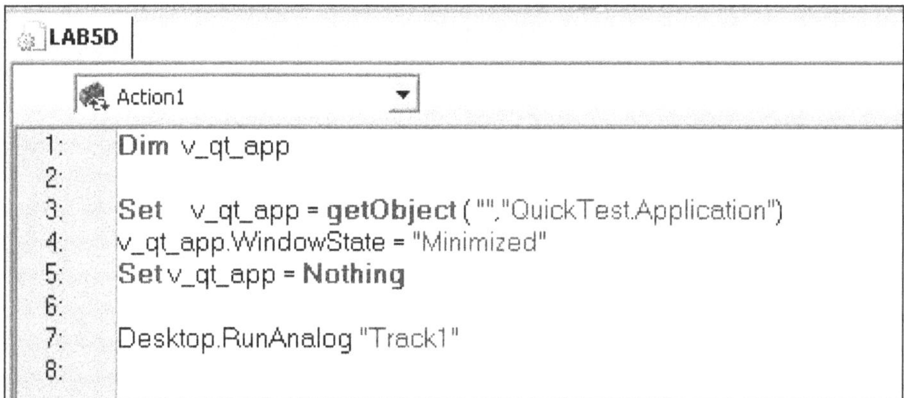

4. Click **Keyword View**

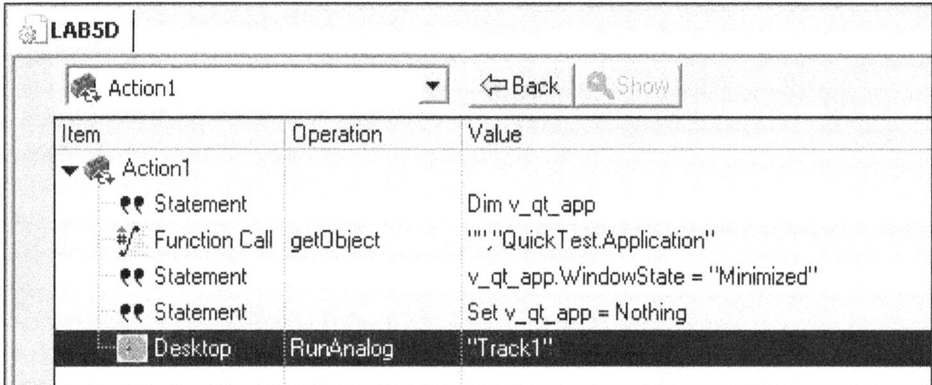

5. Save **LAB5D**

6. Playback → Verify the results →

7. It the test *fails* , it may be a synchronization issue !

** **END OF LAB 5D** ***

RETRIEVING & SETTING TEST OBJECT PROPERTY VALUES

1. Test Object Properties are the set of properties ***defined*** by HP QuickTest for each object
2. You can **set** and **retrieve** a test objects property value
3. You can **retrieve** the values of test object properties from a ***run-time*** object

GetROProperty (Recommended)

1. Returns the value of the property for the object in the AUT at ***run-time*** (execution)

   Format : Object(description).GetROProperty (Property)

   Example 9 : Get the **text** from the two objects in the Flight Reservation

   a. **vs_text = Window("Flight Reservation").WinButton("FLIGHT").GetROProperty("text")**

   b. **vs_name = Window("Flight Reservation").WinEdit("Name:").GetROProperty("text")**

GetTOProperty (Only if the ***RO*** Fails or does not work !)

2. Returns the value of the property from the ***temporary*** test object description

   Format : Object(description).GetTOProperty (Property)

   Example 10 : Get the **text** from the Login window

   a) **vs_text = Dialog("Login").GetTOProperty ("text")**

   b) **vs_enabled = Dialog("Login").WinEdit("Agent Name:").GetTOProperty ("enabled")**

3. **SetTOProperty**

   a) *Sets* the value of the property from the ***temporary*** test object description (at *design-time*)
   b) Valid only for the *properties* captured in the Object Repository
   c) It has *NO* effect on the Active Screen or the values saved in the Object Repository
   d) **DO NOT, *DO NOT*** use this *method* to populate information on the screen/page

   Format : Object(description).SetTOProperty Property, Value

   Example 11 : Set the values for the following object :

   e) **Window("Flight Reservation").WinEdit("Name:").SetTOProperty "text", "John Doe"**

4. **SET Method : ** Sets the *value* to an editable field on the screen (populate fields onto the screen)

   Format : Object(description).Set [value/variable]

   Example 12 : **Window ("Flight Reservation").WinEdit ("Name:").Set "George Washington"**

General Rule (Methods)

1. Parentheses within a **method** implies that a ***value*** is being returned !
2. Exceptions are specific to the *particular* method
3. Review the on-line syntax in the HP QuickTest Professional Help

EXAMPLES using : GetROProperty , SetTOProperty & SET statements

Requirements :

1. Enter User-ID & Password (invalid)
2. Customize a message by printing the values entered by the user
3. Set a new value for the User-ID

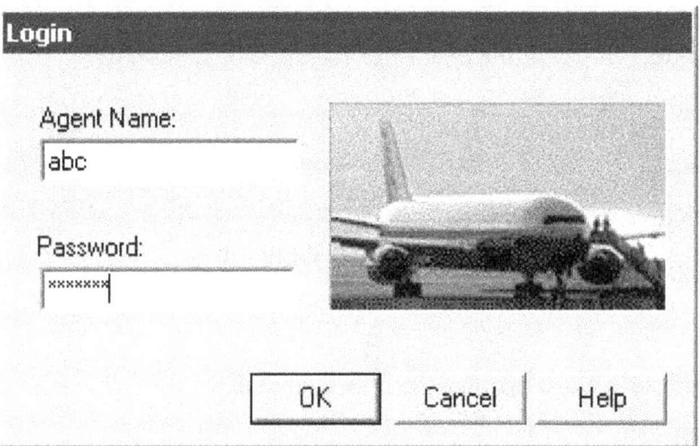

Sample Script :

```
01. DIM  vs_user

02. SystemUtil.Run "C:\Program Files\HP\QuickTest Professional\samples\flight\app\flight4a.exe"

03. Dialog("Login").WinEdit("Agent Name:").Set "abc"
04. Dialog("Login").WinEdit("Password:").SetSecure "477c743994e1a4b7903867e4c09475f5"

05. vs_user = Dialog("Login").WinEdit("Agent Name:").GetROProperty ("text")
06. Reporter.ReportEvent  micDone, "User ID", "User = " & vs_user

07. " Dialog("Login").WinEdit("Agent Name:").SetTOProperty  "text" ,  vs_user
08.   Dialog("Login").WinEdit("Agent Name:").Set   vs_user

09. Dialog("Login").WinButton("OK").Click
10. Dialog("Login").Dialog("Flight Reservations").WinButton("OK").Click          'Login Error Message
11. Dialog("Login").WinEdit("Agent Name:").Set "John"

12. Dialog("Login").WinButton("OK").Click
13. Window("Flight Reservation").WinMenu("Menu").Select "File;Exit"
```

Explanation :

Line 05 :	Retrieves the user's input into **vs_user** variable
Line 06 :	Prints the contents of **vs_user** variable to the **Log** (Test Results)
Line 07 :	Not the recommended method, most likely will NOT work and observe the line is commented !
Lines 5 thru 9 :	Generates an AUT error because it fails the business rule
Line 11 :	Recommended method for populating or *setting* data values into input fields

LAB 5E: ExpertMode

PART I : Test Requirements : Verify the values from the AUT

a) Save objects in ***Local Repository*** (default mode)
b) Use an existing order , Retrieve the values and compare the results, Update the *name*
c) Insert ***VBScript*** functions (GetROProperty)
d) Scenario:

Start the AUT → Login → Open an existing order → Update *Passenger Name* → Exit the AUT

e) It should fail ! Where is the error ? Is it a *QTP* or *AUT* error ?

Steps: (Part 1) :

1. **File** (menu bar) → **New** (Create a *new* script) → **Test** → Click **Keyword View** tab
2. **Automation** (menu bar) → **Record F3**
3. Check **Record and run test on any open Window-based application** → Click **OK**
4. Start the AUT (on your own) → Login with ***yourname*** and use the password ***mercury*** → Click **OK**
5. **File → Open Order** → Select an existing order → Click ***OK***
6. Change the *name* (on your own) → Click **Update Order**
7. **File → Exit** (Close the AUT)
8. **Automation → Stop F4** (Stop Recording)
9. Review the **Keyword View** script → Remove/Delete any *unwanted* actions
10. Click the **Expert View** tab → **Tools** (menu bar) → **View Options...**
11. Click **General** tab → Check **Show Line Numbers** → Click **OK**
12. Script maybe similar to the following :

```
01. SystemUtil.Run "C:\Program Files\HP\QuickTest Professional\samples\flight\app\flight4a.exe"

02  Dialog("Login").WinEdit("Agent Name:").Set "tommy"
03. Dialog("Login").WinEdit("Password:").SetSecure "4123a2ca042e1a8a7286f8"
04. Dialog("Login").WinButton("OK").Click

05. Window("Flight Reservation").WinMenu("Menu").Select "File;Open Order..."
06. Window("Flight Reservation").Dialog("Open Order").WinCheckBox("Order No.").Set "ON"
07. Window("Flight Reservation").Dialog("Open Order").WinEdit("Edit").Set "7"
08. Window("Flight Reservation").Dialog("Open Order").WinButton("OK").Click

09. Window("Flight Reservation").WinEdit("Name:").Set "Jane Doe 2"

10. Window("Flight Reservation").WinButton("Update Order").Click
11. Window("Flight Reservation").WinMenu("Menu").Select "File;Exit"
```

13. Line 9 will vary with the *name* you specify (i.e: instead of ***Jane Doe 2***) it may be a *different* name

14. Save the script as **LAB5E** → Playback → You will get the following error : → Do ***NOT*** Click **OK** button !!

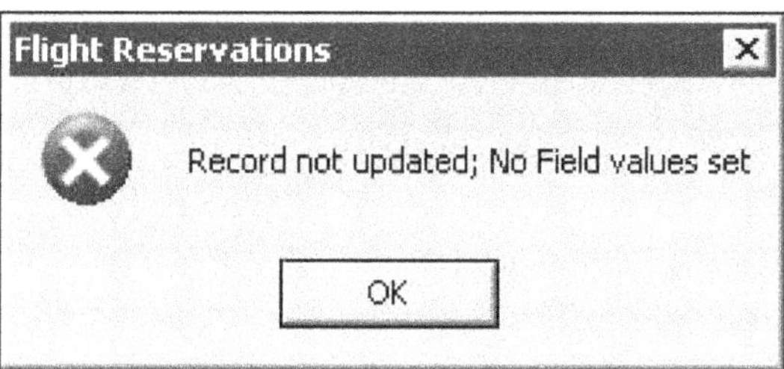

15. Wait for the following error : → Click **Stop** → Review & Close the **Test Results** window

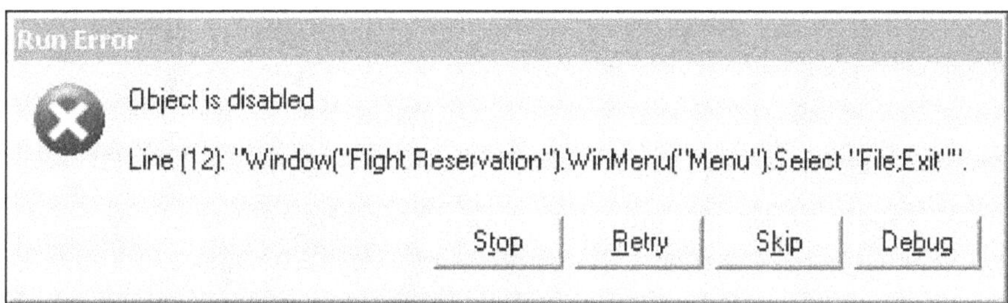

16. Click **OK** (Close the **Flight Reservations** window)

17. Close the AUT

18. Make the following changes :

```
01. DIM vs_name
02. SystemUtil.Run "C:\Program Files\HP\QuickTest Professional\samples\flight\app\flight4a.exe"

03. Dialog("Login").WinEdit("Agent Name:").Set "tommy"
04. Dialog("Login").WinEdit("Password:").SetSecure "4123a2ca042e1a8a7286f8"

05. vs_name = Dialog("Login").WinEdit("Agent Name:").GetROProperty ("text")
06. Reporter.ReportEvent micDone, "Login Process","Name=" & vs_name

07. IF vs_name = "yourname" THEN
08.   Reporter.ReportEvent micPass, "Login ID","Agent Name matches"
09. ELSE
10.   Reporter.ReportEvent micFail, "Login ID","Agent Name Does NOT match"
11. END IF

11. Dialog("Login").WinButton("OK").Click
12. Window("Flight Reservation").WinMenu("Menu").Select "File;Open Order..."
13. Window("Flight Reservation").Dialog("Open Order").WinCheckBox("Order No.").Set "ON"
14. Window("Fli ght Reservation").Dialog("Open Order").WinEdit("Edit").Set "7"
15. Window("Flight Reservation").Dialog("Open Order").WinButton("OK").Click
16. Window("Flight Reservation").WinEdit("Name:").Set "Jane Doe 3"
17. Window("Flight Reservation").WinButton("Update Order").Click
18. Window("Flight Reservation").WinMenu("Menu").Select "File;Exit"
```

19. Change the *name* inside the script (Do NOT use the *same* **name** captured during recording !)

- If you use the *same* name – the AUT may fail !

20. Save the script (**LAB5E**) → **Tools** → **Check Syntax**... **F7** → Fix any errors

21. Playback → Review the *Results* → Here is a sample of the results in **Keyword View**

Item	Operation	Value	Documentation
Action1			
Statement		Dim vs_name	
SystemUtil	Run	"C:\Program Files\HP\QuickTest Pr...	Open the "C:\Program Files\HP\QuickTest Professic
Login			
Agent Name:	Set	"tommy"	Enter "tommy" in the "Agent Name:" edit box.
Agent Name:	Type	micTab	Type micTab in the "Agent Name:" edit box.
Password:	SetSecure	"4c9b58ca042b901666a747e3695...	Enter the encrypted string "4c9b58ca042b901666a7
Agent Name:	GetROProperty	"text"	Retrieve the current value of the "text" property for tl
Reporter	ReportEvent	micDone,"Login Process","Name=:...	Report "Name=:" & vs_name to the report and set th
Statement		vs_name = "tommy"	Check whether (vs_name = "tommy") is true. If so:
Reporter	ReportEvent	micPass,"Login ID","Agent Name M...	Report "Agent Name Matches" to the report and s
Statement			Otherwise:
Reporter	ReportEvent	micfail,"Login ID","Agent Name Do...	Report "Agent Name Does NOT Match" to the rep
OK	Click		Click the "OK" button.
Flight Reservation			

22. Highlight Step **Login Process** → See the right pane

23. Highlight Step **Login Id** → See the right pane

STEPS: (Part 2)

1. Modify the script and change the Agent Name – Do not use the same name from the last run
2. Copy and paste the statement that *sets* the *Agent Name* to **tommy**
3. Comment the original line by put a single quote as the first character on *that* line (it turns green)
4. Change the **Login Id** from ***tommy*** to ***John***

```
' Dialog("Login").WinEdit("Agent Name:").Set "yourname"

Dialog("Login").WinEdit("Agent Name:").Set "John"
```

5. Save the script **LAB 5E→ Tools → Check Syntax F7** → Fix any errors (if any)
6. Playback → Review the scripts → Expand All → It fails
7. Setting the timeout parameter too low – may result in a *synchronization* error (Object Not Found)

***** END OF LAB5E – Part 2 *****

PART 3 (Negative Testing)

Requirements: Enhance the script to:

1. Modify the script and change the Agent Name back to the original
2. Delete the validation for the user id (*agent name*) ***yourname***
3. Delete or *comment* the line that uses "**john**" as the UserId (*agent name*)
4. If the user forgets to enter the agent name:
    a. Issue the appropriate **failed** message !
    b. Insert the action to Click **Cancel** (to exit the LOGIN window)
    c. Exit the run
5. If the user has entered a valid agent name (your name) → issue the appropriate ***pass*** message
6. Add the **Cancel** Object to the repository
7. Include an **IF ELSE** statement to *change* the passengers name back to the original (*Extra*)

STEPS (Part 3)

1. Start the AUT on your own → DO NOT Login !
2. **Resources → Object Repository → Object → Add Objects to Local…**
3. Cursor changes to a hand →Point & click **Cancel** (in the LOGIN window)
4. Verify **WinButton : Cancel** → Click **OK**

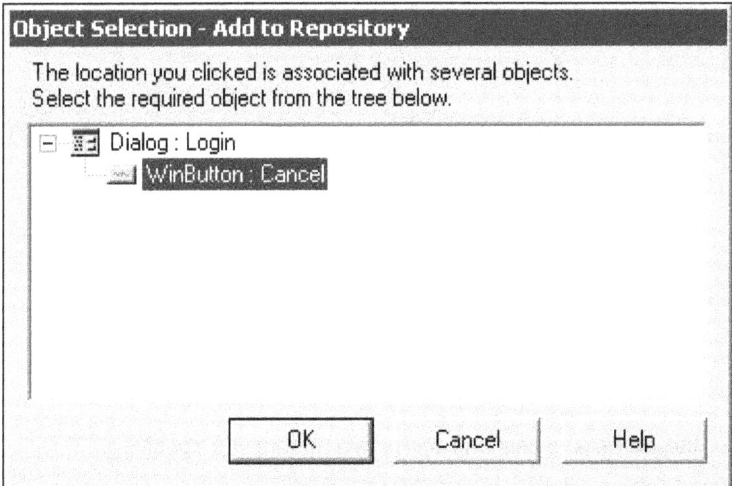

5. **File → Close** (close Object Repository)
6. Save the Script (**LAB 5E**)

7. Modify the script as follows (the changes are in ***bold***):

```
01. DIM vs_name
02. SystemUtil.Run "C:\Program Files\HP\QuickTest Professional\samples\flight\app\flight4a.exe"

03. ' Dialog("Login").WinEdit("Agent Name:").Set "yourname"
04. Dialog("Login").WinEdit("Agent Name:").Set "yourname2"
05. Dialog("Login").WinEdit("Password:").SetSecure "4123a2ca042e1a8a7286f8"

06. vs_name = Dialog("Login").WinEdit("Agent Name:").GetROProperty ("text")

07. IF vs_name > "" THEN
08.    Reporter.ReportEvent micPass, "Login ID","Agent Name OK"
09. ELSE
10.    Reporter.ReportEvent micFail, "Login ID","Missing Agent Name"
11.    Dialog("Login").WinButton("Cancel").Click
12.    ExitTest
13. END IF

14. Dialog("Login").WinButton("OK").Click
15. Window("Flight Reservation").WinMenu("Menu").Select "File;Open Order..."
16. Window("Flight Reservation").Dialog("Open Order").WinCheckBox("Order No.").Set "ON"
17. Window("Flight Reservation").Dialog("Open Order").WinEdit("Edit").Set "7"
18. Window("Flight Reservation").Dialog("Open Order").WinButton("OK").Click
19. Window("Flight Reservation").WinEdit("Name:").Set "Jane Doe 2"
20. Window("Flight Reservation").WinButton("Update Order").Click
21. Window("Flight Reservation").WinMenu("Menu").Select "File;Exit"
```

8. Save the script **LAB 5E**→ Playback → Review the Test Results → Expand All → It should PASS !

9. Modify the script and comment the line for ***tommy.*** (on your own)

    a) You will NOT enter any login ids for this test run

    b) Lines 3 and 4 should be <u>commented</u> (Login ID's). Line 3 is not commented – you must comment it !

10. Save the script **LAB 5E**

11. Playback → Review the Test Results → Expand All → It should fail !

12. Close the *Test Results* window

13. Correct & Fix the problem in the script

<u>**RESOLUTION**</u> :

- Use the ***GetROProperty*** method (to extract the *name* of the passenger from the AUT)
- Insert a conditional statement to check for the ***name*** and use one of the two names you specify.
- Refer to **LAB 5F**

******** **END OF LAB 5D** *******

CUSTOMIZED CHECKPOINT

The following are the steps to create/define your own verifications :

1. Identify the object in the AUT / SUT

   a) Visually locate the field on the *Window / WebPage*
   b) Identify the **Logical Name** of the object (by using the *Active Screen* or *Object Spy*)
   c) Determine the location (or line) in the script where you want to retrieve the *text*
   d) Ideally, the line before the user takes the *final* action (*New, Update, Delete, Fax* , etc …)

2. Create a variable using the **DIM** statement

   a) To hold / store the contents of an object's property in memory
   b) Define the variable at the top of the script

3. Retrieve the object's property value by using one of the following :

   a) **GETROProperty**

   b) **GETTOProperty**

   c) Properties to Request (for an object)

   - text , label, value
   - enabled
   - focused
   - format
   - height
   - width

   d) Use the **Object Spy** to list and review an objects properties

4. Validate the Business Rule (BR) :

   a) You must compare values according to the BR

   b) Use the **IF < expression > THEN**

   c) Compare the values in *memory* by referencing the variables

5. Print the Results of the Comparison (based on the business rules)

   a) Use the **Reporter.ReportEvent**

Question : When should you issue a ***QTP Checkpoint*** or Build a ***Customized*** verification ?

Answer : If the information / data (on the screen) is :

a) *Static* : Issue a QTP Checkpoint

b) *Dynamic* : Create your own customized checkpoint

SAMPLE SCRIPT

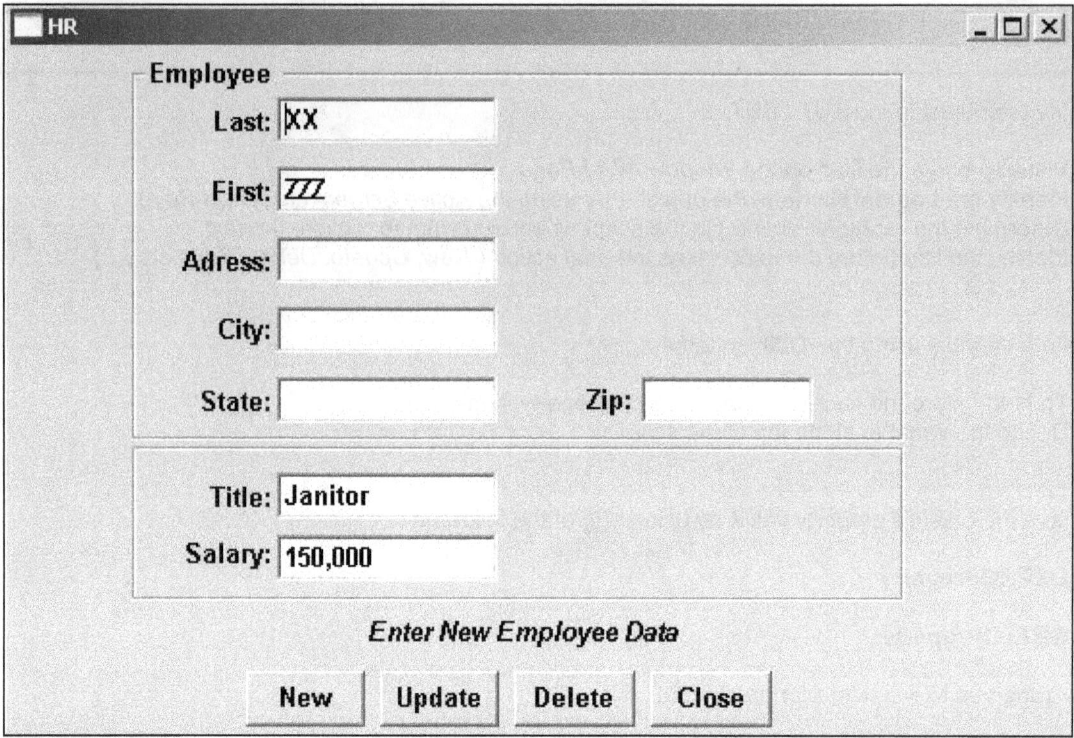


```
DIM  vs_last,  vc_sal
Window ("HR").Check  CheckPoint ("TitleBar")
Window ("HR").WinEdit ("Last:").Set   "XX"
Window ("HR").WinEdit ("First:").Set   "ZZZ"
Window ("HR").WinEdit ("Title:").Set   "Janitor"
Window ("HR").WinEdit ("Salary:").Set   "150,000"
vs_last =  Window ("HR"). WinEdit ("Last:").GetROProperty ( "text" )
IF  vs_last  > ""   THEN
   Reporter.ReportEvent  micPass, "Last", "Name is " &  vs_last
ELSE
   Reporter.ReportEvent  micFail, "Last", "No Name entered  !"
END IF
vc_sal =  Window ("HR"). WinEdit ("Salary:").GetROProperty ( "text" )
IF      vc_sal  >=  50,000
and   vc_sal  <=  90,000    THEN
   Reporter.ReportEvent  micPass, "Janitor Salary", "within range:  " &  vc_sal
ELSE
   Reporter.ReportEvent  micFail, "Janitor Salary", "Not within range:  " &  vc_sal
END IF
Window ("HR").WinButton("New").Click
Window ("HR").AcxMsg("Msg").Check  CheckPoint ("Confirmation Msg")
```

LAB 5F : Test Object & RO Properties

PART I : Test Requirements : Use ***SetTOProperty*** to populate data

a) Use an existing order and **Update** the *name* field. Write down order number : ____________________
b) Populate data by using the **SetTOProperty** method
c) Examine the Test Object properties to evaluate **how** HP QuickTest identifies an object (Explain: *Part II – Step 2*)

STEPS: (Part 1) : Capture the ***TO*** property (Test Object Properties)

1. **File** (menu bar) → **New** → **Test ...** (create a *new* script)
2. **Automation** (menu bar) → **Record... F3**
3. Check **Record and run test on any open Window-based application** → **OK**
4. Start the AUT (on your own) → Login with ***yourname*** and use the password ***mercury*** → Click **OK**
5. **File → Open Order** → Select an *existing* order → Click ***OK***
6. Change the ***name*** → Click **Update Order**
7. **File** → **Exit** (Close the AUT)
8. **Automation** → **Stop F4** (Stop Recording)
9. Click **ExpertView** → Review & Enhance the script to retrieve & display the **text** from the field: ***Name:***

```
DIM vs_ro_name

SystemUtil.Run "C:\Program Files\HP\QuickTest Professional\samples\flight\app\flight4a.exe"

Dialog("Login").WinEdit("Agent Name:").Set "tommy"
Dialog("Login").WinEdit("Agent Name:").Type micTab
Dialog("Login").WinEdit("Password:").SetSecure "41248d5060c56f3ef1811e"
Dialog("Login").WinButton("OK").Click
Window("Flight Reservation").WinMenu("Menu").Select "File;Open Order..."
Window("Flight Reservation").Dialog("Open Order").WinCheckBox("Order No.").Set "ON"
Window("Flight Reservation").Dialog("Open Order").WinEdit("Edit").Set "9"
Window("Flight Reservation").Dialog("Open Order").WinButton("OK").Click
Window("Flight Reservation").WinEdit("Name:").Set "James Brown"

vs_ro_name = Window("Flight Reservation").WinEdit("Name:").GetROProperty ("text")

Reporter.ReportEvent micdone,"Name Check1","RO Name: " & vs_ro_name

Window("Flight Reservation").WinButton("Update Order").Click
Window("Flight Reservation").WinMenu("Menu").Select "File;Exit"
```

- The **Runtime Object** (RO) contains the value in the AUT from memory

10. Change the *name* of the passenger in the script
11. Save the script **LAB5F** → Playback → Write down the Passenger Name : __________________________
12. Close the *Test Results* window

13. Modify the *script* as follows and choose a different *name* in Line 12 :

```
01. DIM vs_ro_name
02. SystemUtil.Run "C:\Program Files\HP\QuickTest Professional\samples\flight\app\flight4a.exe"

03. Dialog("Login").WinEdit("Agent Name:").Set "tommy"
04. Dialog("Login").WinEdit("Password:").SetSecure "41248d5060c56f3ef1811e"
05. Dialog("Login").WinButton("OK").Click

06. Window("Flight Reservation").WinMenu("Menu").Select "File;Open Order..."
07. Window("Flight Reservation").Dialog("Open Order").WinCheckBox("Order No.").Set "ON"
08. Window("Flight Reservation").Dialog("Open Order").WinEdit("Edit").Set "9"
09. Window("Flight Reservation").Dialog("Open Order").WinButton("OK").Click

10. Window("Flight Reservation").WinEdit("Name:").SetSelection 0,20
11. Window("Flight Reservation").WinEdit("Name:").Set "Brown James"
12. Window("Flight Reservation").WinEdit("Name:").SetTOProperty "text","James West"

13. vs_ro_name = Window("Flight Reservation").WinEdit("Name:").GetROProperty ("text")

14. Reporter.ReportEvent micdone,"Name Check1","RO Name: " & vs_ro_name

15. Window("Flight Reservation").WinButton("Update Order").Click
16. Window("Flight Reservation").WinMenu("Menu").Select "File;Exit"
```

14. Save the script **LAB5F** → Playback → It will fail with the following error :

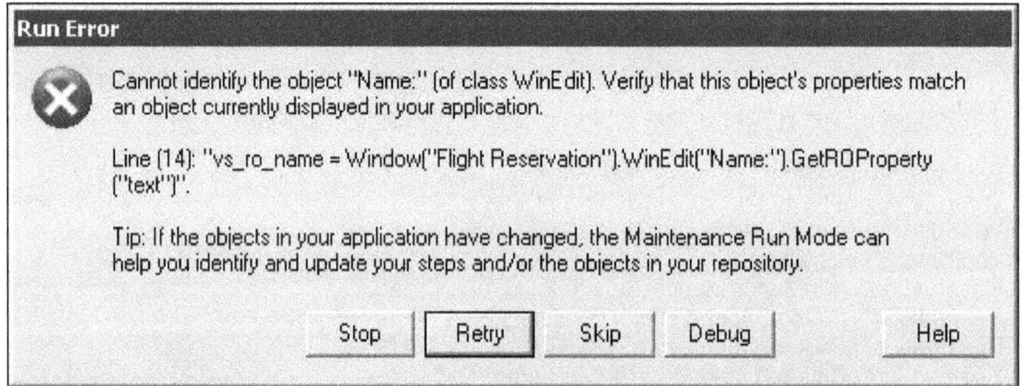

15. Click **Stop** → Review the **Test Results** window → Highlight **Description : Name**

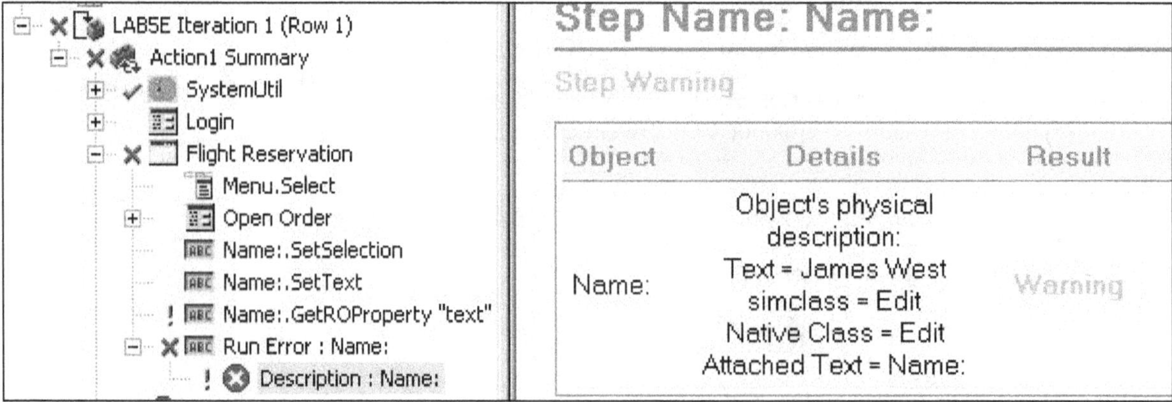

16. The **SetTOProperty** will not work and a warning is generated → Close the **Test Results** window

17. Close the **AUT**

18. Modify the script (*items* in **bold**) & re-arrange the lines as follows :

```
01. DIM  vs_ro_name , vi_length

02. SystemUtil.Run "C:\Program Files\HP\QuickTest Professional\samples\flight\app\flight4a.exe"

03. Dialog("Login").WinEdit("Agent Name:").Set "tommy"
04. Dialog("Login").WinEdit("Password:").SetSecure "41248d5060c56f3ef1811e"
05. Dialog("Login").WinButton("OK").Click

06. Window("Flight Reservation").WinMenu("Menu").Select "File;Open Order..."
07. Window("Flight Reservation").Dialog("Open Order").WinCheckBox("Order No.").Set "ON"
08. Window("Flight Reservation").Dialog("Open Order").WinEdit("Edit").Set "3"
09. Window("Flight Reservation").Dialog("Open Order").WinButton("OK").Click

10. vs_ro_name = Window("Flight Reservation").WinEdit("Name:").GetROProperty ("text")

11. vi_length  =  len ( vs_ro_name )
12. IF  vi_length > 0   THEN
13.     Window("Flight Reservation").WinEdit("Name:").SetSelection 0,vi_length
14. END IF

15. IF   vs_ro_name  =  "James West2"   THEN
16.      vs_ro_name  =  "James West"
17. ELSE
18.      vs_ro_name  = "James West 2"
19. END IF

20. Reporter.ReportEvent micdone,"Name Check1","RO Name: " & vs_ro_name

21. ' Window("Flight Reservation").WinEdit("Name:").Set  "James West"
22.  Window("Flight Reservation").WinEdit("Name:").SetTOProperty "text", vs_ro_name

23. Window("Flight Reservation").WinButton("Update Order").Click
24. Window("Flight Reservation").WinMenu("Menu").Select "File;Exit"
```

a) Add Lines 10 thru 19
b) Comment Line 21
c) Change the name in Line 22

Note: Using the **SetTOProperty** is not the accepted method to populate values onto the screen. It may work based on the AUT.

19. Save the script **LAB 5F**

20. Playback → It should pass ! → Review the *Test Results* window → Verify the passenger Name

21. Close the *Test Results* window

*** **END OF LAB5F – Part 1** ****

LAB 5F - PART II : Test Requirements

Steps: (Part II) : Create additional property identification for the **Name:** field

Logical name:	**Name:**	**Value**	**Action**
Class:	**WinEdit**		
Property	*Nativeclass*	Edit	None
	Attached text	**Name:**	Add
	WindowId	***xxxx***	Add

- This is how QTP identifies the object in the AUT

1. Start AUT → Login → Open an Existing Order (Do NOT Exit the AUT)
2. Swap to QTP → **Tools** (menu bar) → **Object Spy**
3. Re-Align the ***Spy*** & ***AUT*** side by side
4. Click on the **Hand** → cursor changes to a hand
5. Click inside the ***Name*** input field → verify **WinEdit : Name :**

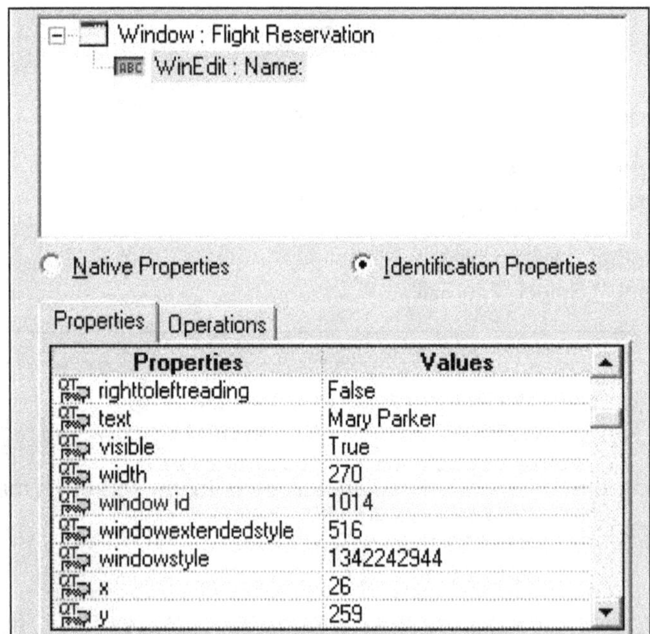

6. Scroll and Highlight **window id** property → Write down the Value : ______________________
7. Click **Close** (Exit the Object Spy)
8. **Resources → Object Repository …**
9. Expand **Flight Reservation** → Highlight **Name:**

10. Click **Green Plus (+) …** button → The **Add Properties** window opens

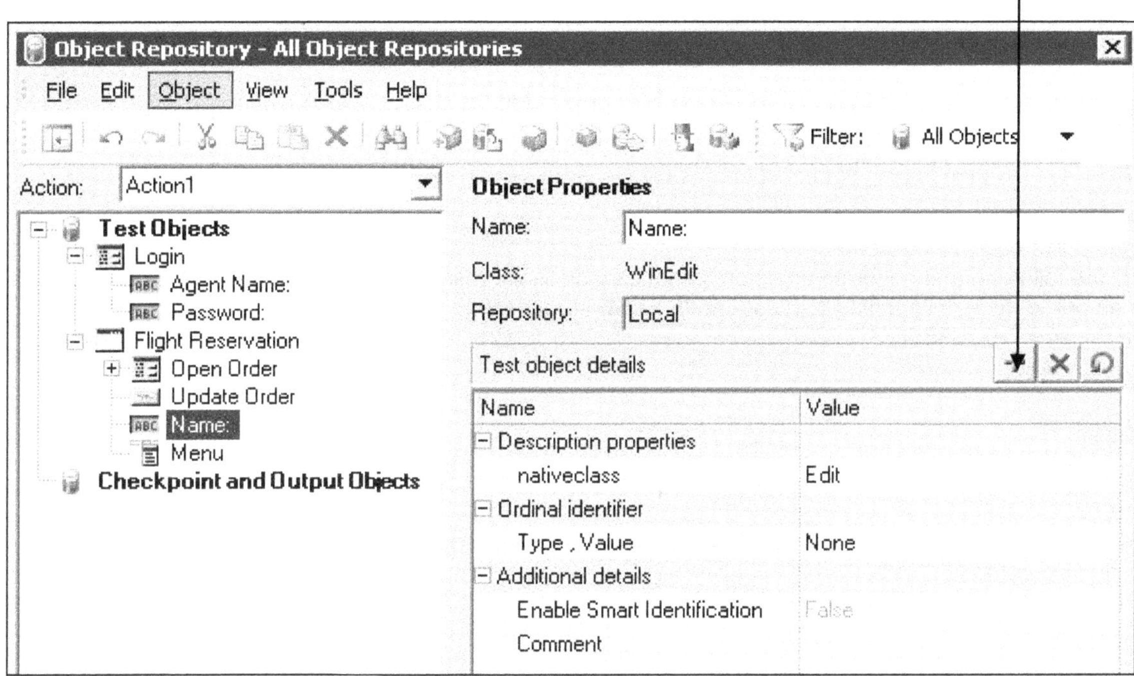

11. Scroll thru and check **attached text Name:**→ Click **OK**

12. Scroll thru and check **window id :**→ Verify the value recorded from the Spy → Click **OK**

13. QTP will identify the **Name:** object by the following properties: ***Nativeclass, attached text,window id***

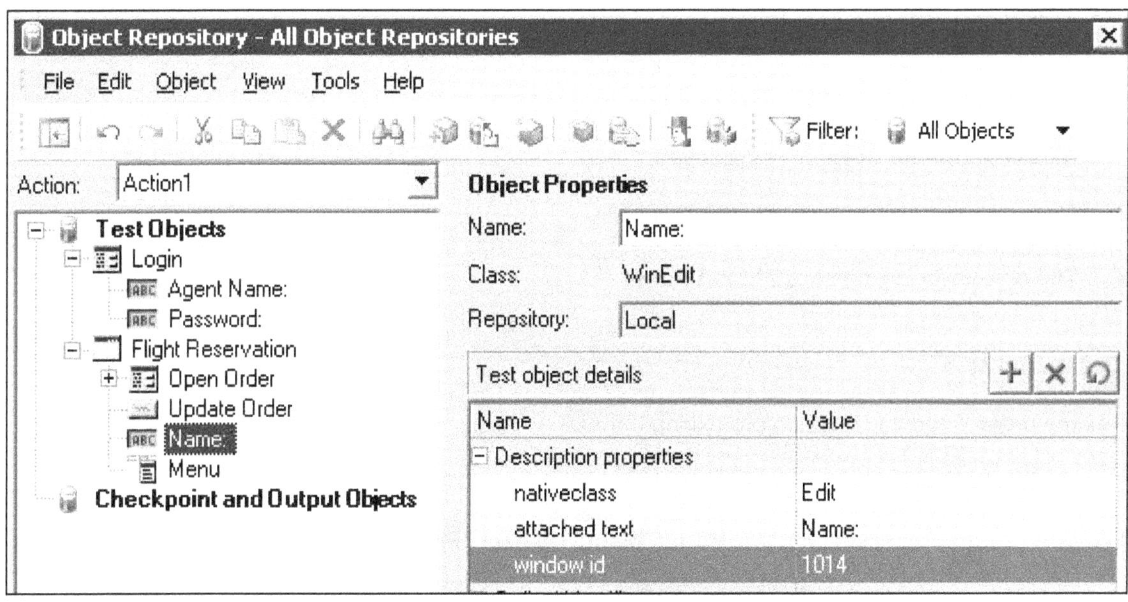

14. **File** → **Close** (Close **Object Repository** window)

15. **File** → **Save** (Save the script) → Close the **AUT** → Playback → It should Pass !

16. Close the **AUT**

*** **END OF LAB 5F – Part 2** ****

LAB 5G : Optional

Test Requirements : On your own

Perform the following Verification by using the ***GetROProperty*** Method

OVERALL

a) Update an existing order
b) Create Customized Checkpoints
c) Verify the following and generate a customized message to the *test results* window
d) Remember to *declare* variables & to use the **IF ELSE** statement
e) Save the script as LAB5F

1. **Login** window

   a) The *title bar* contains the text: **Login**

      Hint : What is the name of *object* for the ***Login*** window ? How can you found out ?

   b) The push button contains the text **OK**

2. **Flight Reservation** window

   Before Update (Initial State)

   a) The *title bar* contains the text: **Flight Reservation**
   b) The *name* field is empty

   After Update

   c) The *name* field is NOT empty (Greater than "") or *spaces*

   d) The message **Update Done...** is displayed

      - Remember to *Add* the object to the repository
      - It does not yet exist
      - The objects *logical* name is **Threed Panel**

3. **Fax Order** window

   a) Fax the order (add the *steps* before exiting the AUT)

   b) Insert a line before :

      Window("Flight Reservation").WinMenu("Menu").Select "File;Exit"

   c) Check for the message **Fax Sent Successfully...**

***** END OF LAB 5G *****

DATA TABLES

1. You can insert Data Table parameters and output values into your test
2. Each test runs on a different set of data called an iteration
3. The data your test uses is stored in the design-time ***Data Table***
4. It is displayed in the Data Table at the bottom of the screen while you create and edit your test

Data Table

D3 Sally Tree

	Flight_Date	Fly_From	Fly_To	Customer	E
1	10/25/2011	Denver	Frankfurt	Mary Heel	
2	12/15/2012	Phila	Seattle	Mark Harmon	
3	05/23/2012	Frankfurt	NYC	Sally Tree	
4					
5					
6					

Global / Action1

5. The Data Table has the characteristics of a **Microsoft Excel** *spreadsheet*

6. You can also create & execute *mathematica*l formulas within the cells
7. When you run a test, HP QuickTest creates a ***run-time*** Data Table
8. It is a *live* version of the Data Table stored with your test (A copy of it)

9. During the test *run*, HP QuickTest displays the run-time data in a table so that you can see any changes to the Data Table as they occur.

10. After the test completes - the:

    a) *run-time* **Data Table** closes
    b) Data Table again displays the *design-time* Data Table that is stored with your test
    c) run-time Data Table is displayed in the *Run-Time* **Data Table** in the Test Results window

11. You can also save the ***resulting*** data from the *run-time* **Data Table** to a file

12. Data can also be *imported* or *exported* using the following methods :

    a) **DataTable.Import**

    b) **DataTable.Export**

13. After running a parameterized test, the ***Run-time*** Data table displays those *passed* values in the **Test Results** window

SUMMARY:

1. You can supply a list of possible values for a parameter by creating a DataTable *parameter*

2. Data Table parameters allow you to create a **Data Driven Test (DDT)** that runs several times using the *data* you supply

3. In each repetition or *iteration* , HP QuickTest substitutes the **constant** value with a different value from the DataTable

GLOBAL & ACTION Sheets

1. There are **two** (2) types of sheets within the Data Table—*Global* and *Action*

2. You can access the different sheets by clicking the appropriate tabs below the Data Table

<u>Global</u> :

- You store data in the ***Global*** tab when you want it to be available to all ***actions*** in your test

  Example: You may want to *pass* parameters from one ***action*** to another

- Global Sheet contains the *data* that replaces parameters in each iteration

<u>Action</u> :

- You store data in the *Action's* tab when you want it to be available to only one action in your test
- Each time you ***add*** a *new* action to the test , a new action sheet is created in the Data Table
- Action sheets are automatically labeled with the exact name of the corresponding action
- The data *contained* in an **action sheet** is relevant for the corresponding *action* only

Editing & Saving the Data Table

1. You can enter data directly into the DataTable

2. The DataTable can be visible or *hidden* at the bottom of the screen

3. The DataTable contains the <u>**values**</u> that HP QuickTest substitutes for *parameters* while executing a test

4. Whenever you save a *test* , QTP will save the DataTable as **DEFAULT.XLS** file

5. By default – the DataTable is saved in the current ***test*** folder

6. You can *assign* a **column** name to the sheet by *double-clicking* on the heading

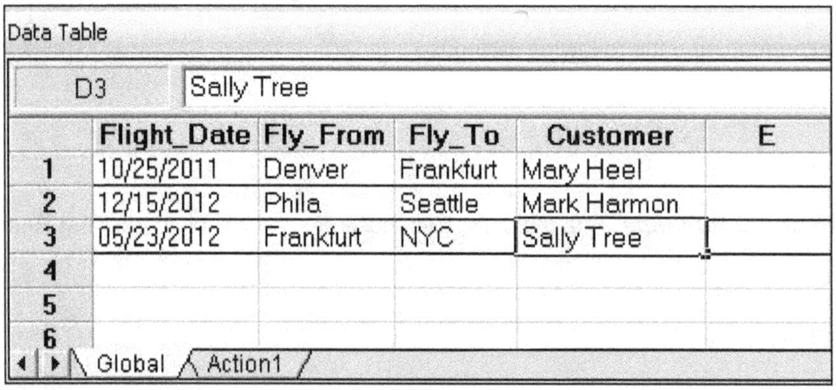

Data Table

D3 | Sally Tree

	Flight_Date	Fly_From	Fly_To	Customer	E
1	10/25/2011	Denver	Frankfurt	Mary Heel	
2	12/15/2012	Phila	Seattle	Mark Harmon	
3	05/23/2012	Frankfurt	NYC	Sally Tree	
4					
5					
6					

Global | Action1

7. The column names must be *unique* (**Flight_Date, Fly_From, Fly_To**)

8. If you change the *name* in the DataTable – you must also make a change to your *script*

9. Each *row* in the table represents the **set** of ***values*** that HP QuickTest submits for <u>parameterization</u> for 1 iteration

10. The number of *iterations* that a test runs is equal to the number of rows in the Global sheet

Review the Run-Time Settings

1. Determine the ***method*** of Data Table *iteration* to be used in the Script. (**File → Settings…**)

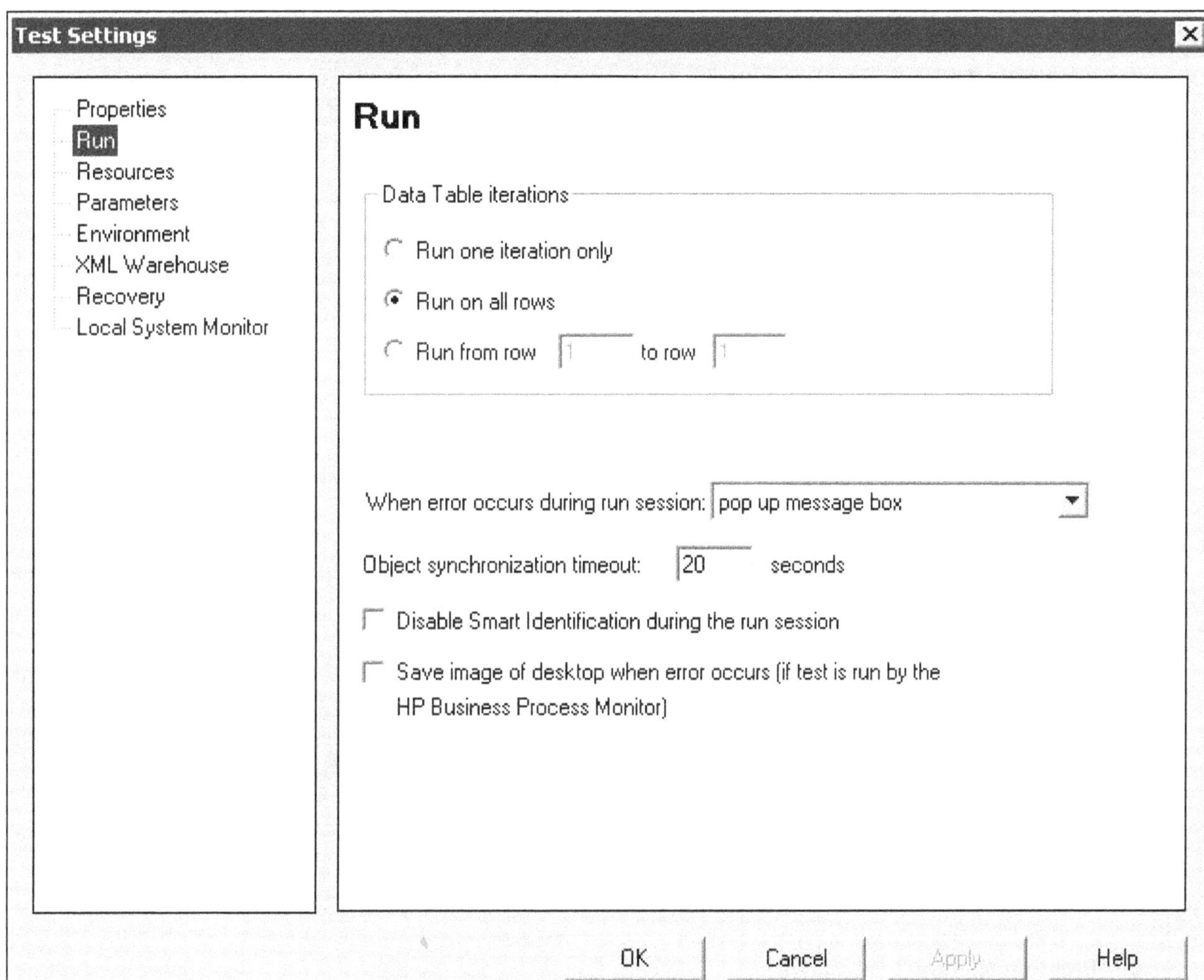

2. The *method* depends on the objective of your test

3. **LAB5A** (QTP Script)

Action0			**Parent Script**
	Action 1		Business Process
		-----------	Step 1
		-----------	Step 2
		-----------	Step *N*
	Action 2		Business Process
		-----------	Step 1
		-----------	Step 2
		-----------	Step *N*
	Action 3		Business Process
		-----------	Step 1
		-----------	Step 2
		-----------	Step *N*

PARAMETERS

1. Are similar to variables and hold values (data) that may change
2. The purpose of a ***parameter*** is to pass or *send* data values to another *process*, *module* or *component*
3. You can define **INPUT** parameters into your business process
4. You can create **OUTPUT** parameters that pass data values to external sources from one step to another step
5. In other words – you use these **parameters** to parameterize both input & output values in steps
6. You can also *parameterize* **steps** & **checkpoints**
7. Four types of parameters :

a) **DataTable** Parameters :

- Enables you to create a Data Driven Test
- With each iteration you can supply different values to complete a business transaction
- DataTable Parameter Names **are** ***Case-Sensitive***

Covered in Level 2 Series

b) **Environment Variable** Parameters :

- Enables you to use variable values from other sources during the run session
- These may be values you supply, or values that HP QuickTest generates for you based on conditions and options you choose.

c) **Random Number** Parameters :

- Enables you to insert random numbers as values in your test
- QTP can generate a random number which can be passed to the Random Number parameter

d) **Test / Action** Parameters :

- Test parameters enable to use values passed from your test
- Action parameters enable you to pass values from other actions in your test

VBScript Methods

1. Sets the encrypted value of an edit field

   Format: Object.**SetSecure** *"encrypted password"*

   Example 13 : **Dialog("Enter Password").WinEdit ("Password:") SetSecure "cvrt55crtty124398"**

   - The text is *encrypted* while **recording** & a statement generated in the script
   - The object must be a secure or *password* field
   - It is *decrypted* during a playback
   - It is ***NOT*** intended to be a secure way to protect password information
   - Each encryption generates a unique value for the same *normal* password
   - No two encrypted passwords are the same , even if the normal password is a constant

2. Encrypts the *values* from a string

   Format : variable = Crypt.Encrypt (*data*)

   Example 14 : **vs_password = "mercury"**

   vs_encrypt = Crypt.Encrypt (vs_password)

   Dialog("Login").WinEdit ("Password:") SetSecure vs_encrypt

3. Imports the specified Microsoft Excel to the run-time Data Table

   Format : DataTable.Import (*filename*)

   Example 15 : **DataTable.Import ("C:\FRS\flights.xls ")**

   - The imported table must match the test
   - The imported table replaces all data in the existing run-time data (including *all* data sheets)
   - The column **names** must ***match*** the parameters in the test
   - The sheet **names** must ***match*** the action names

4. Exports or saves a copy of the run-time Data Table to the specified location

   Format : DataTable.Export (*filename*)

   Example 16 : DataTable.Export ("C:\FRS\flight2.xls ")

5. Extracts Data from the DataTable

   Format : variable = DataTable (column , sheet)

   Example 17 : **vs_password = DataTable ("Password", dtGlobalSheet)**

   - The column **Password** must be in the Data Table

   - ***dtGlobalSheet*** refers to the Global Data Table

LAB 6A : DataTable

PART I : Test Requirements : Test the **LOGIN** process

a) Execute a Data Driven Test (DDT)

- DDT is a volume test that uses different data (*values*) for continuous business transactions in a single run

b) Login using different *user ids* & *passwords*
c) Create Checkpoints to verify the Login Process (Security Function)
d) Save the script twice : **LAB6A, LAB6A_P1**

STEPS : Part 1 : (Create the *initial* script – *prior* to a Data Driven Test)

1. **File → New → Test...** (Create a *new* script) → Click **Keyword View** tab
2. **View** (menu bar) → **Data Table** (Display the Data Table)

   a) The **Data Table** displays – if not Repeat Step 2 again !

3. **Automation → Record F3** (Start recording)
4. Check **Record and run test on any open Window-based application** → Click **OK**
5. Invoke the AUT (on your own)
6. Login with your ***username*** and password **mercury** → Click **OK** → wait for the *Main Window*
7. **Insert → Checkpoint → Standard Checkpoint** ...
8. Point and click on the **title bar** → Verify **Window:Flight Reservation** → Click **OK**
9. The **Checkpoint Property** Window Appears → Uncheck **all** the properties
10. Check only the **text** property and verify it contains **Flight Reservation** → Click **OK**
11. **File → Exit** (Close the AUT) → Stop the Recording (*or* Press **F4**)
12. **Tools → Check for Syntax** → Correct & fix errors
13. Save the script as **LAB6A** (before the DDT script changes)
14. Save the script as **LAB6A_P1** (after the DDT script changes)

Note:

a) The script **now** contains actions for a **Single Business Transaction**

b) A **Single Business Transaction** contains enough information (mandatory fields) for the business process to be completed by a single user

c) Once the script runs (successfully) for a business transaction – it is ready to be **Data Driven**

15. Run the script → It Should Pass ! If not – correct & fix any errors

STEPS: (Part 2 - Add the data & *enhance* the script) :

1. Double-Click the **A** column heading in the *DataTable* cell (located at the bottom *left* of the QTP screen)

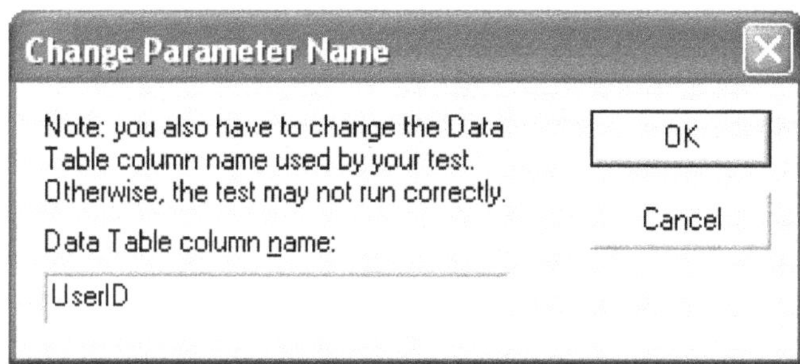

2. Type **UserID** → Click **OK**
3. Double-Click the **B** column heading in the DataTable cell → The **Change Parameter Name** window opens
4. Type **Password** → Click **OK**
5. Click **Keyword View → View** (menu bar) → **Expand All**
6. Highlight **"Agent Name:" Set *"yourname"*** → Move the cursor below the ***Value*** column
7. Click **Configure the value** icon **<#>**

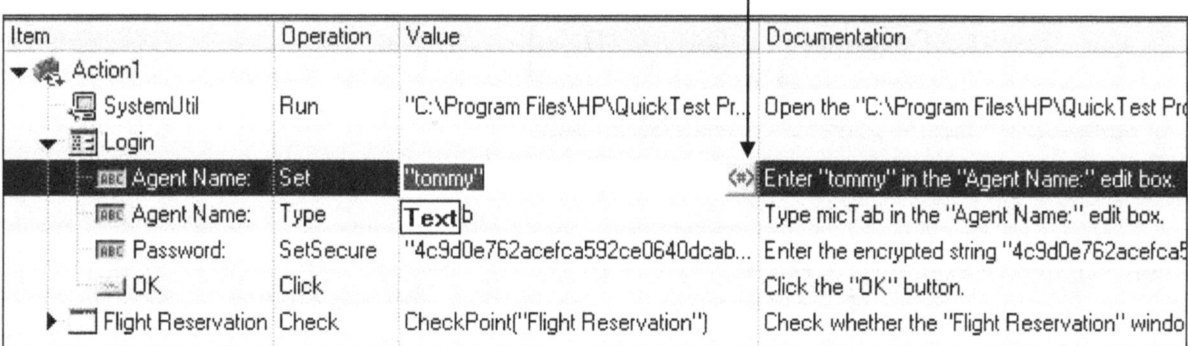

Item	Operation	Value	Documentation
Action1			
SystemUtil	Run	"C:\Program Files\HP\QuickTest Pr...	Open the "C:\Program Files\HP\QuickTest Pr
Login			
Agent Name:	Set	"tommy"	Enter "tommy" in the "Agent Name:" edit box.
Agent Name:	Type	Text b	Type micTab in the "Agent Name:" edit box.
Password:	SetSecure	"4c9d0e762acefca592ce0640dcab...	Enter the encrypted string "4c9d0e762acefca5
OK	Click		Click the "OK" button.
Flight Reservation	Check	CheckPoint("Flight Reservation")	Check whether the "Flight Reservation" windo

8. Check Radio button : **Parameter**

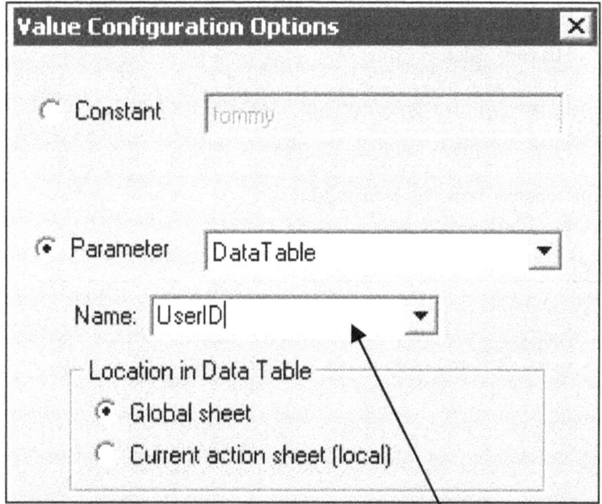

9. In **Parameter Name**: *field* → Select **UserID** (combobox) → Choose **Global sheet** → Click **OK**

10. Highlight **"Password:" SetSecure "43ef53wer334335555"**

11. Click **Configure the value** icon → Check Radio button : **Parameter**

12. In **Parameter Name**: *field* → Select **Password** (combobox) → Choose **Global sheet** → Click **OK**

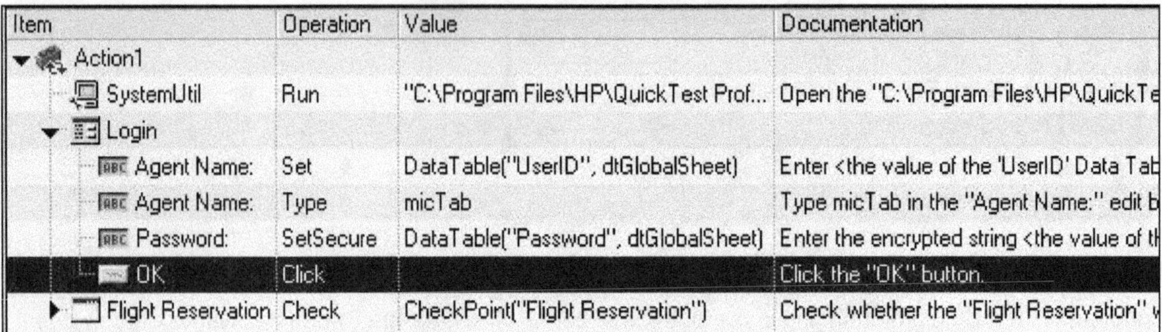

Item	Operation	Value	Documentation
Action1			
SystemUtil	Run	"C:\Program Files\HP\QuickTest Prof...	Open the "C:\Program Files\HP\QuickTe
Login			
Agent Name:	Set	DataTable("UserID", dtGlobalSheet)	Enter <the value of the 'UserID' Data Tab
Agent Name:	Type	micTab	Type micTab in the "Agent Name:" edit b
Password:	SetSecure	DataTable("Password", dtGlobalSheet)	Enter the encrypted string <the value of th
OK	Click		Click the "OK" button.
Flight Reservation	Check	CheckPoint("Flight Reservation")	Check whether the "Flight Reservation" w

13. Click **Expert View** → Click **Keyword View** [Check for any syntax errors]

14. Enter the first row of the transaction into the Data Table

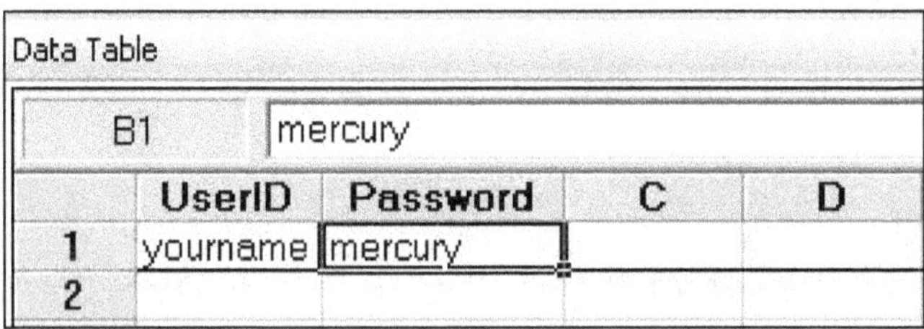

Data Table

B1 mercury

	UserID	Password	C	D
1	yourname	mercury		
2				

15. **File** → **Settings ...** → Click **Run** tab

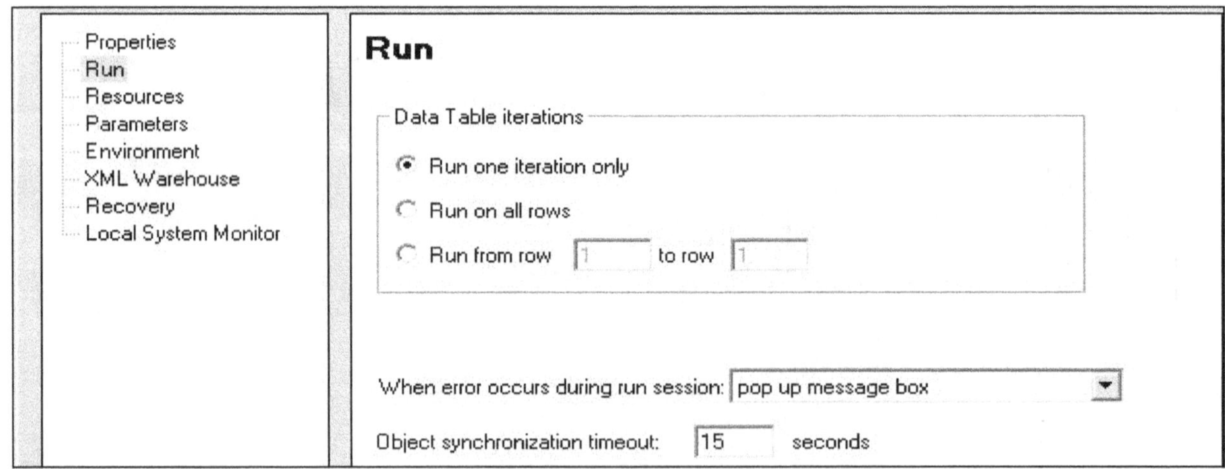

16. Choose the following selections:

    a) **Run one iteration** only
    b) **15** seconds (Object synchronization timeout)

17. Click **OK**

18. Your Script maybe similar to the following:

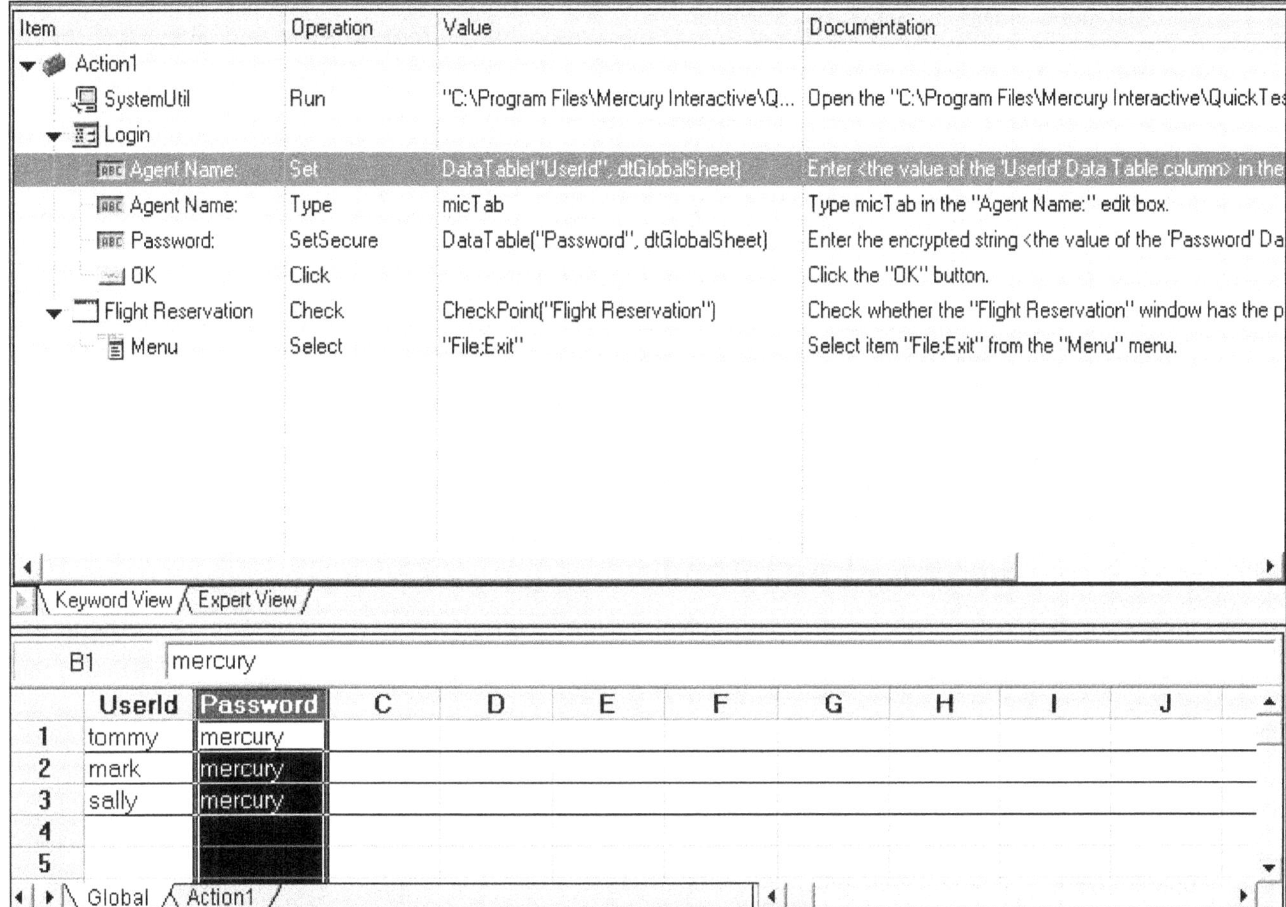

Observe the following:

The following input fields have been parameterized:

a) **Agent Name: Set DataTable ("UserId", dtGlobalSheet)**

b) **Password: SetSecure DataTable ("Password", dtGlobalSheet)**

Note:

If in the *Keyword View* - the above two statements are NOT visible – then :

a) Save the file → Open another QTP script → Then re-open the modified script *OR*
b) Close QTP → Open the modified QTP script

19. Playback the script (on your own !) → It Should Pass !

20. Review the **Test Results** window → **View** → **Expand All**

Why did it run only once ? ______________________

21. Click **Expert View**

22. Modify the Script as follows :

  a) Comment Line 6 - the **SetSecure** statement

  b) Add Line 7 – the **Set** statement

```
01. SystemUtil.Run "C:\Program Files\HP\QuickTest Professional
02.       \samples\flight\app\flight4a.exe",
03.        "", "C:\Program Files\HP\QuickTest Professional\samples\flight\app\", "open"

04. Dialog("Login").WinEdit("Agent Name:").Set DataTable("UserID", dtGlobalSheet)
05. Dialog("Login").WinEdit("Agent Name:").Type micTab

06. ' Dialog("Login").WinEdit("Password:").SetSecure  DataTable ("Password", dtGlobalSheet )
07. Dialog("Login").WinEdit("Password:").Set  DataTable ("Password", dtGlobalSheet )

08. Dialog("Login").WinButton("OK").Click

09. Window("Flight Reservation").Check CheckPoint("Flight Reservation")
10. Window("Flight Reservation").WinMenu("Menu").Select "File;Exit"
```

23. Verify there are no syntax errors by :

  a) Clicking **Keyword View** *or*

  b) **Tools → Check Syntax**

24. Save (**LAB6A_P1**)

25. **File** → **Settings ...** → Click **Run** tab

26. Make the following selections:

  a) **Run on all rows**

27. Click **OK** (close **Test Settings** window)

27. Playback the script

28. Examine the *Test Run* results by *viewing* :

  a) the **Run-Time data** (located near the *top* of the ***left*** pane)

  b) Iteration 1 → Expand *Action1* Summary

  c) Iteration 2 → Expand *Action1* Summary

  d) Iteration 3 → Expand *Action1* Summary

*** **END OF LAB 6A – Part 1** ***

STEPS: (Part 2 - Encrypt the password)

1. Save the script as **LAB6A_P2** → Click **Expert View** → Modify the Script as follows :

```
01. DIM vs_password, vs_encrypt

02. SystemUtil.Run "C:\Program Files\HP\QuickTest Professional\samples
03.     \flight\app\flight4a.exe", "", "C:\Program Files\HP\\QuickTest Professional\samples\flight\app\", "open"

04. Dialog("Login").WinEdit("Agent Name:").Set DataTable("UserID", dtGlobalSheet)
05. Dialog("Login").WinEdit("Agent Name:").Type micTab
06. 'Dialog("Login").WinEdit("Password:").SetSecure DataTable("Password", dtGlobalSheet)

07. vs_password = DataTable ( "Password", dtGlobalSheet )

08. vs_encrypt  = Crypt.Encrypt ( vs_password )

09. Reporter.ReportEvent micDone, "Encryption","Password:" & vs_password & " Encrypt:"
      & vs_encrypt

10. Dialog("Login").WinEdit("Password:").SetSecure  vs_encrypt

11. Dialog("Login").WinButton("OK").Click

12. Window("Flight Reservation").Check CheckPoint("Flight Reservation")

13. Window("Flight Reservation").WinMenu("Menu").Select "File;Exit"
```

Explanation :

a) Line 1 : Declare two variables : **vs_password** , **vs_encrypt**

b) Line 7 : Extract the ***value*** from the Data Table and save it into the variable ***vs_password***

c) Line 8 : Encrypt the *normal* password into the variable **vs_encrypt**

d) Line 9 : Generate a Customize message to the Test Results window

- The statement should be within a single line
- Display the value *read* from the data table (normal password)
- Display the encrypted password *generated* by the method **Crypt.Encrypt**

e) Line 11 : Populate the new encrypted password onto the screen

2. **Keyword View** Sample

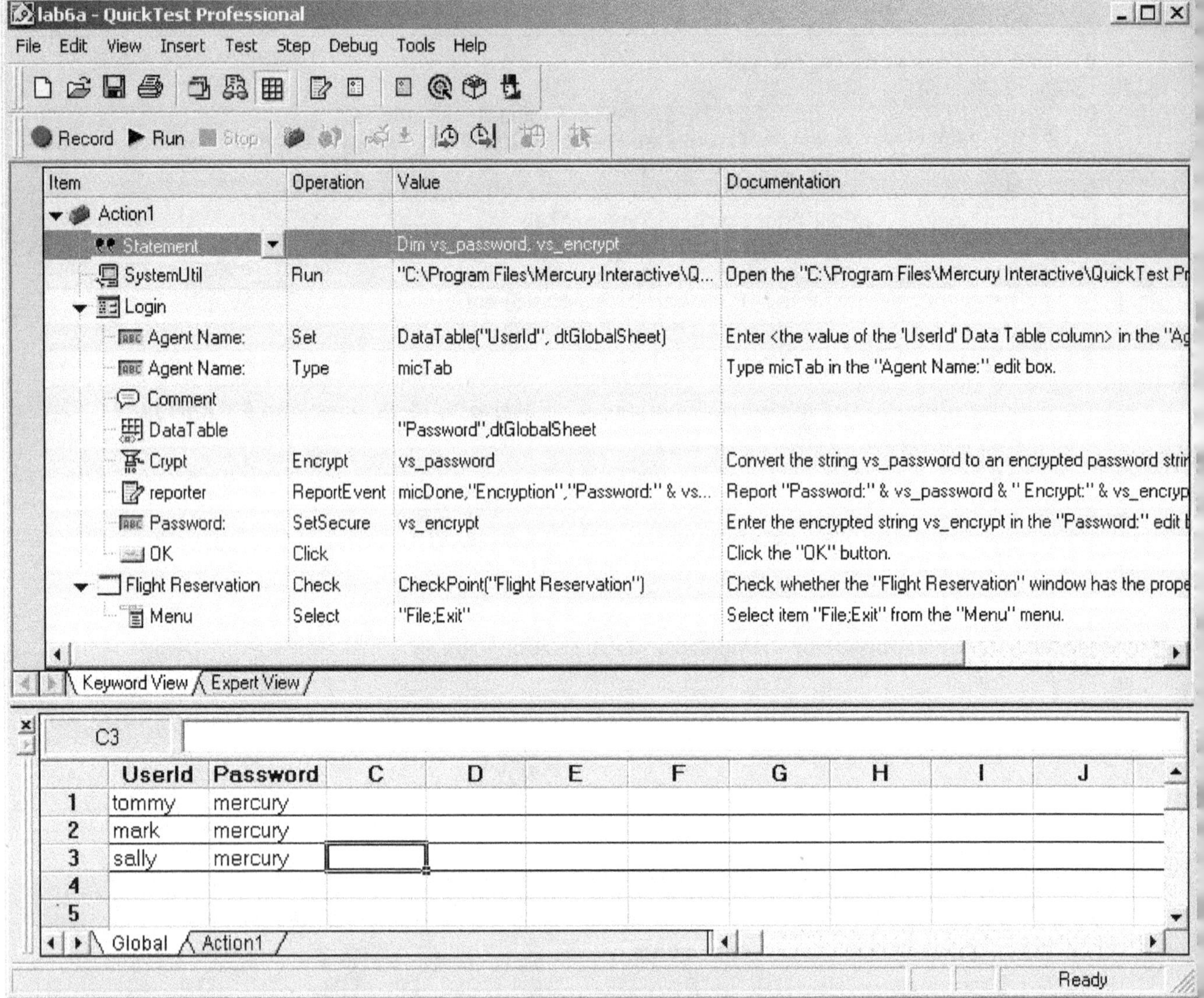

Observe: The ***double quote*** at the top of the Keyword View (Below *Action1*)

3. Save the Script as **LAB6A_P3**

4. Perform any pre-conditions (if required) → Playback the script

5. Review the *results*

   a) Review each iteration

   b) Look for the custom messages which display both : *Password* & *Encrypted* password

   c) The encrypted values vary for each iteration !

**** **END OF LAB 6A** ****

RDBMS – Steps to export information from a database

Guidelines

1. Identify the RDBMS (Oracle, MS SQL Server, Sybase, DB2, etc …) , Tables & Relationships

2. Write a SQL query to confirm the data exists and that it conforms to your needs:

   a) Create your own data (if you have been granted permission)
   b) Make a copy of the tables (from the original source)
   c) Insert / Update columns from your newly created tables for your test scenario
   d) Write a query from a single table or Joins based on multiple tables

3. Use **VIEWS**

   a) Review your application and identify all the input fields and sequence
   b) Create a view based on a SQL Query
   c) The sequence of the view (columns) is based on the required input fields in your application

4. Export Data based on the VIEWS created earlier (Use the vendor provided utility to export data) :

   - BCP (MS SQL Server & Sybase)
   - EXP80 (Oracle)

   a) Exporting is based on your ***VIEW*** and an ASCII text file will contain the data

5. Import ASCII text file into MS-Excel

6. Modify you VB Script to handle the different date formats: **m/d/yyyy mm/d/yyyy m/dd/yyyy**

```
.. VB Script STatements
vs_date = DataTable ( "DOF", dtGlobalSheet )        # mm/dd/yyyy
vi_len   = len ( vs_date );          # Get the length of the date field

IF ( vi_len < 10 )  THEN
     IF  mid (vs_date,2,1) = "/"  and  mid (vs_date,4,1) = "/"  THEN
          vs_date = "0" &  mid (vs_date,1,2) & "0" &  mid (vs_date,3,6)
     END IF

     IF  mid (vs_date,3,1) = "/"  and  mid (vs_date,5,1) = "/"  THEN
          vs_date =  mid (vs_date,1,3) & "0" &  mid (vs_date,4,4)
     END IF

     IF  mid (vs_date,2,1) = "/"  and  mid (vs_date,5,1) = "/"  THEN
          vs_date = "0" &  mid (vs_date,1,2) &  mid (vs_date,1,7)
     END IF
END IF

' vs_date : should contain the correct format

Window ("New Employee").WinEdit ("Date of Birth:"). Set  vs_date
..... Remaining VB statements
```

Examples: 1/1/1960 10/1/1961 3/15/1975

LAB 6B - Part I : Data Driven test using a Text file (from a database)

Requirements:

1. Generate data from the **agents** table in SQL Server
2. Import data into an excel spreadsheet → Run a Data-driven test

STEPS

1. Copy script **LAB6A_P1** and save it is **LAB6B** (Data driven test) :
2. Review the table structure and Appendix A

   agents (schema)   database: **qa_dev1** user: **itm** password: **itm** Server: **training-srvr02**

agent_id	smallint
agent_name	varchar (20)
agent_password	varchar (30)
Dept	char (02)
hire_date	smalldatetime

3. Create a view of the selected columns. Replace ***itm*** (v_*itm*_login_ddt) with your initials

```
create view v_itm_login_lab6b_ddt
As
    select '0UserID' Username, 'Password'  password  from  itm.agents
    UNION
    select   agent_name, agent_password  password from itm.agents
    where    agent_password = "mercury"
```

   - The **union** combines the result set from *each* query into a ***single*** Result Set without duplicates

4. Export the data out to a text file: (MS-DOS or command line prompt) → **Start** → **Run** → **CMD** → **OK**

   C:\bcp\bcp qa_dev1..v_itm_login_lab6b_ddt out C:*yourname*\QTP10\LVL1\logins.txt -c -U*itm* -P*itm* -Straining-srvr02

   Note: The Table Names are case-sensitive !

5. Import the data into excel spreadsheet:

   a) Open MS Excel (on your own) → **Data** → **Import External Data** → **Import Data...**

   b) Point to C:\yourname\QTP10\LVL1\logins.txt → Click **Open**

   c) Check **Delimited** → **Next**

   d) In Delimiters: Check **Tab** → Check **Space**

   e) Click **Next** → **Finish**

   f) Click **OK** (Where do you want to put the data ?) =A1

6. Rename *Sheet1* to **Global**

7. Rename *Sheet2* to **Action1**

8. Replace the current MS Excel spreadsheet in : **C:*yourname*\QTP10\LVL1\LAB6B \Default** → Click **Save**

9. In the ***Global*** sheet change the column ***0UserId*** *to* ***UserID***

10. Click **Yes** (Do you want to replace the existing file ?)

11. Close **MS Excel** → Close **MS-DOS** window

12. Swap to QTP → **File** → **Save** (Save LAB6B)

13. **File** → **Open** → **Test** → ***LAB6B***

14. Click **Yes** (to Revert back to Saved test)

15. Observe the Data is Imported into the excel spreadsheet

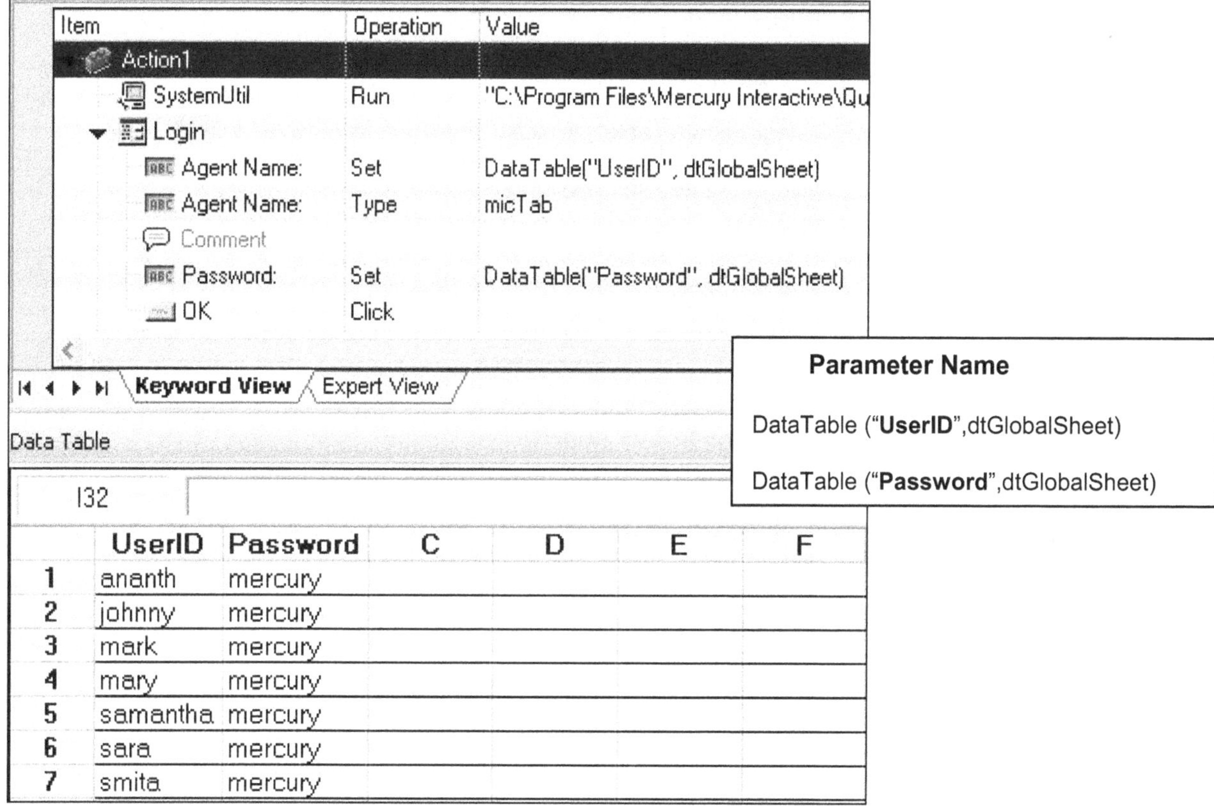

a) The parameter names must match the Column Heading Names

16. Save your script & Playback the test

17. Review & Close the **Test Results** window

*** **END OF LAB 6B** ****

VBScript Statements

1. **WITH** Statement

   a) Makes your script more concise & easier to read
   b) Groups consecutive statements with the same parent hierarchy
   c) Available only in the **expert** mode

   Format : With object
   statement 1
   statement 2
   statement *x*
   End With

   Example 18: Current Script :

   Before ***With***

```
Window("Flight Reservation").WinComboBox("Fly From:").Select "London"

Window("Flight Reservation").WinComboBox("Fly To:").Select "Los Angeles"

Window("Flight Reservation").WinButton("FLIGHT").Click

Window("Flight Reservation").Dialog("Flights Table").WinList("From").Select "19097 LON "

Window("Flight Reservation").Dialog("Flights Table").WinButton("OK").Click
```

   After **With**

```
With Window ("Flight Reservation")
        .WinComboBox("Fly From:").Select "London"
        .WinComboBox("Fly To:").Select "Los Angeles"
        .WinButton("FLIGHT").Click
         With .Dialog("Flights Table")
                 .WinList("From").Select "19097 LON "
                 .WinButton("OK").Click
         End With 'Dialog("Flights Table")
End With 'Window("Flight Reservation")
```

2. Two methods to generate a **WITH** Statement

   a) General Option Settings:

   - Tools → Options → Check Automatically generate 'With' statements after recording

   b) Edit (menu bar) → Advanced → Apply 'With' to script

   - The entire script will be modified (not just specific line numbers)

FOR NEXT Statement

- Repeats a group of *statements* a specified <u>number</u> of times

Format:

```
FOR counter = start TO end [ STEP step ]
    [ Statements ]
    [ EXIT FOR ]
    [ Statements ]
NEXT
```

<u>Arguments</u> :

Argument Name	DataType	Description
counter	Integer	Variable name used as a counter
start	Integer	Initial value of the counter
end	Integer	Ending value of the counter
step	Integer	Increment the *counter* value after the loop ends
statements	VBScript	A valid VB Statement

Example 19 :

```
1. DIM X , vs_order

2. FOR X = 1 to 10

3.     vs_order = DataTable.( "OrderNumber ", dtGlobalSheet )

4.     IF vs_order = "" THEN
           EXIT FOR
5.     END IF

6.     Reporter.ReportEvent micDone, "Order No:" , vs_order

7.     [ VB Script Statements ]

8. NEXT

9. [ VB Script Statements ]
```

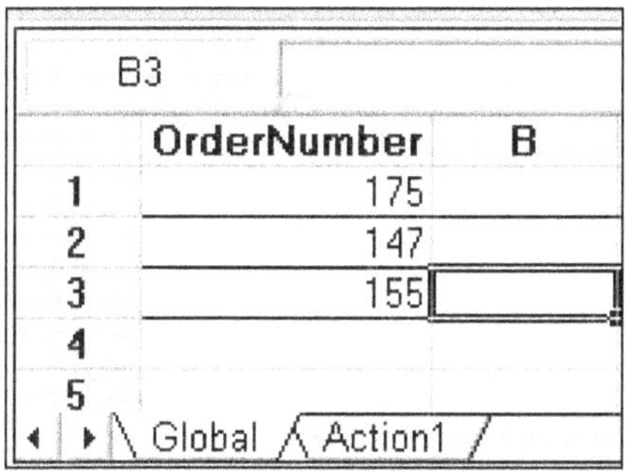

B3

	OrderNumber	B
1	175	
2	147	
3	155	
4		
5		

Global | Action1

GetRowCount Method

- Returns the **total** number of rows in the longest column in the *Global* or *Specified* Data Sheet
- Applies to the Run-Time DataTable

Format: DataTable.GetRowCount Returns : Number

Example 20: Return the total number of rows in the **Global** datasheet

```
DIM vi_max
vi_max = DataTable.GetSheet ( "Global" ).GetRowCount
```

Example 21: Return the total number of rows in the **Action1** datasheet

```
DIM vi_max
vi_max = DataTable.GetSheet ( "Action1" ).GetRowCount
Reporter.ReportEvent 2, "There are " & vi_max, "rows in the data sheet "
```

SetCurrentRow Method

- Sets the specified row as the current (active) row in the run-time DataTable
- You can only set a row that contains at least one value

Format: DataTable.SetCurrentRow (rownumber)

Example 22: Set the current row to the second row

```
DataTable.SetCurrentRow ( 2 )
```

Example 23: Create a loop and extract data from the ***Global*** datatable

```
DIM   vi_row, vi_max
DIM   vs_order, vs_name

vi_max = DataTable.GetSheet ( "Global" ).GetRowCount

FOR vi_row = 1 TO vi_max

    DataTable.SetCurrentRow ( vi_row )

    vs_order = DataTable ( "OrderNo", dtGlobalSheet )
    vs_name = DataTable ( "CustName", dtGlobalSheet )

    Reporter.ReportEvent micDone, "Name", vs_name
    Reporter.ReportEvent micDone, "Order", vs_order
NEXT
```

	OrderNo	CustName	C
1	2912	Abe Lincoln	
2	1929	Sally Green	
3	138	Paul Cricket	
4			
5			

Global | Action1

SetNextRow Method

- Sets the row **after** the current (active) row as the new current row in the run-time DataTable
- You can only set a row that contains at least one value
- If the current row is the last row – then it will set it to the first row

Format: DataTable.SetNextRow

Returns : Number

Example 24: Move to the next row in the **Global** datasheet

```
DataTable.SetNextRow
```

SetPrevRow Method

- Sets the row **above** the current (active) row as the new current row in the run-time DataTable
- You can only set a row that contains at least one value
- If the current row is the first row – then it will set it to the las row

Format: DataTable.SetPrevRow

Example 25: Set the current row to the second row

```
DataTable.SetPrevRow
```

DataTable.Value Method (Saving Values in the Run-Time Data Table)

- Retrieves or Sets the value of the cell in the specified parameter and the current row in the Data Table
- If it is a computed cell – it returns the *computed* value

Format 1: Find the value DataTable.Value (ParameterID [, Sheet ID])

Sheet ID = Any one of the following: **Sheet Name, Index, dtLocalSheet** or **dtGlobalSheet**

Format 2: Set the value DataTable.Value (ParameterID [, Sheet ID]) = *new value*

Example 26: Examples of Setting or *Saving* the **data value** in EXCEL spreadsheet

```
01. DataTable.Value ( “DOF” ,”Action1”) = ‘12/12/10”        “ DOF parameter    Action1 = SheetName

02. DataTable.Value ( 2, 3 ) = “New Jersey”                 “ Second Parameter in the Third Sheet

03. DataTable ( “FlyFrom” , dtLocalSheet ) = “ Denver”      “ Parameter FlyFrom in the LocalSheet
```

Example 27: Create a loop, Extract data from the ***Global*** Datatable in a sequential order

```
DIM   x , vi_row
DIM   order, name

x = DataTable.GetSheet ( dtGlobalSheet ).GetRowCount

FOR vi_row = 1 TO x    STEP 1

    order  = DataTable ( "OrderNo", dtGlobalSheet )

    name = DataTable ( "CustName", dtGlobalSheet )

    Reporter.ReportEvent micDone, "Name",  name

    Reporter.ReportEvent micDone, "Order",  order

    DataTable.SetNextRow

NEXT
```

Example 28: Create a loop and extract data from the ***Global*** datatable in a reverse order. Compute the salary (*hourly rate* **x** *hours*)

```
DIM   x , vi_row , emp_no
DIM   salary, hours, rate

x = DataTable.GetSheet ( dtGlobalSheet ).GetRowCount

DataTable.SetCurrentRow ( x )

FOR vi_row = x  TO 1    STEP -1

    emp_no  = DataTable ( "EmpNum", dtGlobalSheet )

    hours = DataTable ( "Hours", dtGlobalSheet )

    rate   = DataTable ( "PayRate", dtGlobalSheet )

    salary =  hours * rate

    Reporter.ReportEvent micDone, "Salary",  salary

    DataTable.Value ( "New Sal", dtGlobalSheet ) = salary

    DataTable.SetPrevRow

NEXT
```

LAB 6C : Importing Data from MS EXCEL

PART I : Test Requirements : Importing Data

a) Open **LAB6A** as **LAB6C**
b) Create & Use (import) the **C:\QTP10\LVL1\ LAB6C.XLS** data file
c) Display the **UserID** and the encrypted ***password*** in the test results window

STEPS:

1. Open **LAB6A** → Save As **LAB6C** → Click **Keyword view**

2. In the DataTable change the column headings

   - A → **UserID**
   - B → **Password**

3. Create the two parameters (**Agent Name** & **Password**) : [Refer to LAB 6A for help]

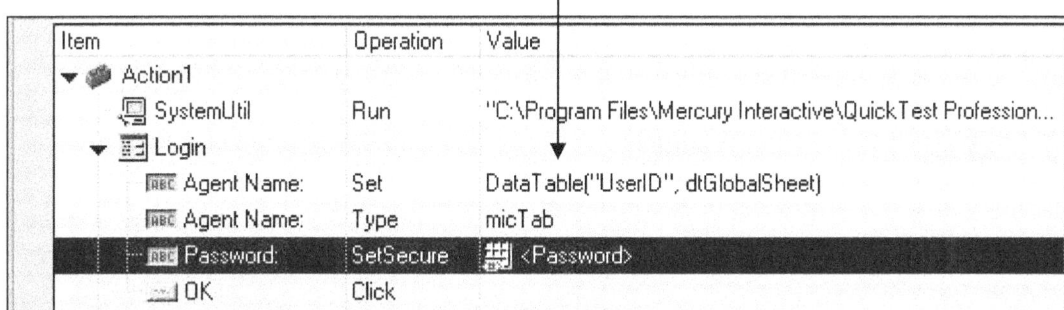

Item	Operation	Value
Action1		
SystemUtil	Run	"C:\Program Files\Mercury Interactive\QuickTest Profession...
Login		
Agent Name:	Set	DataTable("UserID", dtGlobalSheet)
Agent Name:	Type	micTab
Password:	SetSecure	<Password>
OK	Click	

4. Modify the script as follows

```
DIM  vs_password, vs_encrypt, vs_name, vi_max

Datatable.Import ("C:\QTP10\LVL1\LAB6C.XLS")
vi_max  =  DataTable.GetSheet ( "Global" ).GetRowCount

Reporter.ReportEvent micDone,"Total", "Rows:  " &  vi_max

SystemUtil.Run "C:\Program Files\HP\QuickTest Professional
     \samples\flight\app\flight4a.exe", "","C:\Program Files\HP\QuickTest Professional\samples\flight\app\",
"open"

Dialog("Login").WinEdit("Agent Name:").Set DataTable( "UserID", dtGlobalSheet )
Dialog("Login").WinEdit("Agent Name:").Type micTab
' Dialog("Login").WinEdit("Password:").SetSecure DataTable( "Password", dtGlobalSheet )

vs_name     = DataTable("UserID", dtGlobalSheet)
vs_password = DataTable("Password", dtGlobalSheet)
vs_encrypt  = Crypt.Encrypt (vs_password)
Reporter.ReportEvent micDone,"Users", "User ID:  " &  vs_name &  "   Encrypt:"  &  vs_encrypt

Dialog("Login").WinEdit("Password:").SetSecure vs_encrypt
Dialog("Login").WinButton("OK").Click
Window("Flight Reservation").Check CheckPoint("Flight Reservation")
Window("Flight Reservation").WinMenu("Menu").Select "File;Exit"
```

5. Save the script **LAB6C**

6. Create the following : **C:\QTP10\LVL1\LAB6C.XLS**

UserID	Password
itm01	mercury
itm02	mercury
itm03	mercury
itm04	mercury
itm05	mercury

7. Run with 1 iteration only → Playback & Review the results

8. Observe the Run-Time Data-Table

9. Run with All Iterations → Playback & Review the results

   - There is a Problem !! - For all five (iterations) - it is <u>always</u> referring to the first row only !

10. Enhance the script as follows :

```
DIM  vs_password, vs_encrypt, vs_name, vi_max,  vi_row

Datatable.Import ("C:\QTP10\LVL1\itm_lab6b.xls")
vi_max = DataTable.GetSheet ("Global").GetRowCount
Reporter.ReportEvent micDone,"Total", "Rows:  " &  vi_max

FOR vi_row = 1 TO vi_max

   DataTable.SetCurrentRow ( vi_row )

   SystemUtil.Run "C:\Program Files\HP\QuickTest Professional
        \samples\flight\app\flight4a.exe", "","C:\Program Files\HP
        \QuickTest Professional\samples\flight\app\", "open"
   Dialog("Login").WinEdit("Agent Name:").Set DataTable( "UserID", dtGlobalSheet )
   Dialog("Login").WinEdit("Agent Name:").Type micTab
   ' Dialog("Login").WinEdit("Password:").SetSecure DataTable( "Password", dtGlobalSheet )

   vs_name     = DataTable("UserID", dtGlobalSheet)
   vs_password = DataTable("Password", dtGlobalSheet)
   vs_encrypt  = Crypt.Encrypt (vs_password)
   Reporter.ReportEvent micDone,"Users", "User ID:  " &  vs_name &  "   Encrypt:"  &  vs_encrypt

   Dialog("Login").WinEdit("Password:").SetSecure vs_encrypt
   Dialog("Login").WinButton("OK").Click
   Window("Flight Reservation").Check CheckPoint("Flight Reservation")
   Window("Flight Reservation").WinMenu("Menu").Select "File;Exit"

NEXT
```

11. Save **LAB6C**

12. Run with 1 iteration only → Playback & Review the results

****** END OF LAB 6C ******

LAB 6D : Processing Transactions in Reverse

Test Requirements :

1. Use LAB6C
2. Process Information (Data) from the DataTable in the reverse order
3. Modify the starting position in the loop

Steps:

1. Open **LAB6C** → Save As **LAB6D**

2. Modify the script as follows :

```
DIM  vs_password, vs_encrypt, vs_name, vi_max,  vi_row, vs_msg

Datatable.Import ("C:\QTP10\LVL1\itm_lab6c.xls")
vi_max = DataTable.GetSheet ("Global").GetRowCount
Reporter.ReportEvent micDone,"Total", "Rows:  " &  vi_max

FOR vi_row =  vi_max  TO  1   STEP  -1

   DataTable.SetCurrentRow ( vi_row )

   SystemUtil.Run "C:\Program Files\HP\QuickTest Professional
        \samples\flight\app\flight4a.exe", "","C:\Program Files\HP
        \QuickTest Professional\samples\flight\app\", "open"
   Dialog("Login").WinEdit("Agent Name:").Set DataTable( "UserID", dtGlobalSheet )
   Dialog("Login").WinEdit("Agent Name:").Type micTab
   ' Dialog("Login").WinEdit("Password:").SetSecure DataTable( "Password", dtGlobalSheet )

   vs_name     = DataTable("UserID", dtGlobalSheet)
   vs_password = DataTable("Password", dtGlobalSheet)
   vs_encrypt  = Crypt.Encrypt (vs_password )

   vs_msg      = "User ID:  " &  vs_name &  "   Encrypt:"  &  vs_encrypt

   ' Reporter.ReportEvent micDone," Users", "User ID:  " &  vs_name &  "   Encrypt:"  &  vs_encrypt
    Reporter.ReportEvent   micDone,  " Row "  &  vi_row ,  vs_msg

   Dialog("Login").WinEdit("Password:").SetSecure vs_encrypt
   Dialog("Login").WinButton("OK").Click
   Window("Flight Reservation").Check CheckPoint("Flight Reservation")
   Window("Flight Reservation").WinMenu("Menu").Select "File;Exit"

NEXT
```

3. Save **LAB6D**

4. Playback and Review the Results

5. Verify the Transactions were processed in the reverse order

****** END OF LAB 6D ******

LAB 6E : Open & Verify an existing Order

PART I : Test Requirements :

1. Open an existing order number (Positive test)
2. Add objects to the Object Repository
3. Create a Data Driven Test to read from the DataTable
4. Verify if it matches with the order number on the screen

Steps:

1. **File → New → Test…** (Create a *new* script) → Click **Keyword View** tab

2. **Test → Record F3** (Start recording)

3. Check **Record and run test on any open Window-based application** → Click **OK**

4. Invoke the AUT (on your own)

5. Login with your ***username*** and password **mercury** → Click **OK**

6. Create a checkpoint to verify the **Flight Reservation** (on your own – use *any* method …)

7. **File → Open Order** → Check **Order Number** → Type **1** → Click **OK**

8. Create a checkpoint to verify that order number **1** is displayed on the screen

9. **File → Exit** (Close the AUT)

10. Stop Recording → Save the script as **LAB6E**

Add the **Order No:** *object* to the Repository (Which object and window ? ______________________________)

11. Click **Keyword View** mode → Rearrange the **Active Screen** (so you can view the snapshot of the window)
12. Move cursor over the ***Order No*** field (where the Order Number would be displayed)
13. Right-Click → **View/Add Object**

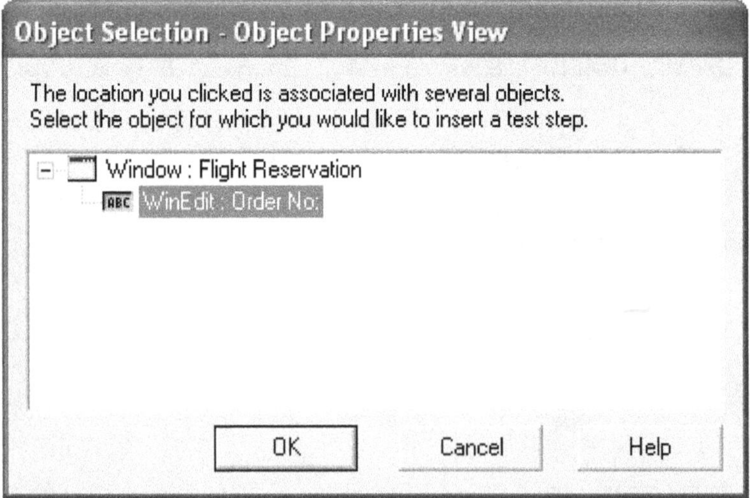

14. Click **OK**

15. Click **Repository>>** → Verify that the Object was added (Visually)

16. Click **OK** (Close **Object Repository** window)

29. Double-Click the **A** column heading in the DataTable cell

30. The **Change Parameter Name** window opens → Type **OrderNo** → Click **OK**

31. Highlight **"Edit" Set "1** → Click **"1"** → Click **Configure the value** icon

32. Check Radio button **Parameter**

33. In **Parameter name**: field → Select **OrderNo** → Click **OK**

17. Add the following three *orders* to the DataTable

1
2
3

18. Click **ExpertView**

19. Modify the script as follows :

```
Dim  vs_ddt_order,  vs_order
SystemUtil.Run "C:\Program Files\HP\QuickTest Professional
   \samples\flight\app\flight4a.exe", "", "C:\Program Files\HP\QuickTest Professional\samples\flight\app\",
"open"

Dialog("Login").WinEdit("Agent Name:").Set "tommy"
Dialog("Login").WinEdit("Password:").SetSecure "412885e5be1ef616772e26"
Dialog("Login").WinButton("OK").Click
Window("Flight Reservation").Check CheckPoint("Flight Reservation")
Window("Flight Reservation").WinMenu("Menu").Select "File;Open Order..."
Window("Flight Reservation").Dialog("Open Order").WinCheckBox("Order No.").Set "ON"

vs_ddt_order  =  DataTable ( "OrderNo", dtGlobalSheet )
Window("Flight Reservation").Dialog("Open Order").WinEdit("Edit").Set   vs_ddt_order

Window("Flight Reservation").Dialog("Open Order").WinButton("OK").Click

vs_order = Window("Flight Reservation").WinEdit("Order No:").GetROProperty ("text")

IF  vs_ddt_order = vs_order THEN
    Reporter.ReportEvent micPass,"MATCH"," Found Order: " & vs_order
ELSE
    Reporter.ReportEvent micFail,"NO MATCH","Expected: " &vs_ddt_order & " Found:" &  vs_order
END IF

Window("Flight Reservation").WinMenu("Menu").Select "File;Exit"
```

20. Save the script → Playback the *script*

21. Verify here are *three* (3) iterations

****** **END OF LAB 6E - Part I** ******

LAB 6E : PART II : Test Requirements : Enhance the script

a) Login in to the AUT → Open an existing order number (Positive test)
b) Verify if it matches with the order number on the screen
c) Repeat the process for as many *orders* exist in the DataTable
d) Exit the AUT
e) Identify the starting & ending steps required to perform the action of Verifying the order

1. Modify the Script as follows :

```
01. Dim vs_ddt_order,vs_order, vs_msg, vi_max, vi_x

02. vi_max = DataTable.GetSheet ("Global").GetRowCount
03. Reporter.ReportEvent micDone,"DataTable","Rows: " & vi_max

04. SystemUtil.Run "C:\Program Files\HP\QuickTest Professional
       \samples\flight\app\flight4a.exe", "", "C:\Program Files\HP
       \QuickTest Professional\samples\flight\app\", "open"
05. Dialog("Login").WinEdit("Agent Name:").Set "tommy"
06. Dialog("Login").WinEdit("Password:").SetSecure "412885e5be1ef616772e26"
07. Dialog("Login").WinButton("OK").Click
08. Window("Flight Reservation").Check CheckPoint("Flight Reservation")
09. ' Start Loop
10. FOR vi_x = 1 TO vi_max
11.    DataTable.SetCurrentRow ( vi_x )

12.    Window("Flight Reservation").WinMenu("Menu").Select "File;Open Order..."
13.    Window("Flight Reservation").Dialog("Open Order").WinCheckBox("Order No.").Set "ON"

14.    vs_ddt_order = DataTable( "OrderNo", dtGlobalSheet )

15.    Window("Flight Reservation").Dialog("Open Order").WinEdit("Edit").Set vs_ddt_order
16.    Window("Flight Reservation").Dialog("Open Order").WinButton("OK").Click
17.    vs_order = Window("Flight Reservation").WinEdit("Order No:").GetROProperty ("text")
18.    IF vs_ddt_order = vs_order THEN
19.       Reporter.ReportEvent micPass,"MATCH"," Found Order: " & vs_order
20.    ELSE
21.       vs_msg = "Expected: " &vs_ddt_order & " Found:" & vs_order
21.       Reporter.ReportEvent micFail,"NO MATCH", vs_msg
22.    END IF
23. NEXT
24. Window("Flight Reservation").WinMenu("Menu").Select "File;Exit"
```

2. **Test → Settings ... Run** (tab) → Check **Run one iteration only** → Click **OK**

3. Save the Script → Playback → Review the results

Explanation:

1. Line 02: **vi_max** contains the total number of rows in the DataTable
2. Line 10: Begins the loop starting from One to **vi_max**
3. Line 11: Moves the row position (in the spreadsheet) to the *current row*
4. Line 14: Extracts the value from the **DataTable** and stores it in ***vs_ddt_order***

Note: Between Lines 16 & Line 17 – you may need to insert a synchronization statement : **Wait (2)**

****** **END OF LAB 6E** ******

NOTES

CHAPTER 4

Objectives of this section

- **Understanding the Data Driver Wizard (DDW)**
- **LAB 7A : Using the DDW Test to validate successful LOGIN's**
- **Basic Validation Methods (*Functions*)**
- **LAB 7B : Create Orders : Basic Validations**
- **RDBMS Import Data**
- **LAB 7C : Import Data for Data Driven Test**
- **Introduction to Action Files**
- **LAB 7D : Creating Multiple Action Files**
- **Batch Tests**
- **LAB 7E : Running Batch Tests**
- **LAB 7F : Mini – Project 1**
- **LAB 7G : Mini – Project 2**
- **LAB 7H : Mini – Project 3**

DATA DRIVER WIZARD

1. Allows you to quickly parameterize your QTP script
2. Existing data values can be parameterized or substituted
3. The data values can be all or any combination of :

   a) CheckPoints
   b) Method Arguments
   c) Constant Values (Data)

4. You can replace all values (with a search & replace all) *or*
5. Step-By-Step : However – you must choose the step to parameterize
6. QTP displays the value and the *number* of occurrences of that value
7. Review the following Test (Add a *New* Reservation) :

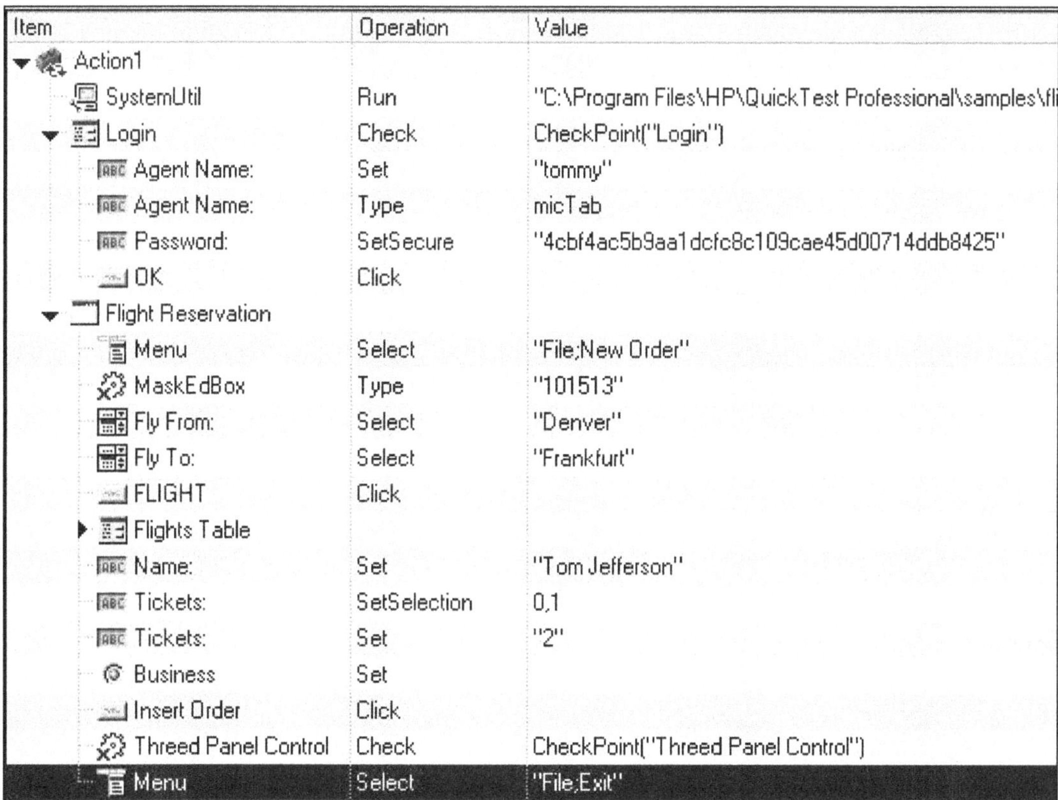

Item	Operation	Value
Action1		
SystemUtil	Run	"C:\Program Files\HP\QuickTest Professional\samples\fl
Login	Check	CheckPoint("Login")
Agent Name:	Set	"tommy"
Agent Name:	Type	micTab
Password:	SetSecure	"4cbf4ac5b9aa1dcfc8c109cae45d00714ddb8425"
OK	Click	
Flight Reservation		
Menu	Select	"File;New Order"
MaskEdBox	Type	"101513"
Fly From:	Select	"Denver"
Fly To:	Select	"Frankfurt"
FLIGHT	Click	
Flights Table		
Name:	Set	"Tom Jefferson"
Tickets:	SetSelection	0,1
Tickets:	Set	"2"
Business	Set	
Insert Order	Click	
Threed Panel Control	Check	CheckPoint("Threed Panel Control")
Menu	Select	"File;Exit"

8. Before displaying the Data Driver window - QTP will scan the current ***Script*** looking for constants !

   a) Identify All Input Values

   b) Wizard will search for values between ***Beginning*** & ***Ending*** quotes

   c) How many Input Fields ?

9. It scans for exact match & performs a case-sensitive search (Creating a Reservation)

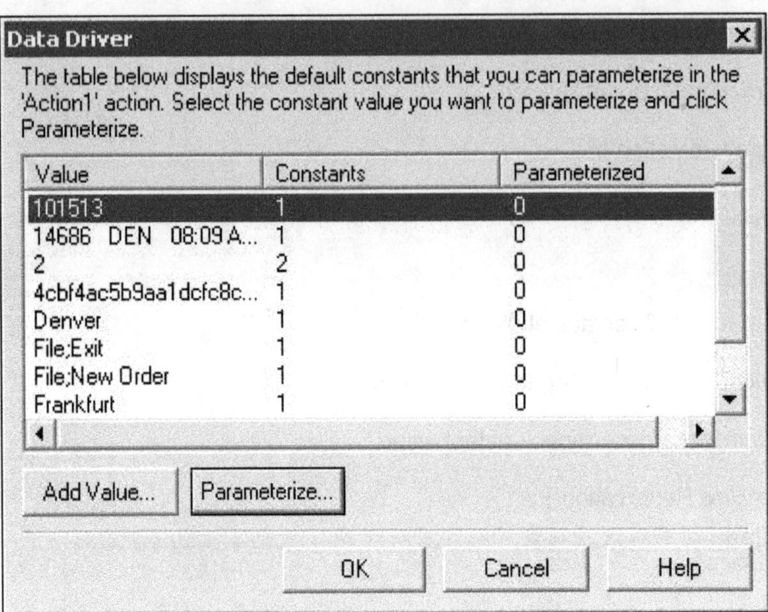

10. When you finish – the Data Driver Wizard displays the total :

    a) How many values you have selected to parameterize

    b) How many remain constants

TIPS

1. Understand the business process (functional requirements)
2. Review the screen flow
3. Visit each window and identify all the input fields and data
4. Determine the source of data
5. Define the acceptance criteria

Note: QTP does NOT parameterize the values of arguments for user-defined methods or *VB Script functions*

LAB 7A : Data Driver Wizard

PART I : Test Requirements :

a) Record a New Script
b) Validate the Login Process
c) Generate *appropriate* messages to confirm the LOGIN was a success
d) Create a Data Driven Test to *LOGIN*
e) Modify column headings in Global Data Table
f) Process specific rows via programming (not using the ***run-time*** settings)

STEPS: (Part 1 : Record the Login Process)

1. **File → New → Test...** (Create a *new* script) → Click **Keyword View** tab
2. **Test → Record F3**
3. Check **Record and run test on any open Windows-based application** → Click **OK**
4. Invoke the AUT (on your own)
5. **Insert → Checkpoint → Standard Checkpoint** ...
6. Point and click on the **title bar** → the **Checkpoint Property** Window Appears
7. Verify **Dialog : Login** → Click **OK**
8. Verify the **text** property contains **Login** → UnCheck the other properties → Click **OK**
9. Login with your ***username*** and password **mercury** → Click **OK**
10. **Insert → Checkpoint → Standard Checkpoint** ...
11. Point and click on the **title bar** → the **Checkpoint Property** Window Appears
12. Verify **Window : Flight Reservation** → Click **OK**
13. Verify the **text** property contains **Flight Reservation** → UnCheck the other properties → Click **OK**
14. **File →Exit** (Close the AUT)
15. Stop the recording → Review the script
16. Save the Script **LAB7A**
17. Playback and verify the script is working
18. Review the **Test Results** window and verify that there are two (2) Checkpoints that passed !
19. Close the **Test Results** window

***** END OF LAB7A – Part 1 *****

STEPS : (Part 2 : Build a Data Driven Test (DDT) by using the ***Data Driver Wizard***)

1. **Tools** (menu bar) → **Data Driver…**

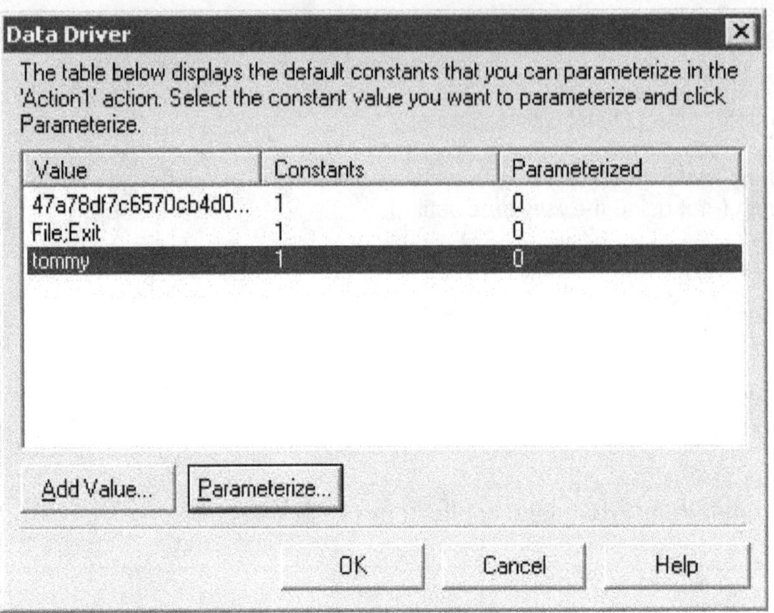

2. Highlight **UserId** (*tommy*) → Click **Parameterize…**

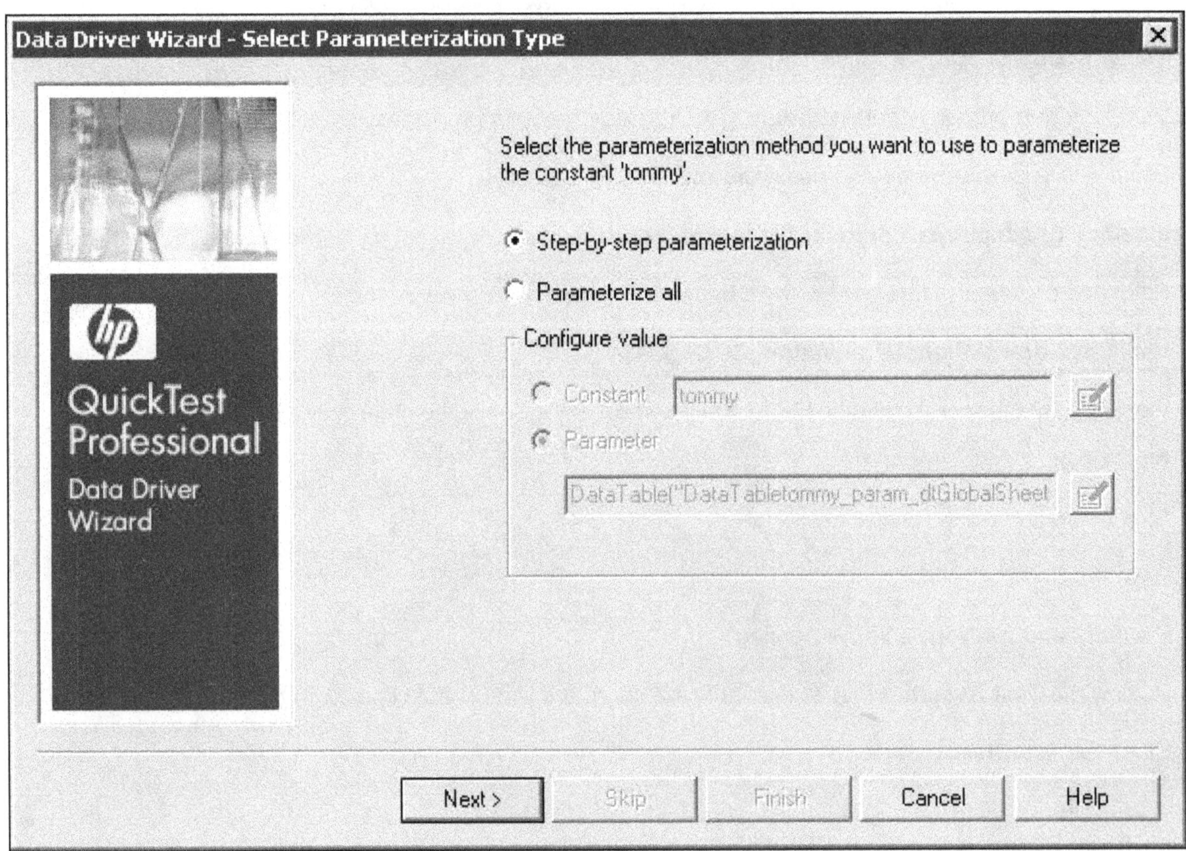

3. Choose **Step-by-step parameterization** → Click **Next >**

4. Verify the following is Highlighted ***Agent Name: Set***

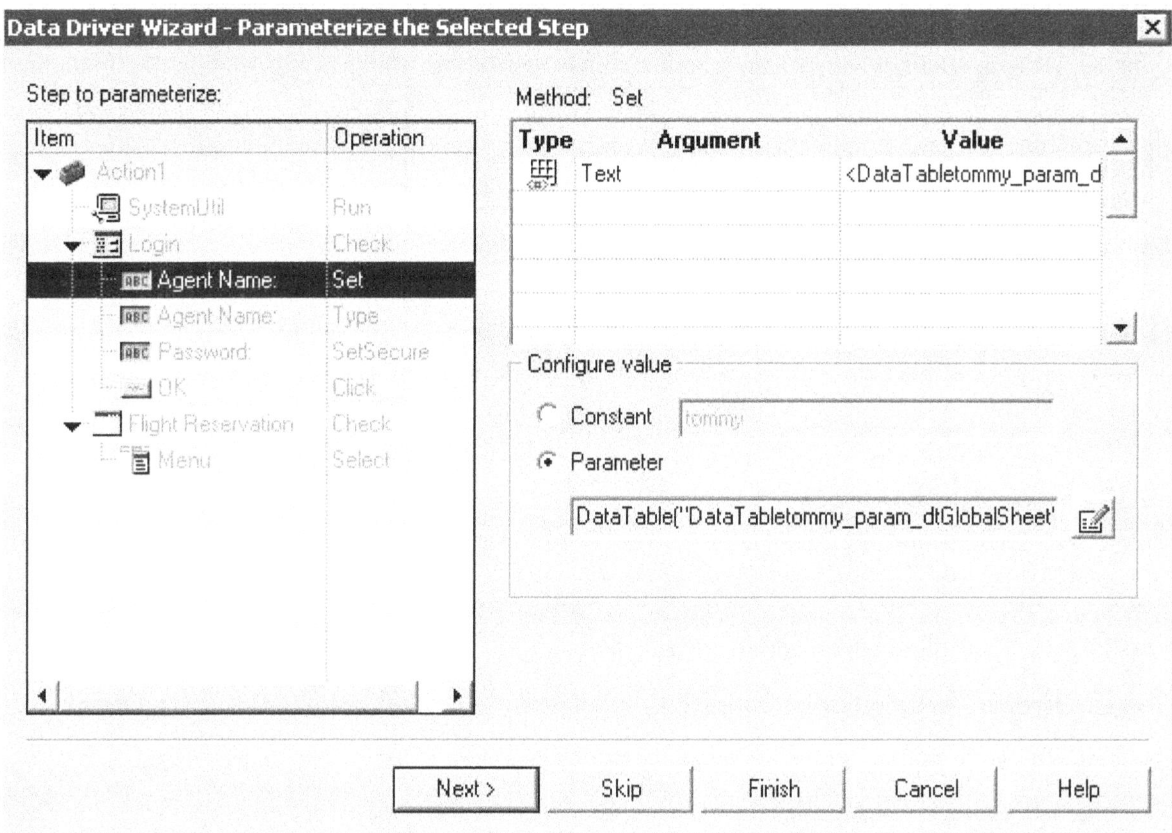

5. Check **Parameter** radio button

6. Verify the input field is correct → Click **Next >** → Click **Finish**

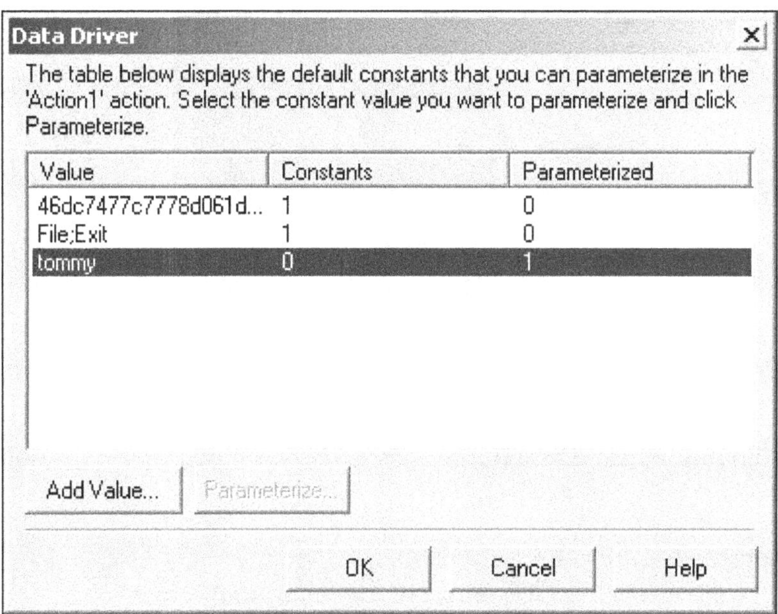

- Observe the **Login Id** is parameterized with a count of **1** (for *tommy*)

7. Highlight the encrypted password → Click **Parameterize...**

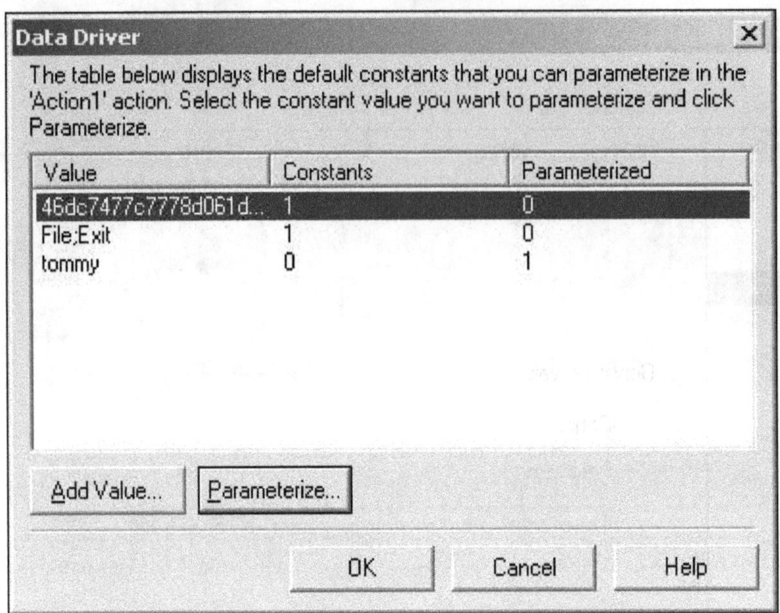

8. Choose **Step-by-step parameterization** → Click **Next >**
9. Verify the following is Highlighted ***Password: SetSecure*** → Check **Parameter** radio button
10. Click **Next >** Click **Finish** (*Congratulations*)
11. Click **OK** (close **Data Driver** window)
12. **View** (menu bar) → **DataTable**
13. Observe the Data Table (Global) contains the data value from your script

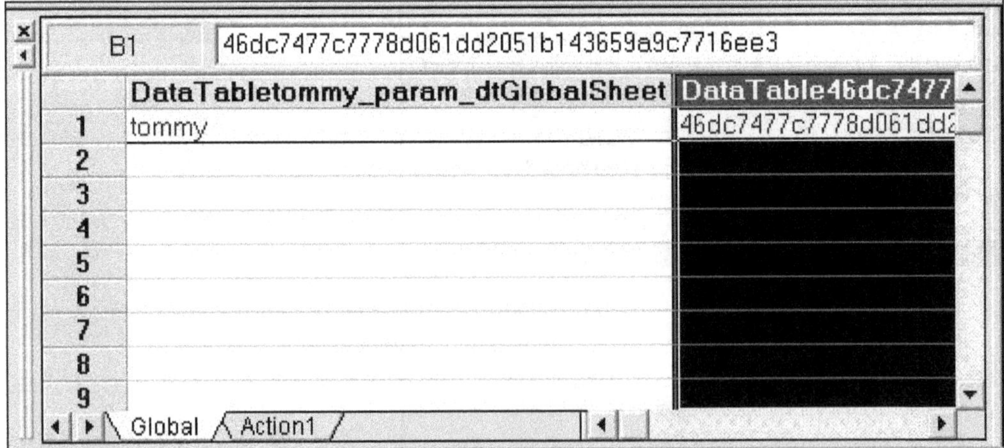

14. Save **LAB7A**
15. Playback and verify the results (No errors) → Correct & Fix any errors

****** END OF LAB7A - PART 2 ******

STEPS : (**Part 3 :** Create Data & Execute the Test)

16. Add the following data to the Global Data Sheet :

B5 | 4ac7bccbe9cd032ac219df4fdb3b65b5a0e5207f

	DataTabletommy_param_dtGlobalSheet	DataTable4ac7bccbe9
1	tommy	4ac7bccbe9cd032ac219df
2	mark	4ac7bccbe9cd032ac219df
3	james	4ac7bccbe9cd032ac219df
4	MARK	4ac7bccbe9cd032ac219df
5	sally	4ac7bccbe9cd032ac219df
6		

17. Save **LAB7D** → Playback & verify that all four (5) rows ran successfully

Modify Run-Time Settings

18. **File** → **Settings…** → **Run** (tab) → Choose **Run one iteration only** → Click **OK**

19. Playback and verify that only the first row (containing *tommy*) ran

20. **File** → **Settings…** → **Run** (tab) → Type **Run from row 2 to row 3** → Click **OK**

21. Playback and verify that rows ***2*** & ***3*** ran (*mark* & *james*)

22. Change the column headings (by double-clicking) to the following in the Global Data Table:

C5

	Users	Password	C
1	tommy	4ac7bccbe9cd032ac219df4fdb3b65b5a0e5207f	
2	mark	4ac7bccbe9cd032ac219df4fdb3b65b5a0e5207f	
3	james	4ac7bccbe9cd032ac219df4fdb3b65b5a0e5207f	
4	MARK	4ac7bccbe9cd032ac219df4fdb3b65b5a0e5207f	
5	sally	4ac7bccbe9cd032ac219df4fdb3b65b5a0e5207f	
6			

23. Playback → It should fail with the following: → Click **Stop** → Close **Test Results** window → Close the **AUT**

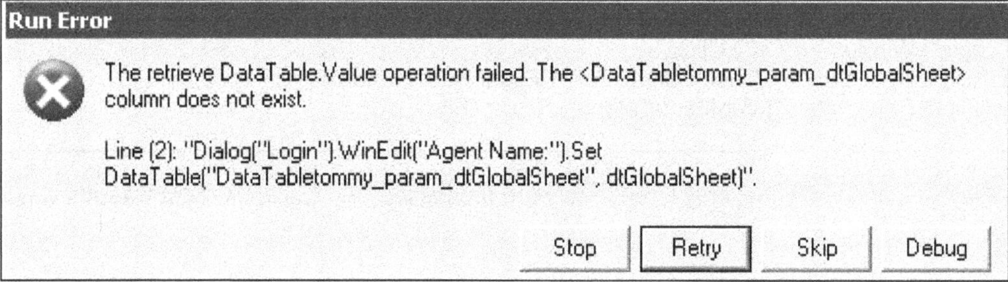
Run Error
The retrieve DataTable.Value operation failed. The <DataTabletommy_param_dtGlobalSheet> column does not exist.
Line (2): "Dialog("Login").WinEdit("Agent Name:").Set DataTable("DataTabletommy_param_dtGlobalSheet", dtGlobalSheet)".
Stop
Retry
Skip
Debug

Why ? ____

24. Add the **Cancel** to the object repository for the Login window (on your own)

    Hint: Use the Active Screen

25. Click **Expertview** → Make the following changes in bold

```
SystemUtil.Run "C:\Program Files\HP\QuickTest
Professional\samples\flight\app\flight4a.exe","","C:\Program Files\HP
\QuickTest Professional\samples\flight\app\","open"

Dialog("Login").Check CheckPoint("Login")
Dialog("Login").WinEdit("Agent Name:").Set DataTable("Users", dtGlobalSheet)
Dialog("Login").WinEdit("Agent Name:").Type micTab
Dialog("Login").WinEdit("Password:").SetSecure DataTable("Password", dtGlobalSheet)
Dialog("Login").WinButton("OK").Click
Window("Flight Reservation").Check CheckPoint("Flight Reservation")
Window("Flight Reservation").WinMenu("Menu").Select "File;Exit"
```

26. Click **KeyWord View** → Save **LAB7A**

27. **File** → **Settings...** → **Run** (tab) → Choose **Run on all rows** → Click **OK**

28. Playback and verify and verify that all five rows ran successfully

Process all rows except for ***mark*** *or* ***MARK*** (row 2 & row 4)

29. Click **Expertview** → Make the following changes in bold

```
Dim  vs_name
SystemUtil.Run "C:\Program Files\HP\QuickTest
Professional\samples\flight\app\flight4a.exe","","C:\Program Files\HP
\QuickTest Professional\samples\flight\app\","open"
Dialog("Login").Check CheckPoint("Login")
vs_name =  DataTable ( "Users",  dtGlobalSheet )

If  ucase ( vs_name ) = "MARK"   THEN
    Reporter.ReportEvent  micFail, "Bypass Transaction", "User : "  &  vs_name
    Dialog("Login").WinButton("Cancel").Click
    ExitActionIteration("Skipping : " & vs_name)
End If

Dialog("Login").WinEdit("Agent Name:").Set DataTable("Users", dtGlobalSheet)
Dialog("Login").WinEdit("Agent Name:").Type micTab
Dialog("Login").WinEdit("Password:").SetSecure DataTable("Password", dtGlobalSheet)
Dialog("Login").WinButton("OK").Click
Window("Flight Reservation").Check CheckPoint("Flight Reservation")
Window("Flight Reservation").WinMenu("Menu").Select "File;Exit"
```

30. Save **LAB7A** → Playback and verify that only three rows were processed → Close the **Test Results** window

Question :

1. Why was nothing displayed for user **mark** in the Test Results window ?

***** **END OF LAB 7A** ******

BASIC VALIDATION Functions

IsDate

- Returns a Boolean value indicating wheter an expression can be converted to a date
- The argument can be any date ***expression*** or ***string*** expression recognizable a **date** or **time**

Format: IsDate (*expression*)

Example 25 : Store the dates in the array & check to see if they contain valid dates

```
DIM   vs_date , vb_found
vs_date = Array ("February 1, 2004","04/15/2004 12:35:40","04/15/2004","04/32/2004")

FOR EACH item IN vs_date
   vb_found  = IsDate ( item )
   IF  vb_found = True  THEN
       Reporter.Reportevent micPass, "Date" , item
   ELSE
       Reporter.Reportevent micFail, "InValid Date" , item
   END IF
NEXT
```

IsEmpty

- Returns a Boolean value indicating whether a variable has been initialized
- Typically used to see if *variables* have assigned <u>any</u> value (data)

Format: IsEmpty (*expression*)

Example 26 : Check to see if the declared variables have been initialized

```
DIM  vs_name, vs_city, vs_check
vs_city = "Edison, NJ"
vs_check = IsEmpty ( vs_city )        ' Returns  False
vs_check = IsEmpty ( vs_name )        ' Returns  True
```

IsNull

- Returns a Boolean value indicating whether an expression contains no valid data (Null)
- Null is **not** the same as *empty*
- Some columns from database table may contain nulls and displayed on the screen via SQL Query

Format: IsNull (*expression*)

Example 27 : Check to see if a declared variables contain NULLS

```
DIM  vs_name, vs_city, vs_check
vs_city = NULL
vs_check = IsNull ( vs_city )        ' Returns True
vs_check = IsNull ( vs_name )        ' Returns  False
```

IsNumeric

- Returns a Boolean value indicating whether an expression can be evaluated as a number

Format: IsNumeric (*expression*)

Example 28 : Check to see if a declared *variable* is numeric

```
DIM  vs_number1, vs_number2, vs_number3
vs_number1  = 173
vs_number2  = "265"
vs_number3  = "X123"
vs_check = IsNumeric ( vs_number1 )        ' Returns True
vs_check = IsNumeric ( vs_number2 )        ' Returns True
vs_check = IsNumeric ( vs_number3 )        ' Returns False
```

ItemsCount

- Returns the number of items in the object list

Example 29 : Return a total count of the items for the object : Refer to online help for *other* objects

Object	Description
ActiveX Combo Box	vi_cnt = Window ("Flight Reservations").AcxComboBox ("Fly From").ItemsCount
VB Combo Box	vi_cnt = VBWindow ("frmMain").VBComboBox ("cmbTO").ItemsCount
WinCombo Box	vi_cnt = Window ("Flight Reservations").WinComboBox ("Fly To").ItemsCount
WinList	vi_cnt = Dialog ("Open").WinList ("Files").ItemsCount

Select

- Selects an *item* from the object's list

Example 30 : Selects the item. The first item is *zero*

Object	Description
VB Combo Box	VbWindow ("frmMain").VBComboBox ("cmbFrom").Select "New York"
or	VbWindow ("frmMain").VBComboBox ("cmbFrom").Select 7
WinCombo Box	Window ("Notepad").Dialog("Font").WinComboBox("Size:").Select "22"
or	Window ("Notepad").Dialog("Font").WinComboBox("Size:").Select 3
WinMenu	Window ("Notepad").WinMenu("Menu").Select "File;Open... Ctrl+O"
or	Window ("Notepad").WinMenu("Menu").Select "<Item 1>;<Item 2>"
WinRadioGroup	Window("Flight Reservation").WinRadioGroup("Order Information:").Select "First "

<u>Note</u>: Other types of checking can be accomplished via (Regular Expressions – covered in *QTP – Level 2* book)

Date & Time : Returns the current date and current time

Format: *Date* Format: *Time*

Example 31 :

```
DIM    vs_date , vs_time
vs_date  = Date       ' Returns the current Date    Format: mm/dd/yyyy
vs_time  = Time       ' Returns the current Time    Format: hh:mm:ss [ AM/PM ]
```

DateAdd : Returns a date to which a specified time interval has been added

Format: DateAdd (interval, number, date)

Interval (yyyy, q=quarter, , m=month, d=day, y=year, h=hour, n=minute s=seconds)

Example 32A :

```
DIM  vs_date, vs_date1 , vs_date2
vs_date  = "04/05/2004"
vs_date1  = DateAdd ( "yyyy" , 1 , vs_date )   ' Add 1 year to date
vs_date2  = DateAdd ( "m" , 5 , vs_date )      ' Add 5 months to date
```

DateDiff : Returns the number of intervals between two dates

Format: DateDiff (interval, date1, date2)

Interval (yyyy, q=quarter, , m=month, d=day, y=year, h=hour, n=minute s=seconds)

Example 32B :

```
DIM  vs_date1, vs_date2 , vs_date3, vi_days
vs_date1  = "12/01/2008"
vs_date2  =  "12/31/2008"
vs_date3  =  "11/01/2008"
vi_days    = DateDiff ( "d" ,vs_date1 , vs_date2 )     ' Returns  30
vi_days    = DateDiff ( "d" ,vs_date2 , vs_date3 )     ' Returns  -60
```

SubString : Returns a specified number of characters are returned

Format: Mid (*variable*, start *pos*, length)

Example 33 :

```
DIM  vs_name, vs_sub1, vs_sub2
vs_name  = "James Brown"
vs_sub1   =  Mid (vs_name , 2, 4)          ' Returns  ames
vs_sub2   =  Mid (vs_name , 7, 5)          ' Returns  Brown
```

Len : Returns the number of characters in a string or the number of bytes

Format: Len (*variable*)

Example 34 :

```
DIM  vs_name1, vs_name2
vs_name1  = "Sally"
vs_name2  = "John Kennedy"
vs_sub1   =  Len (vs_name1)          ' Returns  5
vs_sub2   =  Len (vs_name2)          ' Returns  12
```

LAB 7B : Basic Validation : Create Orders

PART I : Test Requirements :

a) Record a New Script & Save it twice as:

- LAB7 : Prior to Data Driven
- LAB7B : After Data Driven

b) Create a Data Driven Test to ***Add*** new reservations
c) Check for empty fields
d) Bypass transactions that contain an invalid date and print a warning message in the Log
e) Generate *appropriate* messages to confirm the reservation
f) Delete the Reservation (Perform a recording to the existing script)

STEPS : (Part 1 : Record the initial script for one (1) business transaction)

1. **File → New → Test...** (Create a *new* script) → Click **Keyword View** tab
2. **Automation → Record F3**
3. Check **Record and run test on any open Windows-based application** → Click **OK**
4. Invoke the AUT (on your own)
5. **Insert → Checkpoint → Standard Checkpoint** ...
6. Point and click on the **title bar** → the **Checkpoint Property** Window Appears
7. Verify **Dialog : Login** → Click **OK**
8. Uncheck all the properties → Check & verify the **text** property contains **Login** → Click **OK**
9. Login with your ***username*** and password **mercury** → Click **OK**
10. **Insert → Checkpoint → Standard Checkpoint** ...
11. Point and click on the **title bar** → the **Checkpoint Property** Window Appears
12. Verify **Window : Flight Reservation** → Click **OK**
13. Uncheck all the properties → Check & verify the **text** property contains **Flight Reservation** → Click **OK**
14. **File → New Order ...**
15. Enter all the *data* into the fields to **add** an order (on your own)
    - Be sure to *select/highlight* a row in the **Flight Schedule** window
16. Click **Insert Order** → *Progress Bar* appears
17. Swap to QTP → **Insert → Checkpoint → Standard Checkpoint** ...
18. Point & Click on message **Insert Done ...**
19. Verify **Window: Flight Reservation ActiveX : Threed Panel Control** → Click **OK**

20. UnCheck all properties → Check **text** property & verify that **Insert Done...** appears in the selection
21. Check **text** property & verify that **Insert Done...** appears in the selection
22. Click **OK** → Write down the Order Number : ___________________
23. Swap to AUT → **File** → **Exit**
24. Stop Recording
25. File → Save → **C:\yourname\QTP10\LVL1\LAB7**
26. File → Save As → **C:\yourname\QTP10\LVL1\LAB7B**
27. Click **Keyword View** → Review the script → Ensure you selected the *time* of the flight

Fly To:	Select	"Paris"	Select the "Paris" item fr
FLIGHT	Click		Click the "FLIGHT" butto
Flights Table			
From	Select	"13494 FRA 12:48 PM PAR 04:18...	Select the "13494 FRA
OK	Click		Click the "OK" button.
Name:	Set	"Thoman Pratt"	Enter "Thoman Pratt" in

28. Remove any miscellaneous *actions* or statements from your script (if any)
29. Start AUT (Flight 4A) → Login with yourname & password
30. File → Open Order → Open the Order Number from Step 18: ______________ → Click **OK**
31. Highlight **Threed Panel Check** (line above **Menu Select File;Exit**)
32. Press **Record** or **F3** → Click **Delete Order**
33. Click **Yes** (to confirm the **Delete**)
34. Swap to QTP → **Insert** → **Checkpoint** → **Standard Checkpoint** ...
35. Point & Click on message **Delete Done ...**
36. Verify **Window: Flight Reservation ActiveX : Threed Panel Control** → Click **OK**
37. UnCheck all properties
38. Check **text** property & verify that **Delete Done...** appears in the selection → Click **OK**
39. Stop the recording (Press **F4**) → Review the script
40. Save the script: **LAB7B** → Playback & Verify the results

****** END OF Part I : LAB 7B ******

STEPS : (Part 2 : Verify mandatory *fields* in the **Login** window)

Test Requirements

1. If any of the fields (**Login-Id** or **Password** are empty) : Issue a message and *Stop* the test run

Steps: **(Do not use LAB7B)**

1. Open → **C:\yourname\QTP10\LVL1\LAB7** → Click **ExpertView** → Resize the window to view the script
2. Modify the script as follows for Lines 1 thru 15 only – (Review the lines in ***bold***) :

```
01. Dim vs_agent , vs_pass

02. SystemUtil.Run "C:\Program Files\HP\QuickTest\ Professional\samples\flight\
        app\flight4a.exe","","C:\Program Files\HP\QuickTest\
        Professional\samples\flight\app\","open"
03. Dialog("Login").Check CheckPoint("Login")

04. ' Dialog("Login").WinEdit("Agent Name:").Set "yourname"
05. ' Dialog("Login").WinEdit("Password:").SetSecure ""

06. vs_agent = Dialog("Login").WinEdit("Agent Name:").GetROProperty("text")
07. vs_pass = Dialog("Login").WinEdit("Password:").GetROProperty("text")

08. IF vs_agent = "" Then
09.         Reporter.ReportEvent micFail,"User ID","No User ID entered ..."
10.         ExitTest
11. End If
12. IF vs_pass = "" Then
13.         Reporter.ReportEvent micFail,"Password","No Password entered ..."
14.         ExitTest
15. End If
16. Dialog ("Login").WinButton("OK").Click
17. Window("Flight Reservation").Check CheckPoint("Flight Reservation")
18. Window("Flight Reservation").WinMenu("Menu").Select "File;New Order"
19. Window("Flight Reservation").ActiveX("MaskEdBox").Type "101004"
20. Window("Flight Reservation").WinComboBox("Fly From:").Select "Frankfurt"
21. Window("Flight Reservation").WinComboBox("Fly To:").Select "Los Angeles"
22. Window("Flight Reservation").WinButton("FLIGHT").Click
23. Window("Flight Reservation").Dialog("Flights Table").WinList("From").Select "20327  FRA  12:12 AM   LAX
        07:23 PM  AA   $112.20"
24. Window("Flight Reservation").Dialog("Flights Table").WinButton("OK").Click
25. Window("Flight Reservation").WinEdit("Name:").Set "Martha Washington"
26. Window("Flight Reservation").WinButton("Insert Order").Click
27. Window("Flight Reservation").ActiveX("Threed Panel").Check CheckPoint("Insert Done...")
28. Window("Flight Reservation").WinMenu("Menu").Select "File;Exit"
```

3. Remember to **Comment *Lines 4 & 5*** → Save the script **LAB7**
4. Run 1 : Execute the script → Click **OK** → It should fail with : **No User ID entered** → Close the **AUT**
5. Run 2 : Uncomment Line 4 only → Execute the script → Enter **User-ID** (No Password) → Click **OK**
6. It should fail with : **No User ID entered** → Close the **AUT**

**** **END OF Part 2 : LAB 7B** ****

PART 3 : Create a Data Driven Test to *add* new reservations

Test Requirements

1. Parameterize the input fields
2. Create a minimum of five *new* reservations
3. Choose or *select* the first index entry from any *combo boxes* or *tables*
4. Bypass transactions which contain an invalid date *or* the date does not meet the business rule (older than today) and write a message to the Log

Steps: (Do Not use LAB7)

1. Open → **C:\yourname\QTP10\LVL1\LAB7B** → Click **Keyword View**
2. Click **View** (menu bar) → **Collapse All** → Expand ***Action1*** → Expand the first **Flight Reservation**
3. Highlight **"MaskEdBox" Type "*mmddyy*"** → Click **Configure the value** icon
4. Check Radio Button **Parameter** → Type/Replace ***p_KeyboardInput*** *with* **FlightDate** → Click **OK**
5. Highlight **"FlyFrom" Select "*Cityname*"** → Click **Configure the value** icon
6. Check Radio Button **Parameter** → Type/Replace ***p_Item*** *with* **Fly_From** → Click **OK**
7. Highlight **"FlyTo" Select "*Cityname*"** → Click **Configure the value** icon
8. Check Radio Button **Parameter** → Replace ***p_Item*** *with* **Fly_To** → Click **OK**
9. Expand **Flight Tables**
10. Highlight **From Select** "FlightNo city time …." (on the left side of the **Value** column)
11. Click **Configure the value** icon
12. Check Radio Button **Constant** → Replace "FlightNo city time …." *with* **1** → Click **OK**
13. Highlight **"Name" Set "*name*** → Click **Configure the value** icon
14. Check Radio Button **Parameter** → Replace ***p_Text*** *with* **Name** → Click **OK**
15. Expand **Flights Table** (if not already) → **Highlight "From" Select "1"**
16. Click **ExpertView** → Resize the window to view the script
17. The cursor should be at the following *line* : (if not locate & *move* cursor to the following line)

    Window ("Flight Reservation").Dialog("Flights Table").WinList("From").Select "1"

18. Remove the **quotes** around the number **one** (if it exists) - It should look like :

    Window ("Flight Reservation").Dialog("Flights Table").WinList("From").Select 1

19. Press **CTRL + F7** → Correct & Fix any syntax errors
20. Save the Script **LAB7B**
21. Click **Keyword View**

22. Make the following changes to the lines in **bold**

a) Create new variables **vs_date , vb_valid , vb_bypass, vi_days**
b) Uncomment & comment the lines to allow a successful Log-in
c) Bypass transactions containing and invalid or a date in the past & issue a warning message

```
DIM  vs_agent,  vs_pass,  vs_date,  vb_valid , vb_bypass, vi_days

SystemUtil.Run "C:\Program Files\Mercury .......\flight4a.exe",",""",""
Dialog("Login").Check CheckPoint("Login")

Dialog("Login").WinEdit("Agent Name:").Set "userid"
Dialog("Login").WinEdit("Agent Name:").Type  micTab
Dialog("Login").WinEdit("Password:").SetSecure "49776783f74b4f999a78de07cb1606ba120a4b47"

' vs_agent = Dialog("Login").WinEdit("Agent Name:").GetROProperty("text")
' vs_pass = Dialog("Login").WinEdit("Password:").GetROProperty("text")

' if  vs_agent = "" Then
'          Reporter.ReportEvent micFail,"User ID","No User ID entered ..."
'          ExitTest
' end  if
' if  vs_pass  = "" Then
'          Reporter.ReportEvent micFail,"Password","No Password entered ..."
'          ExitTest
' end  if

Dialog("Login").WinButton("OK").Click
Wait ( 2 )                              '  A Syncronization may be required
Window("Flight Reservation").Check CheckPoint("Flight Reservation").......
Window("Flight Reservation").WinMenu("Menu").Select "File;New Order"
vb_bypass = False
vi_days  = 0
vs_date  = DataTable("FlightDate", dtGlobalSheet)
vb_valid = IsDate ( vs_date )

if   vb_valid = False  Then
     Reporter.ReportEvent micWarning, "Invalid Flight Date: ",   vs_date
     vb_bypass = True
else
     vi_days = DateDiff ("d", Date, vs_date)       'Determine Difference in Dates
end if

if  vi_days <  0 Then
    Reporter.ReportEvent micWarning, "Past Date :",  vs_date
    vb_bypass = True
end If
if  vb_bypass = True   Then
    Window("Flight Reservation").WinMenu("Menu").Select "File;Exit"
    ExitActionIteration
end  if

Window("Flight Reservation").ActiveX("MaskEdBox").Type   vs_date

Window("Flight Reservation").WinComboBox("Fly From:").Select DataTable("FlyFrom", dtGlobalSheet)
......
```

23. Save script as **LAB7B_P3**

24. Execute the script → It should Pass ! → Close the *Test Results* window

25. Modify the ***DataTable*** and create at **least** *five* transactions :

Row	FlightDate	Fly_From	Fly_To	Name
1	10/10/10	Denver	London	James Smith
2	11/25/10	Frankfurt	Paris	Mike Thomas
3	12/25/09	London	Frankfurt	Sally Green
4	01/05/11	Paris	Frankfurt	Tiger Woods
5	12/32/16	Portland	Seattle	Susan Shops

<u>**NOTE 1**</u>: Possible Date Problems in the DataTable

- You may need to enter valid dates in the **FlightDate** column for your test data

To enter a **fixed** *date* in the column of the datatable, put a single quote followed by the date

Example: To enter March 4, 2009 (03/04/09)

Enter: **'03/04/09** **# 'mm/dd/yy**

26. Save **LAB7B_P3** → Run the Test → It Should Pass with two warning messages !

27. Close the *Test Results* window

28. Save as **LAB7B_P4** (new file)

29. Start the AUT → Login → Delete all five rows that were created

<u>**NOTE 2**</u>: Step 28 Must be completed before **LAB7B - PART 4** before begins

<u>**NOTE 3**</u>: Password *settings* :

a) HP QuickTest *encrypts* normal passwords during recording captures it in the script

b) HP QuickTest also *encrypts* **normal** passwords during playback onto the screen

c) When you retrieve passwords from the screen – HP QuickTest does **NOT** capture the *text*

d) You will ***not*** be able to use *functions* or *methods* to retrieve password

****** END OF LAB7B : Part 3 ******

LAB 7B - PART 4 : Single Action Iteration with a User Controlled Loop

Requirements :

a) Delete the User-Id & Password Verifications
b) Login once and read all the rows in the DataTable
c) Define new variables to check for empty or blank field (Fly From & Fly To)
d) Create a user-controlled loop to process all the transactions & Save the Order Number in the DataTable

1. Open **LAB7B_P4** (Saved in **LAB7B_P3**)

2. Add a New Column in the Data Table **Order_No** (on your own)

3. Modify the Script (your actions & input data maybe different) :

```
DIM  vs_order, vs_from, vs_to, vs_name, vi_max, vi_x
DIM  vb_bypass, vb_valid, vi_days

vi_max = DataTable.GetSheet ("Global").GetRowCount
Reporter.ReportEvent micDone,"DataTable","Rows: " & vi_max

SystemUtil.Run "C:\Program Files\HP\QuickTest Professional\
     samples\flight\app\flight4a.exe", "", "C:\Program Files\HP\
     QuickTest Professional\samples\flight\app\", "open"

Dialog("Login").Check CheckPoint("Login")
Dialog("Login").WinEdit("Agent Name:").Set "tommy"
Dialog("Login").WinEdit("Password:").Set "mercury"
Dialog("Login").WinButton("OK").Click

Wait ( 2 )                          ' Synchronization
Window("Flight Reservation").Check CheckPoint("Flight Reservation")

FOR   vi_x = 1 TO vi_max

   DataTable.SetCurrentRow (vi_x)

   Window("Flight Reservation").WinMenu("Menu").Select "File;New Order"
   vi_days     = 0
   vb_bypass  = False
   vs_date     = DataTable("FlightDate", dtGlobalSheet)
   vb_valid    = IsDate ( vs_date )

   If  vb_valid = False  Then
       Reporter.ReportEvent micWarning, "Invalid Flight Date " , vs_date
       vb_bypass = True
   else
       vi_days = DateDiff ("d", Date, vs_date)
   end If

   If  vi_days < 0  Then
       Reporter.ReportEvent micWarning, "Past Date :"  ,  vs_date
       vb_bypass = True
   end If
```

Script continues on the next page .....

```
  If  vb_bypass  =  False   Then
     Window("Flight Reservation").ActiveX("MaskEdBox").Type  vs_date
     vs_from    = DataTable("Fly_From",dtGlobalSheet)
     vs_to       = DataTable("Fly_To",dtGlobalSheet)
     vs_name   = DataTable("Name",dtGlobalSheet)

     If   vs_from = "" THEN
        Window("Flight Reservation").WinComboBox("Fly From:").Select  1
     else
        Window("Flight Reservation").WinComboBox("Fly From:").Select   vs_from
     end if

     If   vs_to  = "" THEN
         Window("Flight Reservation").WinComboBox("Fly To:").Select  1
     else
         Window("Flight Reservation").WinComboBox("Fly To:").Select   vs_to
     end if

     Window("Flight Reservation").WinButton("FLIGHT").Click
     Window("Flight Reservation").Dialog("Flights Table").WinList("From").Select 1
     Window("Flight Reservation").Dialog("Flights Table").WinButton("OK").Click

     Window("Flight Reservation").WinEdit("Name:").Set  vs_name

     Window("Flight Reservation").WinButton("Insert Order").Click

     Window("Flight Reservation").ActiveX("Threed Panel").Check CheckPoint("Threed Panel Control")

     vs_order = Window("Flight Reservation").WinEdit("Order No:").GetROproperty ("text")

     DataTable ( "Order_No",dtGlobalSheet )  =  vs_order

   end  if
 NEXT

 Window("Flight Reservation").WinMenu("Menu").Select "File;Exit"

 DataTable.Export ( "C:\yourname\QTP10\LVL1\LAB7A_P4.XLS" )
```

34. **Test** → **Settings ...** → Click **Run** tab → Check **Run one iteration only** → Click **OK**

35. Save script **LAB7B_P4** → Run script → It should Fail with the following error message !

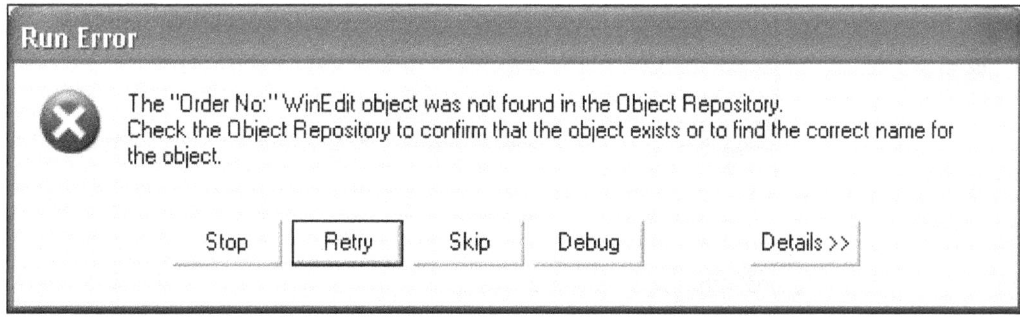
Run Error
The "Order No:" WinEdit object was not found in the Object Repository.
Check the Object Repository to confirm that the object exists or to find the correct name for the object.
Stop
Retry
Skip
Debug
Details >>

36. Click **Details >>** Identify the Line # : _________ Click **Stop** → Review & Close the **Test Results** window

Question:

a) Why was the object **Order No:** not in the Object Repository ?

b) How would you correct / fix the problem ?

37. Go to the **AUT** → Click **Delete Order** → Click **Yes** → Close the **AUT**
38. Click **Keyword View** → Highlight **Threed Panel Control Check** .. Go to the Active Screen

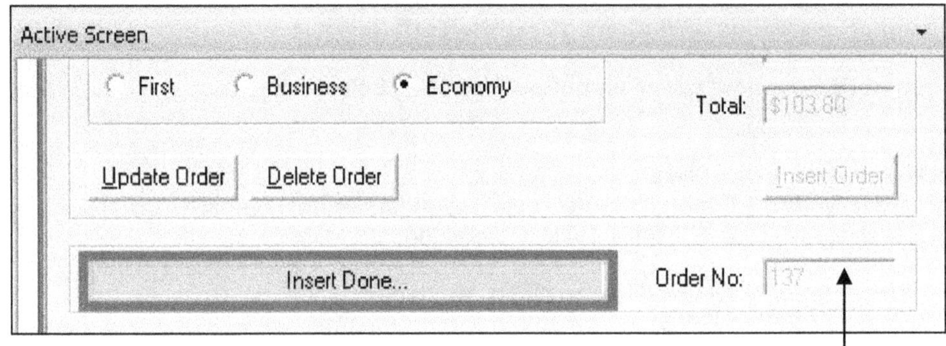

39. Move cursor to the right of **Order No:** and inside the new number
40. Right-Click → Select **View Add / Object ...** → Verify **WinEdit:Order No:** → Click **OK**
41. Click **Add to Repository** → Click **OK**
42. Save **LAB7B_P4**
43. Run script → It should Pass !
44. Review the **Test Results** window
45. Click ***Run-Time Data Table*** → Observe the New Order Number was saved

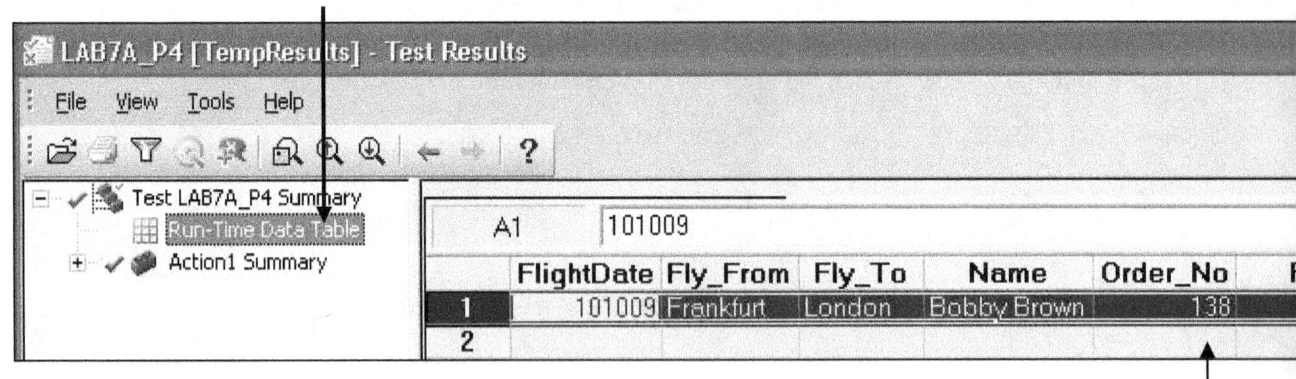

46. Close the **Test Results** window
47. Start the AUT → Login → Open the order that was just created !
48. Delete the Order → Exit the AUT
49. **Test** → **Settings ...** → Click **Run** tab

50. Check **Run on all rows** → Click **OK**

51. Add *more* data to the Data Table (rows 6 & 7) → Save **LAB7A_P4**

Row	**FlightDate**	**Fly_From**	**Fly_To**	**Name**	**Order_No**
1	10/10/10	Denver	London	James Smith	1
2	11/25/10	Frankfurt	Paris	Mike Thomas	2
3	12/25/09	London	Frankfurt	Sally Green	3
4	01/05/11	Paris	Frankfurt	Tiger Woods	4
5	12/32/16	Portland	Seattle	Susan Shops	5
6	**05/10/10**		**Denver**	**Soupy Sales**	
7	**10/25/10**	**Denver**		**Alice Sparks**	

52. **Test** → **Settings ...** → Click **Run** tab → Check **Run one iteration only** → Click **OK**

53. Execute the Test → It should Pass !

54. In *windows explorer* – verify the *excel* spreadsheet file was created :
C:*yourname*\QTP10\LVL1\LAB7B_P4.XLS

55. Open the **LAB7B_P4.XLS** & review the data

56. Close the **LAB7B_P4.XLS** file

****** END OF LAB 7B ******

RDBMS – Steps to export information from a database (Summary)

Guidelines

1. Identify the RDBMS : (ORACLE, MS SQL Server, DB2, SYBASE, INFORMIX , etc …)

2. Review Schema(s)

   a) Data Model (Entity Relationship Diagram)
   b) Table Definition
   c) Identify Relationships (between tables)

3. Write a SQL query to confirm the data exists and that it conforms to your needs:

   a) Create your own data (if you have been granted permission)
   b) Make a copy of the tables (from the original source)
   c) Insert / Update columns from your newly created tables for your test scenario
   d) Write a query from a single table or Joins based on multiple tables

4. Use **VIEWS**

   a) Review your application and identify all the input fields and sequence
   b) Create a view based on a SQL Query
   c) The sequence of the view (columns) is based on the required input fields in your application
   d) The view should include a **UNION** statement (Review LAB 7C)

5. Export Data based on the VIEWS created earlier (Use the vendor provided utility to export data) :

   - BCP (MS SQL Server & Sybase)
   - EXP, IMP (Oracle)

   a) Change the file extension to **XLS** while specifying the output file name in the *export* utility

6. Review the data in MS EXCEL file (**logins.xls**)

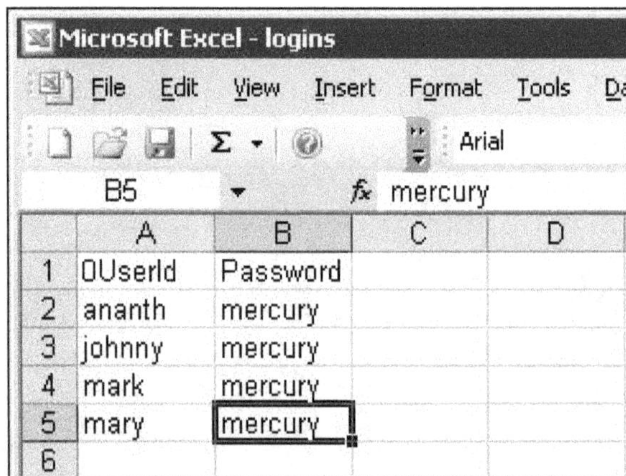

7. Open → MS-Excel → Change the Column Heading to remove the Number Zero (0)

8. QTP uses EXCEL as the default source for the DataTable

LAB 7C : Import Data from RDBMS

Requirements:

1. Prerequisites: LAB7A is completed & working !
2. Create the data from the **agents** table in MS SQL Server
3. The table **agents** should contain both *positive* and *negative* data
4. Export data into an ***excel*** spreadsheet
5. Bypass Transactions that DON'T match the Business Rule
6. Run the Data-Driven test

STEPS

1. Review the table structure and Appendix A

agents (schema) database: **qa_dev1** user: **itm** password: **itm** Server: **training-srvr02**

agent_id	smallint
agent_name	nvarchar (20)
agent_password	nvarchar (30)
dept	nchar (02)
hire_date	smalldatetime

2. Create a view of the selected columns. Replace ***itm*** (v_*itm*_lab7c_ddt) with your initials

```
create view v_itm_lab7c_ddt
As
   select  '0User' , '0Password'  from  itm.agents
   UNION
   select  agent_name,  agent_password from itm.agents
```

- The **union** combines the result set from *each* query into a ***single*** Result Set without duplicates
- In the first **select** statement, the first character is the number zero

3. Export the data out to a EXCEL file: (MS-DOS or command line prompt)

   a) Start → Run → CMD → Click **OK**

   b) bcp qa_dev1..v_itm_lab7c_ddt out C:\yourname\QTP10\LVL1\LAB7C.XLS -c -U*itm* -P*itm* -Straining-srvr02

   Note : if you have created your own tables & views – you can specify your database id -Usql_id -Pwelcome -Straining-srvr02

4. Open & review the **C:\yourname\QTP10\LVL1\LAB7C.XLS** data is correct

5. Change the Column Heading in Excel

   a) Open MS Excel (on your own)

6. **Edit → Replace CTRL + H →**

   a) Find & Replace **0Users** with **Users** → Click **Replace All**

   b) Find & Replace **0Password** with **Password** → Click **Replace All**

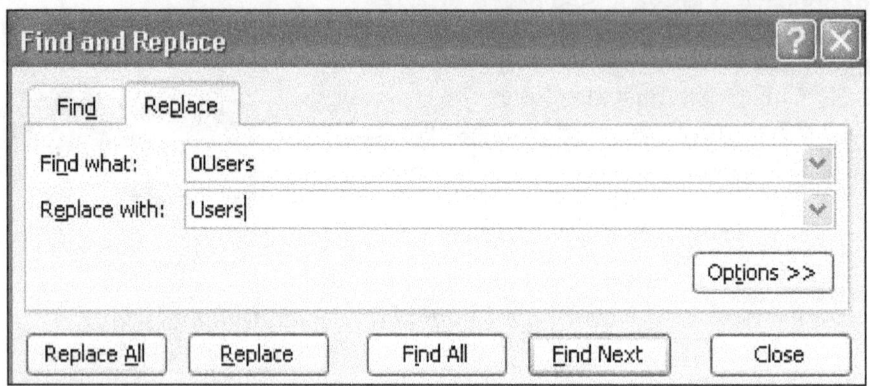

7. Click **Close** (close the *Find and Replace* window)

8. Save the MS Excel spreadsheet: **C:*yourname*\QTP10\LVL1\LAB7C.XLS**

9. Open **LAB7A** (Data Driver Wizard) → Save as **LAB7C**

10. Click **ExpertView** → Make the following changes

```
DIM  vs_name , vi_max, vi_row , vs_pass, vi_len, vb_login

DataTable.Import  ( "C:\YOURNAME\QTP10\LVL1\LAB7C.XLS" )
vi_max = DataTable.GetSheet ("Global").GetRowCount

FOR  vi_row =  1  to  vi_max

     DataTable.SetCurrentRow ( vi_row )
     vs_name = DataTable ("Users",dtGlobalSheet)
     vs_pass  = DataTable ("Password",dtGlobalSheet )

    ' IF ucase (vs_name) = "MARK"  THEN
    '    Dialog("Login").Winbutton("Cancel").Click
    '    ExitActionIteration ("Skipping:" & vs_name )
    ' END IF
   vb_login =  True
    vi_len  = len ( vs_name )
    IF  vi_len  <  1  THEN
         Reporter.ReportEvent  micwarning, "User Name", "Empty "
         vb_login = False
    END IF

    IF  vi_len  <  4  THEN
         Reporter.ReportEvent  micwarning, vs_name,  "InValid Length "  & vi_len
         vb_login = False
    END IF
```

Script continues on the next page

LAB 7C - Script

```
vi_len = len ( vs_pass )
IF vi_len < 1 THEN
    Reporter.ReportEvent micwarning, "Password", "Empty "
    vb_login = False
END IF

IF vs_pass < > "mercury" THEN
    Reporter.ReportEvent micwarning, "Invalid Password", vs_pass
    vb_login = False
END IF

IF vb_login = True THEN
    ' Reporter.reportEvent micpass,"User Id", vs_name
    SystemUtil.Run "C:\Program Files\HP\QuickTest
        Professional\samples\flight\app\flight4a.exe",
        "","C:\Program Files\HP\QuickTest Professional\
        samples\flight\app\","open"
    Dialog("Login").Check CheckPoint("Login")

    Dialog("Login").WinEdit("Agent Name:").Set DataTable("Users", dtGlobalSheet)
    Dialog("Login").WinEdit("Agent Name:").Type micTab
    Dialog("Login").WinEdit("Password:").SetSecure DataTable("Password", dtGlobalSheet)
    Dialog("Login").WinButton("OK").Click

    Window("Flight Reservation").Check CheckPoint("Flight Reservation")
    Window("Flight Reservation").WinMenu("Menu").Select "File;Exit"

END IF
NEXT
```

11. Save **LAB7C** → Playback & Review the Test Results

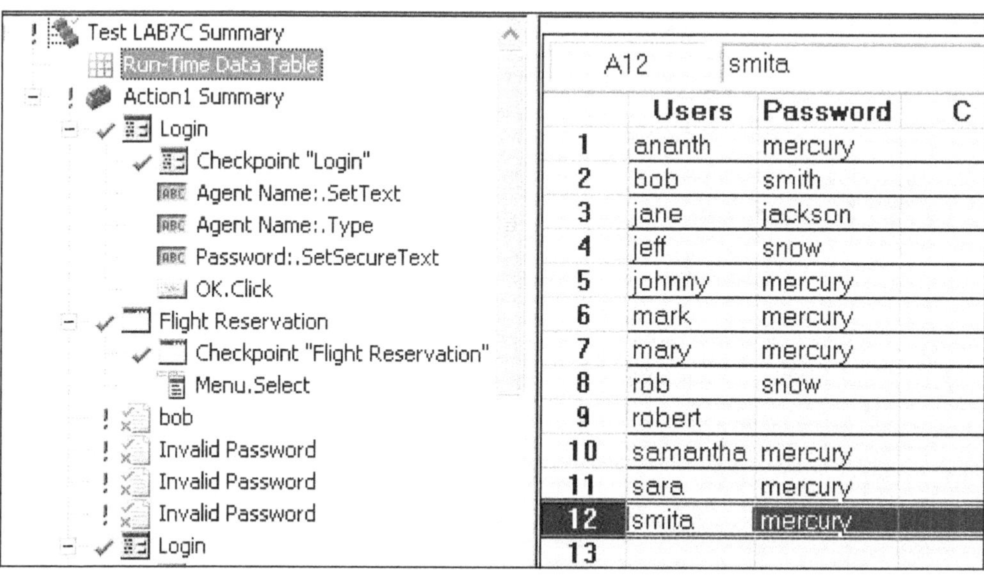

	Users	Password	C
1	ananth	mercury	
2	bob	smith	
3	jane	jackson	
4	jeff	snow	
5	johnny	mercury	
6	mark	mercury	
7	mary	mercury	
8	rob	snow	
9	robert		
10	samantha	mercury	
11	sara	mercury	
12	smita	mercury	
13			

*** **END OF LAB7C** ***

ACTION Files

1. An ***Action*** file represents the steps a user took to complete the business process

2. You can divide your **test** into *actions* to streamline the testing process of your ***Website*** or ***AUT***

3. By dividing your tests into multiple actions, you can design more modular & efficient tests

4. Action ***files*** help divide your test into **logical units** or *modules*

5. An ***action*** consists of a :

   a) Test Tree
   b) Test Script
   c) All *steps* recorded within *that* action
   d) Each having its own Object Repository (assuming *per-action*)
   e) A separate DataTable with ***same*** *name* as the name of the **Action*x***

6. You must choose where the data is stored ? In the Global or Local (Action*x*) sheet ?

Example 1: General Template

Main		Action Name	Script	DataTable
Test 1	Action 0		Parent (File Name)	Global
		Action 1 (Bus Process Name)		
			Action1 : Steps	LocalSheet (Action1)
		Action 2 (Bus Process Name)		
			Action2 : Steps	LocalSheet (Action2)
		Action 3 (Bus Process Name)		
			Action3 : Steps	LocalSheet (Action3)

GENERAL GUIDELINES

1. Each ***Action*** file should be kept to a minimum !

2. All the *steps* represented - should complete *that* process

3. The total *steps* in an ***Action*** should be between 20 – 30 steps (User Actions – Not VB Statements)

4. Maintenance is minimized

5. Understanding the logic and business rules are simplified

6. Debugging is easier

MULTIPLE ACTION Files

1. When you ***create*** a QTP test – it will *always* include one action called ***Action1***
2. You can rename ***Action1*** to a meaningful *business process action*
3. All the steps you *record* and/or *modifications* made to the script are part of a **single** action
4. You can divide your test into multiple actions by creating ***new*** actions or inserting ***existing*** actions

THREE ACTIONS (Via TreeView)

Non-reusable action

- An *action* that can be used only in the test in which it was created

Reusable action

- An *action* that can be **called** *multiple* times by the test in which it was created (the *local* test) as well as by other tests

Default (action)

- ***New*** actions are non-reusable
- You can re-classify or ***mark*** the action as *reusable* or *Non-reusable*
- For each Action (in your test) there is a corresponding ***action sheet*** in the Data Table

Example 2: FRS : Build the Modular Tree to: *Open Existing Reservations*

Main	Action Name	Sheet	Description
LAB7D		(Global)	
	fl_login	fl_login (Action1)	Login Process (Pre-Conditions)
	fl_open_order	fl_open_order (Action2)	Open Order (Functionality)
	fl_exit	fl_exit (Action3)	Exit the AUT (Clean-Up / Reset)

Steps To Build the Multiple Action Files (*Modular Programming*) :

1. Understand the ***Objective*** of the Business Process (or test requirement)
2. Initially - Don't be overly concerned about the ***technical*** design
3. Break-down the Business Process into smaller workable units (screens / windows)
4. Manually (think or) ask yourself - How can I explain (overall) how to conduct the test ?
5. Each module or *Action* will represent the *steps* to complete only that process

Question : Where does it make sense to store the data ? Which Sheet ?

LAB 7D : Multiple Action Files

Test Requirements :

a) Build a Data Driven Test to ***Open*** existing reservations

b) Create three (3) multiple action files (for this functionality)

c) Perform your own Checkpoint (via **GETROProperty**)

- Compare the value on the AUT screen with the current row in the Local sheet
- Display a customized message in the **Log** (*Test Results* window)

d) Enter Positive Data into the **Global** data sheet

PART 1 : Record the initial script for one (1) business transaction

Steps:

1. **File → New → Test...** (Create a *new* script) → Click **Keyword View** tab
2. **Automation → Record F3**
3. Check **Record and run test on any open Windows-based application** → Click **OK**
4. Invoke the AUT (on your own) → Login with your ***username*** and password **mercury** → Click **OK**
5. Issue a Standard Checkpoint on the **Flight Reservation** *title bar* (on your own !)
6. **File → Open Order** ... → Choose **Order No.** check box → Type **1** → Click **OK**
7. **File → Exit** (Close the AUT)
8. Click **Stop** or Press **F4** (Stop the QTP recording)
9. Your script should be similar to :

Item	Operation	Value	Documentation
Action1			
SystemUtil	Run	"C:\Program Files\Mercury Interacti...	Open the "C:\Program File
Login			
Agent Name:	Set	"tommy"	Enter "tommy" in the "Age
Agent Name:	Type	micTab	Type micTab in the "Agent
Password:	SetSecure	"4a15bcf2101de6b049e199422a14...	Enter the encrypted string '
OK	Click		Click the "OK" button.
Flight Reservation	Check	CheckPoint("Flight Reservation")	Check whether the "Flight
Menu	Select	"File;Open Order..."	Select item "File;Open Ord
Open Order			
Order No.	Set	"ON"	Set the state of the "Order
Edit	Set	"1"	Enter "1" in the "Edit" edit
OK	Click		Click the "OK" button.
Menu	Select	"File;Exit"	Select item "File;Exit" from

10. Save as **C:\YOURNAME\QTP10\LVL1\LAB7D_ORIG**

11. Save As **C:\YOURNAME\QTP10\LVL1\LAB7D**

12. Add **Order No:** following object to the repository (using the Active Screen) in the **Flight Reservation** window

    a) Click **Keyword View** → Highlight **Flight Reservation Check CheckPoint("Flight Reservation")**

    b) In the Active Screen – Scroll and mouse over the **Order No:** → Right-Click → **View / Add Object ...**

    c) Verify **WinEdit : Order No:** → Click **OK**

    d) Click **Add to repository** → Click **OK** (close **Object Properties** window)

13. Save **LAB7D**

Part 2 : Build the Data Driven Test

14. Highlight **Edit Set "1"** → In the **Value** Click **Configure the value** icon

15. Check **Parameter** Radio Button → Type & Replace **p_Text** with **Order_No**

16. Check **Current action sheet (local)** radio button → Click **OK**

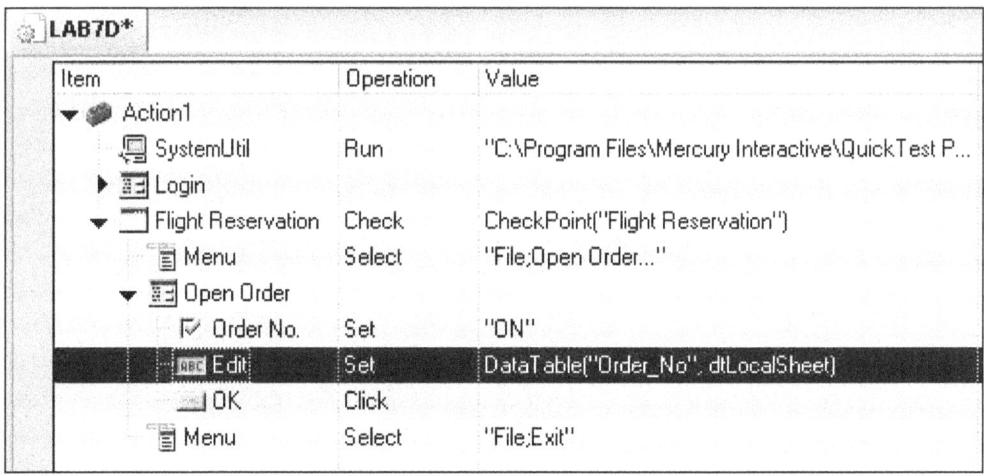

17. Save **LAB7D** → Playback and it should Pass ! → Review & Close the **Test Results** window

***** END OF LAB7D - Part 2 *****

Part 3 : Enhance the script to perform a customized Checkpoint

18. Click **ExpertView** → Modify the script as follows :

```
01. DIM vs_aut_order, vs_dt_order
02. SystemUtil.Run "C:\Program Files\HP\QuickTest Professional\samples\flight\app
        \flight4a.exe","","C:\Program Files\HP\QuickTest Professional\samples\flight\app\","open"

03. Dialog("Login").WinEdit("Agent Name:").Set "tommy"
04. Dialog("Login").WinEdit("Agent Name:").Type micTab
05. Dialog("Login").WinEdit("Password:").SetSecure "4a15bcf2101de6b049e199422a148d413d5c3309"
06. Dialog("Login").WinButton("OK").Click

07. Window("Flight Reservation").Check CheckPoint("Flight Reservation")
08. Window("Flight Reservation").WinMenu("Menu").Select "File;Open Order..."
09. Window("Flight Reservation").Dialog("Open Order").WinCheckBox("Order No.").Set "ON"
10. Window("Flight Reservation").Dialog("Open Order").WinEdit("Edit").Set DataTable("Order_No",
dtGlobalSheet)
11. Window("Flight Reservation").Dialog("Open Order").WinButton("OK").Click

12. vs_aut_order = Window("Flight Reservation").WinEdit("Order No:").GetROProperty ("text")
13. vs_dt_order  = DataTable ( "Order_No",dtLocalSheet )

14. IF  vs_aut_order = vs_dt_order THEN
15.       Reporter.Reportevent micPass, "Match" , "Order:  " & vs_aut_order
16. ELSE
17.       Reporter.Reportevent micFail, "No Match" ,  "DataTable: " & vs_dt_order
18. END IF

19. Window("Flight Reservation").WinMenu("Menu").Select "File;Exit"
```

19. Save **LAB7D** → Run the script → It Should Pass ! → Review & Close the **Test Results** window

20. Add the following data to the Action1 Data Table (Local Sheet)

Data Table

B4

	Order_No	B
1	1	
2	5	
3	3	
4	2	
5		

21. Save **LAB7D** → Run the script → It Should Pass !

22. However, only the first row was processed → Review the Customized messages

23. Each iteration has the same Order Number

24. Close **Test Results** window

25. Click **Keyword View** → Highlight **Action1** → Right-Click → **Action Call Properties...**

26. Select **Run on all rows** → Click **OK** → Click **OK** (to **Hint** window if prompted – else go to next step)

27. Save **LAB7D** → it should Pass → Review the **Test Results** window → All four rows were processed !

Part 5 : Create Multiple Action Files

28. Click **Keyword View** → Highlight **Flight Reservation Check CheckPoint ("Flight Reservation")**

29. Right-Click → **Action** → **Split...**

30. Click **Yes** (to **If you split this action , the run may file ?**)

31. Rename *Action1_1* to **fl_login** → Rename *Action1_2* to **fl_open_order** → Click **OK**

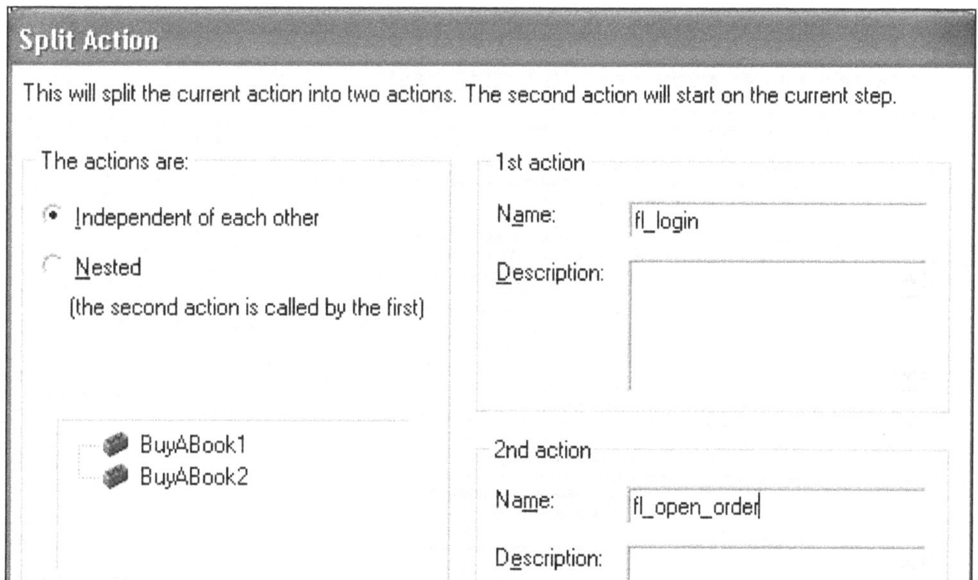

32. Save **LAB7D** → **View** → **Collapse All** → Highlight & Expand **fl_open_order**

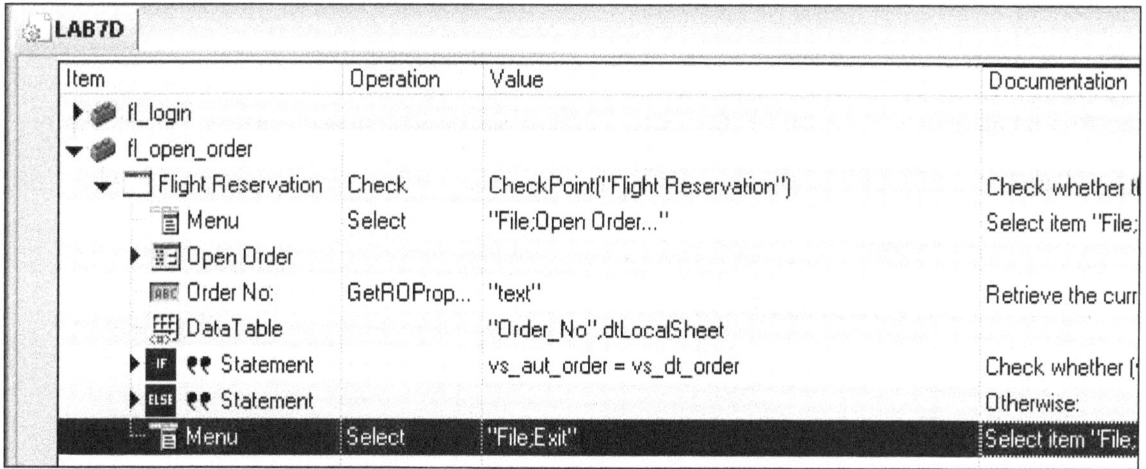

33. Highlight **Menu Select "File;Exit"** → Right-Click →
34. Highlight **Menu Select "File;Exit"** → Right-Click → **Action** → **Split...**

35. Rename *fl_open_order_1* to **fl_open_order**

36. Rename *fl_open_order_2* to **fl_exit** → Click **OK**

37. Save **LAB7D**

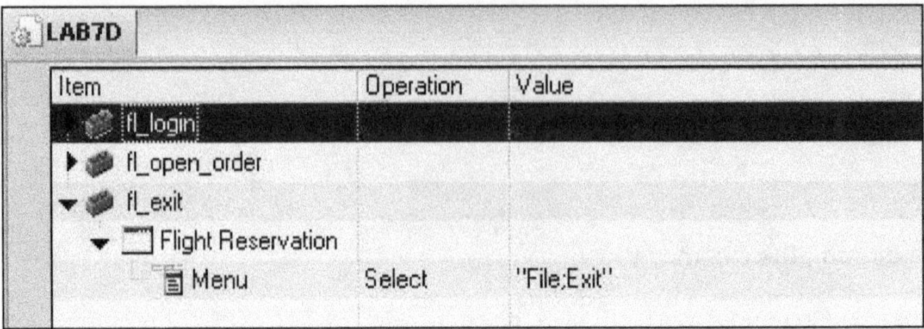

Note : The data is now in all three local sheets ! (It was previously in *Action1*)

38. In the Data Table → Click **fl_login** , **fl_open_order** & **fl_exit** sheet

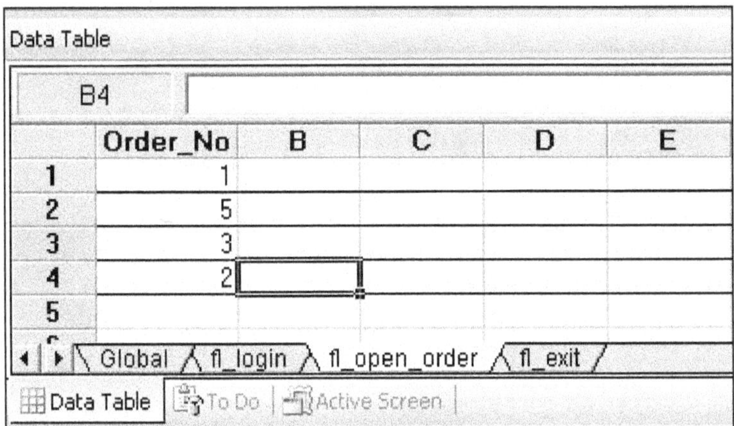

39. Click **fl_login** sheet → Delete all the rows (Do NOT clear or empty the cell – but delete the row !)

40. Click **fl_exit** sheet → Delete all the rows (Do NOT clear or empty the cell – but delete the row !)

41. Highlight **fl_open_order** → Right-Click → → **Action Call Properties…**

42. Select **Run on all rows** → Click **OK** → Click **OK** (to **Hint** window if prompted – else go to next step)

43. Save **LAB7D**

44. Playback → It should pass !

45. Review & Close the **Test Results** window

*** **END OF LAB 7D** ****

BATCH TESTS

1. The **Test Batch Runner** is a separate utility which allows you to run several Tests consecutively

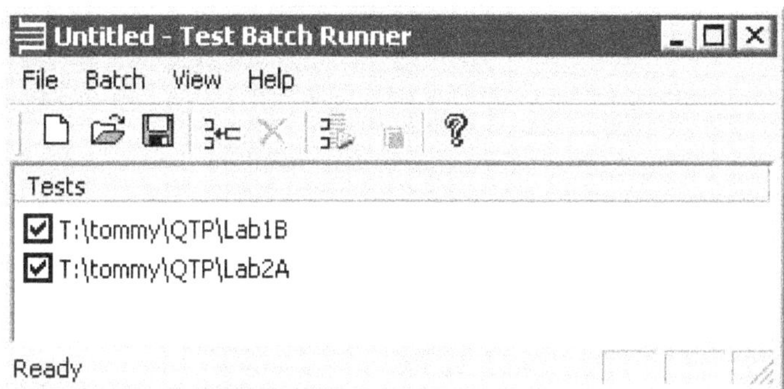

2. If QTP is not open, the *Batch Runner* will open it for you

3. All the tests are saved in a file with an extension of **.mtb** (mercury test batch)

4. You can choose to *include* or *exclude* a QTP script from running during the Batch Execution

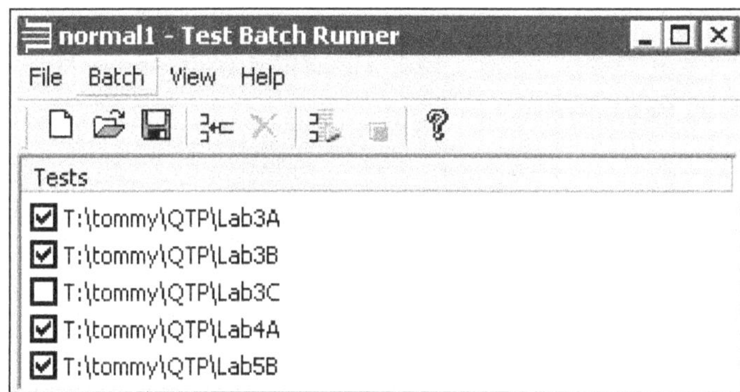

5. The ***Batch Runner*** is valid for scripts that saved in a File System

6. If you want to run Tests that are stored in Quality Center – then download the test into a *folder*

7. The QTP scripts should :

   a) Not contain errors

   b) Require Manual Pre-Conditions to be performed

   c) Have access to any required *shared network , folders and host data file(s)*

   d) Modify the Test Results to be saved into a permanent folder (Not Temporary)

      Example: **C:\YOURNAME\QTP10\LABX\Results**

   e) QTP Options must be set to allow :

      Allow other HP products to run tests and components

LAB 7E : Batch Test Runner

Test Requirements :

a) Identify two or more QTP Scripts to be executed consecutively

Example: ***LAB4A , LAB4B***

b) Add the scripts to the Batch Test Runner

c) Save the Batch Test as **LAB7E.MTB**

d) Execute all the tests in the Batch & Review the Results

Steps:

1. Open **LAB4A** → Change the Location from temporary to **C:\YOURNAME\QTP10\LVL1\LAB4A**
2. Run → Save Results in **C:\YOURNAME\QTP10\LVL1\LAB4A\Results**
3. Review and Close the **Test Results** window
4. Open **LAB4A** → Change the Location from temporary to **C:\YOURNAME\QTP10\LVL1\LAB4B**
5. Run → Save Results in **C:\YOURNAME\QTP10\LVL1\LAB4B\Results**
6. Review and Close the **Test Results** window
7. Close **QTP**
8. **Start → Programs → QuickTest Professional → Tools → Batch Test Runner**
9. **Batch** (menu bar) → **Add.. INS** → Point to **C:\YOURNAME\QTP10\LVL1\LAB4A** → Click **Open**
10. **Batch** (menu bar) → **Add.. INS** → Point to **C:\YOURNAME\QTP10\LVL1\LAB4B** → Click **Open**
11. **File** (menu bar) → **Save** → **C:\YOURNAME\QTP10\LVL1\LAB7E** → Click **Save**
12. **Batch** (menu bar) → **Run F5**
13. Wait for each Test to *Run* → Review the Results for each Run
14. Close and Save the **LAB7E.MTB**

TO RESTORE THE HP QTP MENU Options (If missing)

1. Right Click on the First row of the Toolbar icons
2. Select **Customize...** → Select **Toolbars** tab
3. Click **Restore All**

*** **END OF LAB 7E** ****

LAB 7F - FRS (Flight Reservation System) (Mini Project 1)

REQUIREMENTS :

1. Save LAB7B_P4 as LAB7E

2. Create a Data Driven Test using parameterization

3. Delete the orders that were added in LAB7B

   a) Refer to this statement : **DataTable.Import ("C:\yourname\QTP10\LVL1\LAB7B_P4.XLS")**
   b) The above statement should be one of the first statements in the script

4. Include appropriate:

   a) Display a message to indicate the total rows processed
   b) Checkpoints
   c) Validations
   d) Messages

STEPS

1. Set the Record & Run Settings

2. Perform a recording to record the ***actions*** to Delete an order :

   c) Start Recording
   d) Invoke the AUT
   e) Login
   f) Open an Existing Order
   g) Delete the order
   h) Exit the Application
   i) Stop Recording

3. Verify the Script in **Keyword View** mode

4. Review the Script in **ExpertView**

5. Enhance the script :

   a) Declare variables
   b) Identify the starting and ending statements for the actions that need to be repeated
   c) Create a Loop to read the *cell name* from the DataTable to find the Order

6. Additional Enhancement :

   a) Read *Values* from the Data Table
   b) Do NOT populate fields onto the screen
   c) Instead – verify if the data is valid
   d) If invalid *userid, password*, etc … issue an appropriate message, and read the next row

**** END OF LAB 7F ****

LAB 7G - FRS (Flight Reservation System) (Mini Project 2)

REQUIREMENTS :

1. Requires LAB7F to be completed
2. Save LAB7F as LAB7G
3. Validate the following windows : **Login** & **Flight Reservation**
4. Use the methods (described in the *Validation Functions* section) :
   - IsDate, IsNumeric, DateAdd, Mid, Len, Date, Time, etc ....
5. Business Rules

   Login

   a) Agent Name must contain at least five or more characters (new Rule)
   b) Fields cannot be empty

   New Order

   a) Valid Date (Date of Flight)

   b) New Orders should use the **current date** plus two days

   c) Display the total number of items in **Fly From** and **Fly To** (*report_msg*)

   d) Customer Name (only for this LAB 7G):

   - Last , First
   - Three or more characters
   - Check for a comma

   e) Display appropriate error Messages when required

7. Create your own data :

   a) Positive & Negative

   b) Add 30,60,90 days to the Date of Flight

   c) Use the newly ***computed*** date as the data for *that* iteration

   d) Display appropriate error Messages when required

*** **END OF LAB 7G** ****

LAB 7H - FRS (Flight Reservation System) (Mini Project 3)

REQUIREMENTS : Create a Volume Test to Add New Orders

1. Create a new RDBMS Table

   a) Use the User-ID from the SQL course . DO **NOT** USE itm / itm !

   b) Create the following table :

CREATE TABLE reservation		
(reservation_no	int	primary key
flight_date	smalldatetime	default getdate ( )
depart_from	nvarchar (40)	null
arrival_city	nvarchar (40)	null
name	nvarchar (40)	null
tickets	smallint	default 0
seat	int	default 3
unit_price	money	default 0
total_price	money	default 0
)		

   c) **sp_help reservation** -- To review the schema definition

2. Load the Data into the table **reservation** : (Examples)

```
insert into reservation
      values ( 1, "10/10/10", "Denver", "Frankfurt", "James Madison 1" , 1, 3 , 121.60, 121.60 )

insert into reservation
      values ( 2, "11/14/10", "Frankfurt", "London", "James Madison 2 ", 2, 3 , 103.80, 207.60 )

insert into reservation
      values ( 3, "11/25/10", "Paris", "Sydney", "James Madison 3 ", 3, 1 , 532.41, 1592.23 )
```

Add more data into the reservation table

   a) Ensure the order number (reservation number) matches with existing data in the FRS database

      - Login into FRS → **Analysis** → **Report**
      - The report lists all reservations for the Travel Agent (including *order number*)
      - Instead of Orders 1, 2, 3 start with the highest order number plus one (1)

   b) Your dates may be different (it must conform to the business rules)

   c) The *Name, Number of tickets, unit price & total* may vary

   d) Add at least ten (10) reservations

3. Write a SQL Query to verify the data was loaded

   SELECT * FROM reservation

4. Define a **VIEW** containing the data required for your test

   a) Sample view

```
SELECT  '0ORDER_NO' Order, 'DOF' DOF, 'FROM' From, 'TO' TO, 'NAME' Name, 'QTY' Qty
  FROM  reservation
 UNION
SELECT  CONVERT (NVARCHAR, reservation_no), Convert (NVARCHAR, flight_date),
        depart_from, arrival_city, name , Convert ( nvarchar, tickets)
 FROM  reservation
```

   b) Verify the view was created properly :

   SELECT * FROM v_orders

5. Export Data from SQL Server to an ASCII file. In MS-DOS prompt , type the following *command*

   BCP qa_dev1..v_orders out C:\YOURNAME\QTP10\LVL1\LAB7H.TXT –c
   -Usql_user -Pwelcome -Straining-srvr02

6. Import ASCII file into MS EXCEL → Open MS EXCEL

   a) Rename *sheet1* to **Global**
   b) Rename *sheet2* to **fl_login**
   c) Rename *sheet3* to **fl_exit**
   d) Right-Click on **fl_exit** → **Insert ..** → Highlight **Worksheet** → Click **OK**
   e) Rename *sheet4* to **fl_open_order**
   f) Highlight **fl_open_order**
   g) **Data → Import External Data → Import Data**
   h) Follow steps from earlier lab to complete the load & Save the excel file as **LAB7H.XLS**

7. New reservations should default to :

   a) Select the **first** entry in the flight timings (from the **FLIGHTS** window)
   b) Seat Type: default **economy**

8. Create a New script :

   a) Record all the actions required to add a single reservation → Save as **LAB7H**

   b) Data Drive the Test (any method you choose) → Save **LAB7H** → Playback & Verify the results

   Note: Use the same column headings as created in the VIEW

   c) Create the following action files :

LAB7H

Parent	Action File	DataSheet	Description
Action 0		Global	Parent
	fl_login	Local	No Data
	fl_new_order	Local	EXCEL File from Step 7
	fl_exit	Local	No Data

   d) In the **fl_login** remember to import the EXCEL file : DataTable.Import ("LAB7H.XLS")

***** END OF LAB 7H ******

APPENDIX A : MS SQL Server Schema Defintion

4. Create the **agents** table

agent_id	smallint
agent_name	varchar (20)
agent_password	varchar (30)
dept	char (02)
hire_date	smalldatetime

5. Load the following data (12 rows)

1	johnny	mercury	02	05-02-2000
5	sara	mercury	02	05-04-2000
10	ananth	mercury	02	04-22-2000
15	mary	mercury	02	12-12-1999
20	smita	mercury	02	03-15-2001
25	samantha	mercury	03	03-15-2001
30	jeff	snow	03	03-15-2001
35	bob	smith	03	03-15-2002
40	jane	Jackson	03	03-15-2001
45	robert		03	03-15-2001
50	rob	snow	03	04-25-2002
55	mark	mercury	02	05-02-2001

6. Write the query to confirm the data is loaded

select * from agents

www.ingramcontent.com/pod-product-compliance
Lightning Source LLC
Chambersburg PA
CBHW080402060326
40689CB00019B/4099

* 9 7 8 1 4 3 2 7 5 3 4 0 5 *